Laboratory Manual

Biology

MILLER • LEVINE

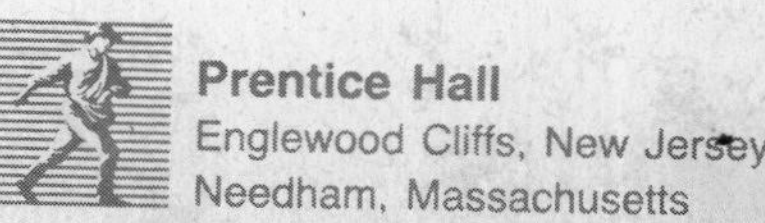

Prentice Hall
Englewood Cliffs, New Jersey
Needham, Massachusetts

BIOLOGY
Miller • Levine

Laboratory Manual

Designed and produced by
Function Thru Form Inc.

ISBN 0-13-803073-1

10 9 8 7 98 97 96

Prentice Hall
A Paramount Communications Company
Englewood Cliffs, New Jersey 07632

Contents

Safety in the Biology Laboratory

Working in the biology laboratory can be interesting, exciting, and rewarding. But it can also be quite dangerous if you are not serious and alert and if proper safety precautions are not taken at all times. You are responsible for maintaining an enjoyable, instructional, and safe environment in the biology laboratory. Unsafe practices endanger not only you but the people around you as well.

Read the following information about safety in the biology laboratory carefully. Review applicable safety information before you begin each Laboratory Investigation. If you have any questions about safety or laboratory procedures, be sure to ask your teacher.

SAFETY SYMBOL GUIDE

All the investigations in this laboratory manual have been designed with safety in mind. If you follow the instructions, you should have a safe and interesting year in the laboratory. Before beginning any investigation, make sure you read the safety rules on pages 10–12 of the *Laboratory Manual.*

The eight safety symbols shown on page 10 are used throughout the *Laboratory Manual.* They appear first next to the Safety section of an investigation and then next to certain steps in an investigation where specific safety precautions are required. The symbols alert you to the need for special safety precautions. The description of each symbol indicates the precaution(s) you should take whenever you see the symbol in an investigation.

Glassware Safety

1. Whenever you see this symbol, you will know that you are working with glassware that can be easily broken. Take particular care to handle such glassware safely. And never use broken glassware.
2. Never heat glassware that is not thoroughly dry. Never pick up any glassware unless you are sure it is not hot. If it is hot, use heat-resistant gloves.
3. Always clean glassware thoroughly before putting it away.

Fire Safety

1. Whenever you see this symbol, you will know that you are working with fire. Never use any source of fire without wearing safety goggles.
2. Never heat anything—particularly chemicals—unless instructed to do so.
3. Never heat anything in a closed container.
4. Never reach across a flame.
5. Always use a clamp, tongs, or heat-resistant gloves to handle hot objects.
6. Always maintain a clean working area, particularly when using a flame.

Heat Safety

Whenever you see this symbol, you will know that you should put on heat-resistant gloves to avoid burning your hands.

Chemical Safety

1. Whenever you see this symbol, you will know that you are working with chemicals that could be hazardous.
2. Never smell any chemical directly from its container. Always use your hand to waft some of the odors from the top of the container toward your nose—and only when instructed to do so.
3. Never mix chemicals unless instructed to do so.
4. Never touch or taste any chemical unless instructed to do so.
5. Keep all lids closed when chemicals are not in use. Dispose of all chemicals as instructed by your teacher.
6. Immediately rinse with water any chemicals, particularly acids, off your skin and clothes. Then notify your teacher.

Eye and Face Safety

1. Whenever you see this symbol, you will know that you are performing an investigation in which you must take precautions to protect your eyes and face by wearing safety goggles.
2. Always point a test tube or bottle that is being heated away from you and others. Chemicals can splash or boil out of the heated test tube.

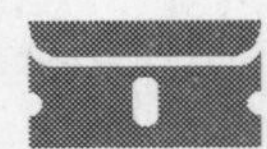

Sharp Instrument Safety

1. Whenever you see this symbol, you will know that you are working with a sharp instrument.
2. Always use single-edged razors; double-edged razors are too dangerous for use.
3. Handle any sharp instrument with extreme care. Never cut any material toward you; always cut away from you.
4. Notify your teacher immediately if you cut yourself or receive a cut.

Electrical Safety

1. Whenever you see this symbol, you will know that you are using electricity in the laboratory.
2. Never use long extension cords to plug in any electrical device. Do not plug too many different appliances into one socket or you may overload the socket and cause a fire.
3. Never touch an electrical appliance or outlet with wet hands.

Animal Safety

1. Whenever you see this symbol, you will know that you are working with live animals.
2. Do not cause pain, discomfort, or injury to any animal.
3. Follow your teacher's directions when handling animals. Wash your hands thoroughly after handling animals or their cages.

One of the first things a scientist learns is that working in the laboratory can be an exciting experience. But the laboratory can also be quite dangerous if proper safety rules are not followed at all times. To prepare yourself for a safe year in the laboratory, read over the following safety rules. Then read them a second time. Make sure you understand each rule. If you do not, ask your teacher to explain any rules you are unsure of.

DRESS CODE

1. Many materials in the laboratory can cause eye injury. To protect yourself from possible injury, wear safety goggles whenever you are working with chemicals, burners, or any substance that might get into your eyes. Never wear contact lenses in the laboratory.
2. Wear a laboratory apron or coat whenever you are working with chemicals or heated substances.
3. Tie back long hair to keep your hair away from any chemicals, burners and candles, or other laboratory equipment.
4. Remove or tie back any article of clothing or jewelry that can hang down and touch chemicals and flames. Do not wear sandals or open-toed shoes in the laboratory. Never walk around the laboratory barefoot or in stocking feet.

GENERAL SAFETY RULES

5. Be serious and alert when working in the laboratory. Never "horse around" in the laboratory.
6. Be prepared to work when you arrive in the laboratory. Be sure that you understand the procedure to be employed in any laboratory investigation and the possible hazards associated with it.
7. Read all directions for an investigation several times. Follow the directions exactly as they are written. If you are in doubt about any part of the investigation, ask your teacher for assistance.
8. Never perform activities that are not authorized by your teacher. Obtain permission before "experimenting" on your own.
9. Never handle any equipment unless you have specific permission.
10. Take extreme care not to spill any material in the laboratory. If spills occur, ask your teacher immediately about the proper cleanup procedure. Never simply pour chemicals or other substances into the sink or trash container.
11. Never eat or taste anything in the laboratory unless directed to do so. This includes food, drinks, candy, and gum, as well as chemicals. Wash your hands before and after performing every investigation.
12. Know the location and proper use of safety equipment such as the fire extinguisher, fire blanket, first-aid kit, safety shower, and eyewash station.
13. Notify your teacher of any medical problems you may have, such as allergies or asthma.
14. Keep your laboratory area clean and free of unnecessary books, papers, and equipment.

FIRST AID

15. Report all accidents, no matter how minor, to your teacher immediately.
16. Learn what to do in case of specific accidents such as getting acid in your eyes or on your skin. (Rinse acids off your body with lots of water.)
17. Become aware of the location of the first-aid kit. But remember that your teacher should administer any required first aid due to injury. Or your teacher may send you to the school nurse or call a physician.
18. Know where and how to report an accident or fire. Find out the location of the fire extinguisher, phone, and fire alarm. Keep a list of important phone numbers such as the fire department and school nurse near the phone. Report any fires to your teacher at once.

HEATING AND FIRE SAFETY

19. Never use a heat source such as a candle or burner without wearing safety goggles.
20. Never heat a chemical you are not instructed to heat. A chemical that is harmless when cool can be dangerous when heated.
21. Maintain a clean work area and keep all materials away from flames.
22. Never reach across a flame.
23. Make sure you know how to light a Bunsen burner. (Your teacher will demonstrate the proper procedure for lighting a burner.) If the flame leaps out of a burner toward you, turn the gas off immediately. Do not touch the burner. It may be hot. And never leave a lighted burner unattended.
24. Point a test tube or bottle that is being heated away from you and others. Chemicals can splash or boil out of a heated test tube.
25. Never heat a liquid in a closed container. The expanding gases produced may blow the container apart, injuring you or others.

26. Never pick up a container that has been heated without first holding the back of your hand near it. If you can feel the heat on the back of your hand, the container may be too hot to handle. Use a clamp, tongs, or heat-resistant gloves when handling hot containers.

USING CHEMICALS SAFELY

27. Never mix chemicals for the "fun of it." You might produce a dangerous, possibly explosive, substance.
28. Never touch, taste, or smell a chemical that you do not know for a fact is harmless. Many chemicals are poisonous. If you are instructed to note the fumes in an investigation, gently wave your hand over the opening of a container and direct the fumes toward your nose. Do not inhale the fumes directly from the container.
29. Use only those chemicals needed in the investigation. Keep all lids closed when a chemical is not being used. Notify your teacher whenever chemicals are spilled.
30. Dispose of all chemicals as instructed by your teacher. To avoid contamination, never return chemicals to their original containers.
31. Be extra careful when working with acids or bases. Pour such chemicals over the sink, not over your work bench.
32. When diluting an acid, pour the acid into water. Never pour water into the acid.
33. Rinse any acids off your skin or clothing with water. Immediately notify your teacher of any acid spill.

USING GLASSWARE SAFELY

34. Never force glass tubing into a rubber stopper. A turning motion and lubricant will be helpful when inserting glass tubing into rubber stoppers or rubber tubing. Your teacher will demonstrate the proper way to insert glass tubing.
35. Never heat glassware that is not thoroughly dry. Use a wire screen to protect glassware from any flame.
36. Keep in mind that hot glassware will not appear hot. Never pick up glassware without first checking to see if it is hot.
37. If you are instructed to cut glass tubing, fire polish the ends immediately to remove sharp edges.
38. Never use broken or chipped glassware. If glassware breaks, notify your teacher and dispose of the glassware in the proper trash container.
39. Never eat or drink from laboratory glassware. Clean glassware thoroughly before putting it away.

USING SHARP INSTRUMENTS

40. Handle scalpels or razor blades with extreme care. Never cut material toward you; cut away from you.
41. Be careful when handling sharp, pointed objects such as scissors, pins, and dissecting probes.
42. Notify your teacher immediately if you cut yourself or receive a cut.

HANDLING LIVING ORGANISMS

43. No investigations that will cause pain, discomfort, or harm to mammals, birds, reptiles, fishes, and amphibians should be done in the classroom or at home.
44. Treat all living things with care and respect. Do not touch any organism in the classroom or laboratory unless given permission to do so. Many plants are poisonous or have thorns, and even tame animals may bite or scratch if alarmed.
45. Animals should be handled only if necessary. If an animal is excited or frightened, pregnant, feeding, or with its young, special handling is required.
46. Your teacher will instruct you as to how to handle each species that may be brought into the classroom.
47. Treat all microorganisms as if they were harmful. Use antiseptic procedure, as directed by your teacher, when working with microbes. Dispose of microbes as your teacher directs.
48. Clean your hands thoroughly after handling animals or the cage containing animals.
49. Wear gloves when handling small mammals. Report animal bites or stings to your teacher at once.

END-OF-INVESTIGATION RULES

50. When an investigation is completed, clean up your work area and return all equipment to its proper place.
51. Wash your hands after every investigation.
52. Turn off all burners before leaving the laboratory. Check that the gas line leading to the burner is off as well.

SAFETY CONTRACT

Once you have read all the safety information on pages 10–12 in the *Laboratory Manual* and are sure you understand all the rules, fill out the safety contract that follows. Signing this contract tells your teacher that you are aware of the rules of the laboratory. Return your signed contract to your teacher. You will not be allowed to work in the laboratory until you have returned your signed contract.

Safety Contract

I, ______________________________, have read the

Safety in the Biology Laboratory section on pages 9–12 in the *Biology Laboratory Manual.* I understand its contents completely, and agree to follow all the safety rules and guidelines that have been established in each of the following areas:

(please check)

____ Dress Code
____ General Safety Rules
____ First Aid
____ Heating and Fire Safety
____ Using Chemicals Safely
____ Using Glassware Safely
____ Using Sharp Instruments
____ Handling Living Organisms
____ End-of-Investigation Rules

Signature ______________________________ Date ____________

How to Use the Laboratory Manual

This is probably the most exciting time in history to be studying biology. The science of biology is progressing at a rapid rate; scientists are learning more about the nature of life and the living world than they ever have before. Biology is directly related to many of the most important news stories reported on television and radio and in the newspapers and magazines. The Alaskan oil spill; the AIDS epidemic; animal rights; genetic fingerprinting; acid rain; the war against drugs; and efforts to save whales, elephants, pandas, and other endangered species are but a few of the current events and issues that can be better understood and evaluated by a person familiar with the principles of biology.

In order to gain a working knowledge of biological principles, you need to understand some of the processes that biologists and other scientists use to study and find answers to problems. The Laboratory Investigations and activities in the *Biology Laboratory Manual* provide an opportunity to learn about and practice processes and techniques used by scientists in their quest to increase human knowledge.

In each Laboratory Investigation, your objective is to solve a problem or problems using organized common sense, which is formally known as the scientific method. Each Laboratory Investigation follows a standard outline that will help you tackle the problem or problems in a systematic and organized manner.

Pre-Lab Discussion The Pre-Lab Discussion provides background information you will need to complete the investigation and ties the Laboratory Investigation to specific concepts discussed in the textbook. The Pre-Lab Discussion corresponds to an initial step in any scientific endeavor—gathering information about the topic you wish to study so that you can develop a hypothesis.

Problem This section presents a problem in the form of a question or series of questions. Your job during the Laboratory Investigation is to solve the problem based on your observations.

Materials A list of all materials required to conduct the investigation appears at the beginning of the investigation. The quantity of material for each investigation is indicated for individual students, pairs of students, or groups of students.

Safety The Safety section warns you of potential hazards before the investigation is begun and tells you about precautions you should take to decrease the risk of accidents. The safety symbols that are relevant to the Laboratory Investigation appear next to the title of the Safety section. They also appear next to certain steps of the Procedure.

Procedure This section provides a detailed step-by-step outline of the Laboratory Investigation procedure. Diagrams are included where necessary to further explain a technique or illustrate an experimental setup. The Procedure corresponds to the step of the scientific method known as testing the hypothesis.

Make sure you read the entire procedure carefully before you begin the investigation. Look for safety symbols and special instructions. If safety symbols appear next to a certain step in the Procedure, you should begin to follow the corresponding safety precaution(s) for that step and all following steps. **CAUTION** statements within the steps of the Procedure warn of possible hazards and indicate how accidents can be avoided. **Notes** in the Procedure provide special directions or explain techniques.

Observations In keeping with the traditional scientific method, you are asked to record your data in Observations in each investigation. Observations often include filling in data tables, graphing data, labeling diagrams, and drawing observed structures, as well as answering general questions related to the investigation.

Analysis and Conclusions Two steps of the traditional scientific method—analyzing data and forming a conclusion—are represented in this section. Using data gathered during the investigation and knowledge gained from the textbook and Pre-Lab Discussion, you are to analyze and interpret your experimental results.

Critical Thinking and Application This section challenges you to use critical-thinking skills to answer a variety of questions based on the Laboratory Investigation and your textbook reading.

Going Further Each Laboratory Investigation concludes with a section entitled Going Further. This section suggests additional activities for you to pursue on your own. Some of these are brief extensions of the Laboratory Investigation. Others involve library research. Still others suggest experimental research that you might perform with your teacher's permission.

Presenting Data

To seek answers to problems or questions they have about the world, scientists typically perform many experiments in the laboratory. In doing so, they observe physical characteristics and processes, select areas for study, and review the scientific literature to gain background information about the topic they are investigating. They then form hypotheses, test these hypotheses through controlled experiments, record and analyze data, and develop a conclusion about the correctness of the hypotheses. Finally, they report their findings in detail, giving enough information about their experimental procedure so that other scientists are able to replicate the experiments and verify the results.

The Laboratory Investigations in the *Biology Laboratory Manual* provide an opportunity for you to investigate scientific problems in the same manner as that of a typical scientist. As you perform these investigations, you will employ many of the techniques and steps of the scientific method a working scientist does. Some of the most important skills you will acquire are associated with the step of the scientific method known as recording and analyzing data. Three of these skills are creating and filling in data tables, making drawings, and finding averages. Another set of skills useful in presenting data is examined in the Laboratory Skills activity titled Using Graphing Skills.

It is important to record data precisely—even if the results of an investigation appear to be wrong. And it is extremely important to keep in mind that developing laboratory skills and data analysis skills is actually more valuable than simply arriving at the correct answers. If you analyze your data correctly—even if the data are not perfect—you will be learning to think as a scientist thinks. And that is the purpose of this laboratory manual and your experience in the biology laboratory.

DATA TABLES

When scientists conduct various experiments and do research, they collect vast amounts of information: for example, measurements, descriptions, and other observations. To communicate and interpret this information, they must record it in an organized fashion. Scientists use data tables for this purpose.

You will be responsible for completing data tables for many of the Laboratory Investigations. Each column in a data table has a heading. The column headings explain where particular data are to be placed. The completed data tables will help you interpret the information you collected and answer the questions found at the end of each Laboratory Investigation.

EXERCISE 1

Given the following information, complete Data Table 1. Then interpret the data and answer the five questions that follow.

Information: The following hair colors were found among three classes of students:

Class 1: brown—20
black—1
blond—4

Class 2: brown—18
black—0
blond—6

Class 3: brown—15
black—4
blond—5

Data Table 1

Hair Color	Class 1	Class 2	Class 3	Total
Brown				
Black				
Blond				

1. What type of information is being gathered? ______________________________

____

2. Which hair color occurs most often? ______________________________

3. From the information in the Data Table, can you give the number of boys with black hair? __________ What information can you give about the number of students with black hair?

____

4. Which class has the most blond students? ______________________________

5. How many students made up the entire student population? ______________________________

EXERCISE 2

Given the following information, organize the data into a table. Use the blank table provided in Figure 1 to draw in the necessary columns and rows. Then interpret the data and answer the questions that follow.

Information: On an expedition around the world, several scientists collected the venom of various snakes. One of the tests that the scientists conducted determined the toxicity of the venom of each snake. Other data obtained by the scientists included the mortality percentage, or relative death rate, from the bites of various snakes.

The snakes observed were the (1) southern United States copperhead, (2) western diamondback rattlesnake, (3) eastern coral snake, (4) king cobra, (5) Indian krait, (6) European viper, (7) bushmaster, (8) fer-de-lance, (9) black-necked cobra, and (10) puff adder.

The mortality percentage of people bitten by the snakes varied from 100% to less than 1%. The scientists noted the mortality percentage for each of the snakes was (1) less than 1%, (2) 5–15%, (3) 5–20%, (4) greater than 40%, (5) 77%, (6) 1–5%, (7) usually 100%, (8) 10–20%, (9) 11–40%, and (10) 11–40%.

Figure 1

1. Which snake's venom has the highest mortality rate. ______
2. Which snake's venom has the lowest mortality rate? ______
3. From the information recorded, can you determine the snake whose venom works the most rapidly? The least rapidly? ______
4. Which two snakes' venom have the same mortality rate? ______
5. How many snakes were observed? ______

DRAWINGS

Laboratory drawings can be made using several methods, depending on a particular Laboratory Investigation. Some drawings are made in circles that represent the viewing field of a microscope or another type of magnifier. When completing these drawings, be sure to include the magnification at which you viewed the object. Other laboratory drawings are representative of entire organisms or parts of organisms. These drawings show the relative size, shape, and location of anatomical structures. When completing representative drawings, make the structures as clear and as accurate as possible.

Most laboratory drawings are labeled. Use the following guidelines to help make your laboratory drawings clear and legible.

- Use a ruler to draw label lines.
- Label lines should point to the center of the structure being labeled.
- Do not write on the label lines.
- Print all labels horizontally.
- Label the right-hand side of the drawing, if possible.
- Do not cross label lines.

EXERCISE 3

The following laboratory drawing was completed without using the guidelines for laboratory drawings. Circle those parts of the drawing that do not follow the guidelines. Then, on the lines provided, explain how these parts of the drawing should be done properly.

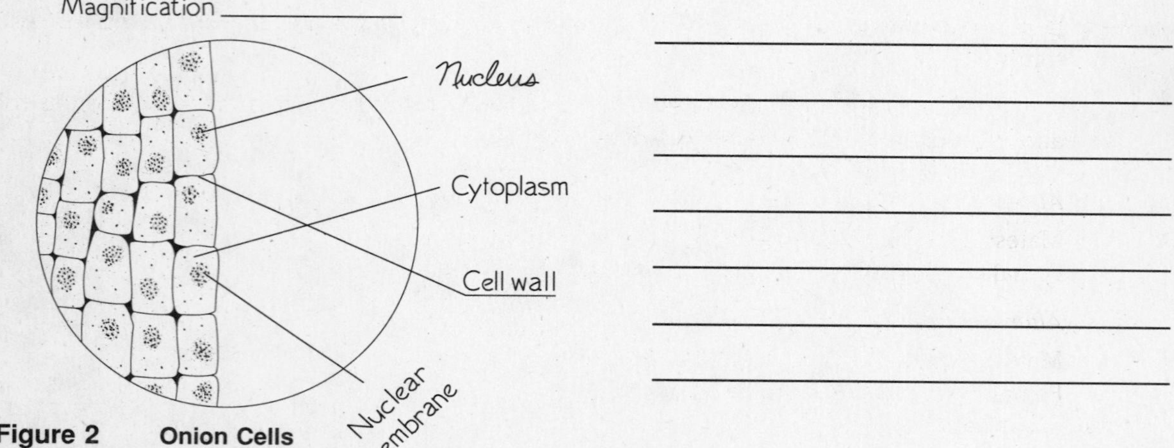

Figure 2 **Onion Cells**

AVERAGES

Occasionally you will be required to find the average of data gathered from an investigation. To find an average, add the items in the group together and then divide the total by the number of items. For example, if there were five students of different ages—12, 13, 14, 17, and 19—how would you find the average age of the group? Add the five ages together and divide the total by 5, which is the number of items (students) in the group. What is the average age of this group of students? Your answer should be 15 years old.

EXERCISE 4

In a garden the heights of six sunflowers are 135 cm, 162.5 cm, 180 cm, 235 cm, 185 cm, and 167.5 cm. What is the average height of the sunflowers?

EXERCISE 5

Find the average for the following group of data. Then use the results to answer the questions that follow.

In an experiment on plant growth and overcrowding, plants of the following heights are in three equal-sized containers.

Flowerpot 1: 20 cm and 18 cm
Flowerpot 2: 12 cm, 10.8 cm, 11.2 cm, and 12.4 cm
Flowerpot 3: 7.5 cm, 8 cm, 6 cm, 6.2 cm, 5.8 cm, and 7.3 cm

1. What is the average height of the plants in each flowerpot? ______________________

____

2. In which flowerpot did the plants grow the tallest? Explain. ______________________

____

____

____

EXERCISE 6

Find the averages for the following group of data. Express your answers to the nearest tenth.

In a sample group of students, the number of breaths per minute was taken at rest and after exercise. The results were as follows:

At rest
Males: 10.1, 13, 12.5, 10.2, 13.1, 11.8
Females: 10.4, 13.0, 12.1, 11.9, 10.5, 12.8

After exercise
Males: 18.9, 23.7, 22.6, 21.3, 19.2, 20.6
Females: 25, 26.7, 29, 35.3, 33.1, 31.7

1. What is the average number of breaths per minute for males at rest? ______________

Females at rest? ______________________

2. What is the average number of breaths per minute for males after exercise? ______________

Females after exercise? ______________________

3. How many students make up the sample group? ______________________

4. What is the average number of breaths per minute for the entire group at rest? ______________

After exercise? ______________________

5. Do males or females take more breaths per minute at rest? ______________________

After exercise? ______________________

Name ______________________ Class ____________ Date ____________

LABORATORY SKILLS 1

Recognizing Laboratory Safety

Pre-Lab Discussion

An important part of your study of biology will be working in a laboratory. In the laboratory, you and your classmates will learn biology by actively conducting and observing experiments. Working directly with living things will provide opportunities for you to better understand the principles of biology discussed in your textbook or talked about in class.

Most of the laboratory work you will do is quite safe. However, some laboratory equipment, chemicals, and specimens can be dangerous if handled improperly. Laboratory accidents do not just happen. They are caused by carelessness, improper handling of equipment and specimens, or inappropriate behavior.

In this investigation, you will learn how to prevent accidents and thus work safely in a laboratory. You will review some safety guidelines and become acquainted with the location and proper use of safety equipment in your classroom laboratory.

Problem

What are the proper practices for working safely in a biology laboratory?

Materials *(per group)*

Biology textbook
Laboratory safety equipment (for demonstration)

Procedure

Part A. Reviewing Laboratory Safety Rules and Symbols

1. Carefully read the list of laboratory safety rules listed on page 19 of your textbook.
2. Special symbols are used throughout this laboratory manual to call attention to investigations that require extra caution. Use page 19 in your textbook as a reference to describe what each symbol means by completing numbers 1 through 8 of Observations.

Part B. Location of Safety Equipment in Your Biology Laboratory

1. Your teacher will point out the location of the safety equipment in your classroom laboratory. Pay special attention to instructions for using such equipment as fire extinguishers, eyewash fountains, fire blankets, safety showers, and items in first-aid kits. Use the space provided under Observations to list the location of all safety equipment in your laboratory.

Observations

Part A. Reviewing Laboratory Safety Rules and Symbols

1. ______________________________

____

____

2. ______________________________

____

____

3. ______________________________

____

____

4. ______________________________

____

____

5. ______________________________

____

____

6. ______________________________

____

____

7. ______________________________

____

____

8. ______________________________

____

____

Name ________________ Class ________ Date ________

Part B. Location of Safety Equipment in Your Biology Laboratory

____

____

____

____

____

Analysis and Conclusions

Look at each of the following drawings and explain why the laboratory activities pictured are unsafe.

1. ____

2. ____

3. ____

4. ____

Critical Thinking and Application

In each of the following situations, write "yes" if the proper safety procedures are being followed and "no" if they are not. Then give a reason for your answer.

1. Kathy cannot find matches to light her Bunsen burner. The student next to her picks up a lighted burner and says, "Here, you can use my flame to light your burner."

____

____

2. Jim notices that the electrical cord on his microscope is frayed near the plug. He takes the microscope to his teacher and asks for permission to use another one.

____

____

3. The printed directions in the laboratory manual tell a student to pour a small amount of hydrochloric acid into a beaker. Alan puts on safety goggles before pouring the acid into the beaker. ____

____

____

4. It is rather warm in the biology laboratory during a late spring day. Joanne slips off her shoes and walks barefoot to the sink to clean her glassware. ______________________

____

____

5. While washing glassware, Mike splashes some water on Tim. To get even, Tim splashes him back. ____

____

____

Going Further

Many houseplants and some plants found in biology laboratories are poisonous. Use appropriate library resources to do research on several common poisonous plants. You might like to share your research with your classmates. You may prepare a booklet describing common poisonous plants. Use drawings or photographs to illustrate your booklet.

Name ______________________ Class ____________ Date ____________

LABORATORY SKILLS 2

Identifying Laboratory Equipment

Pre-Lab Discussion

Scientists use a variety of tools to explore the world around them. Tools are very important in the advancement of science. The type of tools scientists use depends on the problems they are trying to solve. A scientist may use something as simple as a metric ruler to measure the length of a leaf. At another time, the same scientist may use a complex computer to analyze large amounts of data concerning hundreds of leaves.

In this investigation, you will identify pieces of laboratory equipment likely to be found in a biology laboratory. You will also learn the function of each piece of laboratory equipment.

Problem

What are the names and functions of some of the pieces of laboratory equipment found in a typical biology laboratory?

Materials *(per group)*

Equipment shown in this activity (for inspection and demonstration)

Safety

Handle all glassware carefully. Be careful when handling sharp instruments. Always handle the microscope with extreme care. You are responsible for its proper care and use. Use caution when handling glass slides as they can break easily and cut you. Note all safety alert symbols next to the steps in the Procedure and review the meanings of each symbol by referring to the symbol guide on page 10.

Procedure

Part A. Identifying Laboratory Equipment

1. Look at the drawings of the laboratory equipment. In Observations, write the letter of the drawing next to the name that correctly identifies it.

Part B. Identifying the Function of Certain Types of Laboratory Equipment

1. Carefully inspect the different types of laboratory equipment that have been set out by your teacher.
2. In Observations, identify the function of each piece of laboratory equipment.

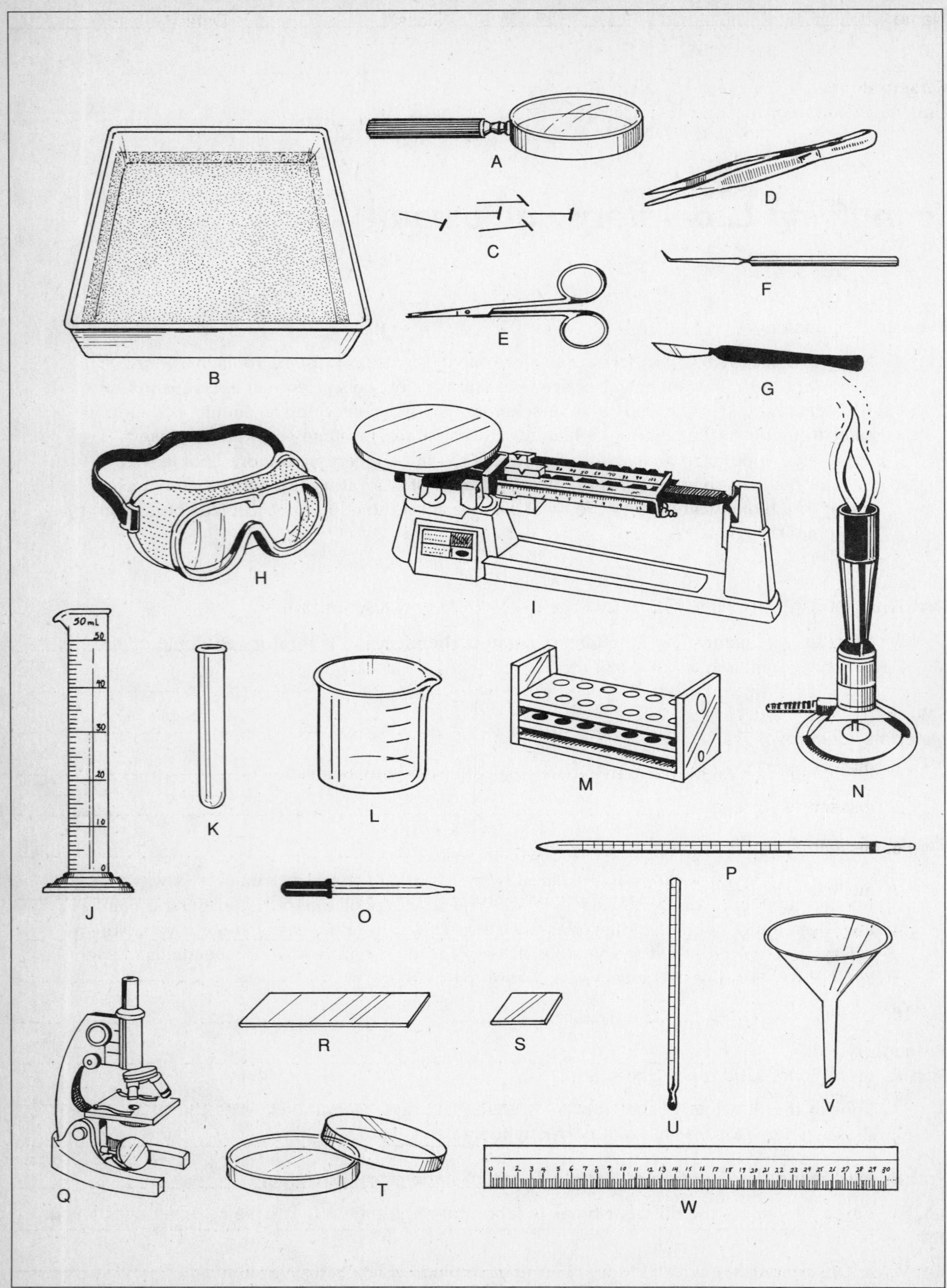

Figure 1

Name ______________________ Class __________ Date __________

Observations

Part A. Identifying Laboratory Equipment

_____ 1. beaker

_____ 2. Bunsen burner

_____ 3. coverslip

_____ 4. dissecting pins

_____ 5. dissecting scissors

_____ 6. dissecting tray

_____ 7. medicine dropper

_____ 8. forceps

_____ 9. funnel

_____ 10. graduated cylinder

_____ 11. hand lens

_____ 12. microscope

_____ 13. glass slide

_____ 14. petri dish

_____ 15. pipette

_____ 16. probe

_____ 17. metric ruler

_____ 18. safety goggles

_____ 19. scalpel

_____ 20. test tube

_____ 21. test tube rack

_____ 22. thermometer

_____ 23. triple-beam balance

Part B. Identifying the Function of Certain Types of Laboratory Equipment

1. beaker ______________________________

2. Bunsen burner ______________________________

3. coverslip ______________________________

4. dissecting pins ______________________________

5. dissecting scissors ______________________________

6. dissecting tray ______________________________

7. medicine dropper ______________________________

8. forceps ______________________________

9. funnel ______________________________

10. graduated cylinder ______________________________

11. hand lens ______________________________

12. microscope ______________________________

13. glass slide ______________________________

14. petri dish ______________________________

15. pipette ______________________________

16. probe ______________________________

17. metric ruler ______________________________

18. safety goggles ______

19. scalpel ______

20. test tube ______

21. test tube rack ______

22. thermometer ______

23. triple-beam balance ______

Analysis and Conclusions

1. Which laboratory tools can be used to magnify small objects so they can be seen more easily?

______

______

2. Which laboratory tools are useful when looking at the internal organs of an earthworm?

______

______

______

Critical Thinking and Application

1. What tool or tools would you use to make each of the following measurements?

   a. amount of milk in a small glass ______

   b. length of a sheet of paper ______

   c. temperature of a swimming pool ______

   d. mass of a baseball ______

2. How do laboratory tools improve the observations made by a scientist?

______

______

Going Further

Examine other types of laboratory equipment that you will be using in the biology laboratory. Try to determine the function of each piece of equipment.

Name ______________________ Class ____________ Date ____________

LABORATORY SKILLS 3

Making Metric Measurements

Pre-Lab Discussion

In many biology investigations, precise measurements must be made before observations can be interpreted. For everyday measuring, we still use English units such as the inch, quart, and pound. For scientific work, and for everyday measuring in most countries, the International System of Units (SI) is used. Eventually our country will use SI units for everyday measuring too.

Like our money system, SI is a metric system. All units are based on the number 10. In the SI system it is easy to change one unit to another because all units are related to one another by a power of 10.

In this investigation, you will review SI units for measuring length, liquid volume, and mass. You will also learn how to use some common laboratory equipment used for measuring.

Problem

How are metric units of measurement used in the laboratory?

Materials *(per group)*

Meterstick
Metric ruler
Small test tube
Rubber stopper
Coin
Triple-beam balance
50-mL beaker
100-mL graduated cylinder

Safety

Handle all glassware carefully. Note all safety alert symbols next to the steps in the Procedure and review the meanings of each symbol by referring to the symbol guide on page 10.

Procedure

Part A. Measuring Length

1. Use the meterstick to measure the length, width, and height of your laboratory table or desk in meters. Record your measurements to the nearest centimeter in Data Table 1.
2. Convert the measurements from meters to centimeters and then to millimeters. Record these measurements in Data Table 1.

3. Use a metric ruler to measure the length of a small test tube and the diameter of its mouth in centimeters. Record your measurements to the nearest millimeter in Data Table 2.

4. Convert the measurements from centimeters to millimeters. Record these measurements in Data Table 2.

Part B. Measuring the Volume of a Liquid

1. Fill the test tube to the top with water. Pour the water into the graduated cylinder.

2. If the graduated cylinder is made of glass, the surface of the liquid will be slightly curved. This curved surface is called a *meniscus.* To measure the volume accurately, your eye must be at the same level as the bottom of the meniscus. This is the mark on the graduated cylinder you must read. See Figure 1. Record the volume of the water from the test tube to the nearest milliliter in Data Table 3.

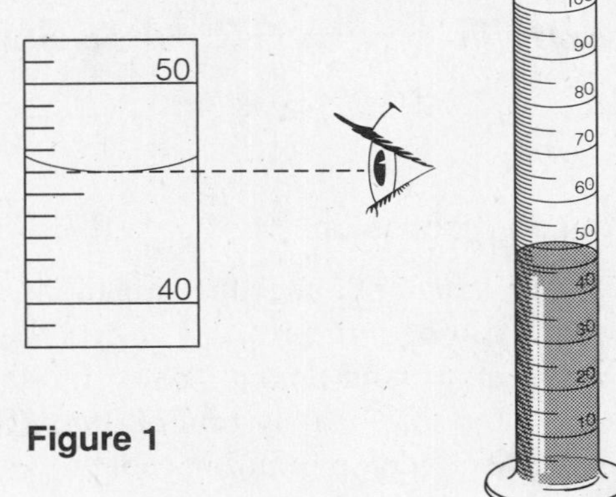

Figure 1

Part C. Measuring Mass

Before beginning this part of the investigation, be sure that the riders on the triple-beam balance are moved all the way to the left and that the pointer rests on zero. See Figure 2.

1. Place the 50-mL beaker on the pan of the balance.

2. Move the rider on the middle beam one notch at a time until the pointer drops below zero. Move the rider back one notch.

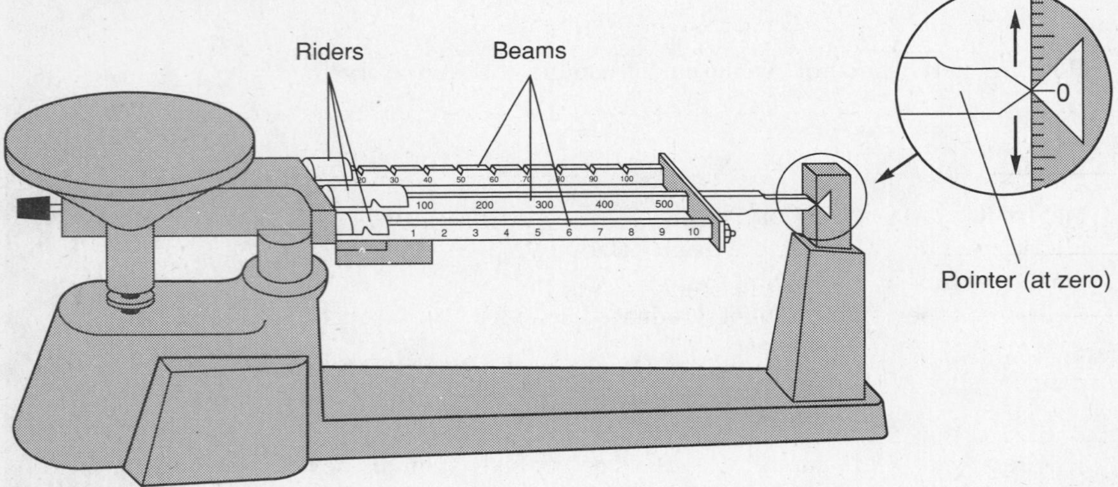

Figure 2

3. Move the rider on the back beam one notch at a time until the pointer again drops below zero. Move the rider back one notch.

4. Slide the rider along the front beam until the pointer stops at zero. The mass of the object is equal to the sum of the readings on the three beams.

5. Record the mass of the beaker to the nearest tenth of a gram in Data Table 4.

6. Remove the beaker and repeat steps 2 through 5 using the rubber stopper and then the coin.

7. Use the graduated cylinder to place exactly 40 mL of water in the beaker. Determine the combined mass of the beaker and water. Record this mass to the nearest tenth of a gram in Data Table 4.

Name ______________________ Class ____________ Date ____________

Observations

Data Table 1

Lab Table Measurements			
Dimension	m	cm	mm
Length			
Width			
Height			

Data Table 2

Test Tube Measurements		
Dimension	cm	mm
Length		
Diameter of mouth		

Data Table 3

Measurement of Volume	
Object	mL
Water in test tube	

Data Table 4

Measurement of Mass	
Object	g
50-mL beaker	
Rubber stopper	
Coin	
50-mL beaker plus 40 mL of water	

Analysis and Conclusions

1. How do you convert meters to centimeters? Centimeters to millimeters?

____

____

2. What is the largest volume of a liquid your graduated cylinder can measure?

____

3. What is the smallest volume of a liquid your graduated cylinder can measure?

____

4. What is the largest mass of an object your balance can measure? ______________

____

5. What is the smallest mass of an object your balance can measure? ______________

____

6. What is the mass of 40 mL of water? ______________

Critical Thinking and Application

1. Why is it easier to convert meters to centimeters or millimeters than it is to convert miles to yards or feet? ______________

____

____

2. How would you find the mass of a certain amount of water that you poured into a paper cup? ______________

____

____

3. In this investigation you found the mass of 40 mL of water. Based on your observations, what is the mass of 1 mL of water? ______________

____

Going Further

If other types of laboratory balances are available, such as an electronic balance or a double-pan balance, use them to find the masses of several different objects. Compare the accuracy of the different balances.

Name ______________________ Class __________ Date __________

LABORATORY SKILLS 4

Applying the Scientific Method

Pre-Lab Discussion

The scientific method is a procedure used to gather information and test ideas. Scientists use the scientific method to answer questions about life and living organisms. Experimentation is an important part of the scientific method. In order to ensure that the results of an experiment are due to the variable being tested, a scientist must have both an experimental setup and a control setup. The experimental setup contains the variable that is being tested. The control setup is exactly like the experimental setup except it does not contain the variable being tested.

In this investigation, you will form a hypothesis, test it, and draw a conclusion based on your observations.

Problem

Is light necessary for the sprouting of a potato?

Materials *(per group)*

1 medium-sized potato
2 plastic bags with twist ties
Knife
2 paper towels

Safety

Put on a laboratory apron if one is available. Be careful when handling sharp instruments. Note all safety alert symbols next to the steps in the Procedure and review the meanings of each symbol by referring to the symbol guide on page 10.

Procedure

1. With the members of your group, discuss whether or not the potato needs light to sprout. Based on your discussion, record your hypothesis in Observations.
2. Carefully cut the potato in half lengthwise. Count the number of eyes on the potato half to be put in the dark and the half to be put in the light. Record this in Data Table 1.
3. Fold each paper towel repeatedly until you have a rectangle about the same size as your potato halves. Moisten the towels with water. Place a folded paper towel in each plastic bag.

4. Place a potato half in each plastic bag with the cut surface on the paper towel. Tie each bag with a twist tie. See Figure 1.

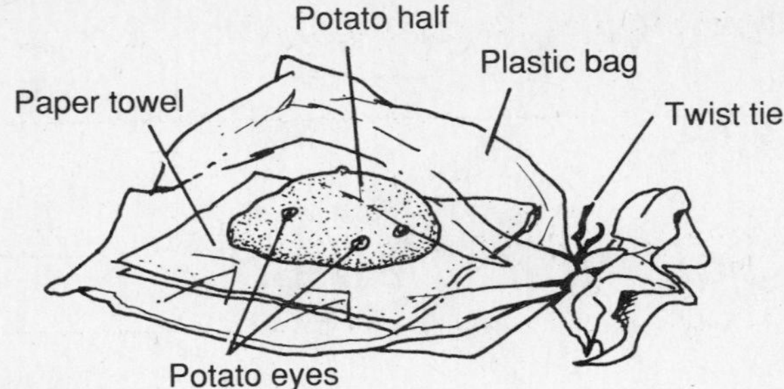

Figure 1

5. Place one of the plastic bags in a place that receives light. Place the other plastic bag in a dark place. Be sure that the potato halves remain on top of the paper towels and that both potato halves are kept at the same temperature.
6. After one week, open each plastic bag and count the number of sprouts. Record this information in Data Table 2.
7. To calculate the percentage of eyes sprouting, divide the number of sprouts by the number of eyes and multiply the result by 100. Record your answers in Data Table 3.
8. Have one person from your group go to the chalkboard to record your group's data in the table that has been drawn by your teacher.

Observations

Hypothesis ____

____

Data Table 1

Number of Eyes	
Potato in dark	
Potato in light	

Data Table 2

Number of Sprouts	
Potato in dark	
Potato in light	

Data Table 3

Percentage of Eyes Sprouting	
Potato in dark	
Potato in light	

Analysis and Conclusions

1. Did the most sprouts grow in the light or in the dark? ______________________

____

Name ______________________ Class __________ Date __________

2. What was the control setup in this investigation? ______________________

3. What was the experimental setup in this investigation? ______________________

4. What conclusion can you draw from this investigation? ______________________

5. How does your hypothesis compare with your results after completing the investigation?

Critical Thinking and Application

1. Why was it important to keep both the control setup and the experimental setup at the same temperature throughout the experiment? ______________________

2. Why were the plastic bags sealed? ______________________

3. Why was it necessary to keep the potato halves on moist paper towels?

Going Further

Devise an experiment to see if another variable, such as temperature or water, affects the number of sprouts a potato produces.

Name ______________________ Class ____________ Date ____________

LABORATORY SKILLS 5

Using a Compound Light Microscope

Pre-Lab Discussion

Many objects are too small to be seen by the eye alone. They can be seen, however, with the use of an instrument that magnifies, or visually enlarges, the object. One such instrument, which is of great importance to biologists and other scientists, is the compound light microscope. A compound light microscope consists of a light source or mirror that illuminates the object to be observed, an objective lens that magnifies the image of the object, and an eyepiece (ocular lens) that further magnifies the image of the object and projects it into the viewer's eye.

Objects, or specimens, to be observed under a microscope are generally prepared in one of two ways. Prepared or permanent slides are made to last a long time. They are usually purchased from biological supply houses. Temporary or wet mount slides are made to last only a short time—usually one laboratory period.

The microscope is an expensive precision instrument that requires special care and handling. In this investigation, you will learn the parts of a compound light microscope, the functions of those parts, and the proper use and care of the microscope. You will also learn the technique of preparing wet-mount slides.

Problem

What is the proper use of a compound light microscope?

Materials *(per group)*

Compound light microscope
Prepared glass slide
Lens paper
Soft cloth (or cheesecloth)
Newspaper
Glass slide
Coverslip
Dissecting needle
Medicine dropper
Scissors

Safety

Put on a laboratory apron if one is available. Always handle the microscope with extreme care. You are responsible for its proper care and use. Use caution when handling glass slides as they can break easily and cut you. Never use direct sunlight as a light source for a compound light microscope. The sunlight reflecting off the mirror up through the microscope could damage your eye. Be careful when handling sharp instruments. Observe proper laboratory procedures when using electrical equipment. Note all safety alert symbols next to the steps in the Procedure and review the meanings of each symbol by referring to the symbol guide on page 10.

Procedure

Part A. Care of the Compound Light Microscope; Parts of the Compound Light Microscope

1. Figure 1 shows the proper way to carry a microscope. Always carry the microscope with both hands. Grasp the arm of the microscope with one hand and place your other hand under the base. Always hold the microscope in an upright position so that the eyepiece cannot fall out. Take a microscope and place it on your worktable or desk at least 10 cm from the edge. Position the microscope with the arm facing you.

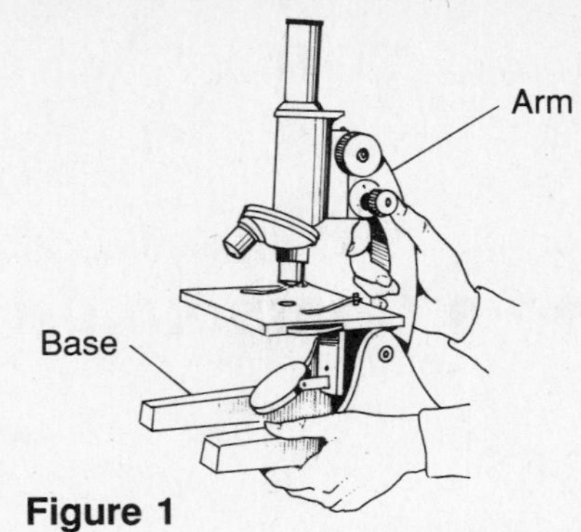

Figure 1

2. Study the labeled drawings of the microscopes in Figure 2. Identify the following parts on your microscope: eyepiece, arm, coarse adjustment, fine adjustment, revolving nosepiece, low-power objective, high-power objective, stage, stage clips, stage opening, diaphragm, light source (mirror or lamp), and base. Review the function of each part of the microscope in Figure 2. Do not proceed with this investigation until you can identify all the parts of your microscope and describe the function of each part. **Note:** *Tell your teacher at once if you find any parts of the microscope missing or damaged.*

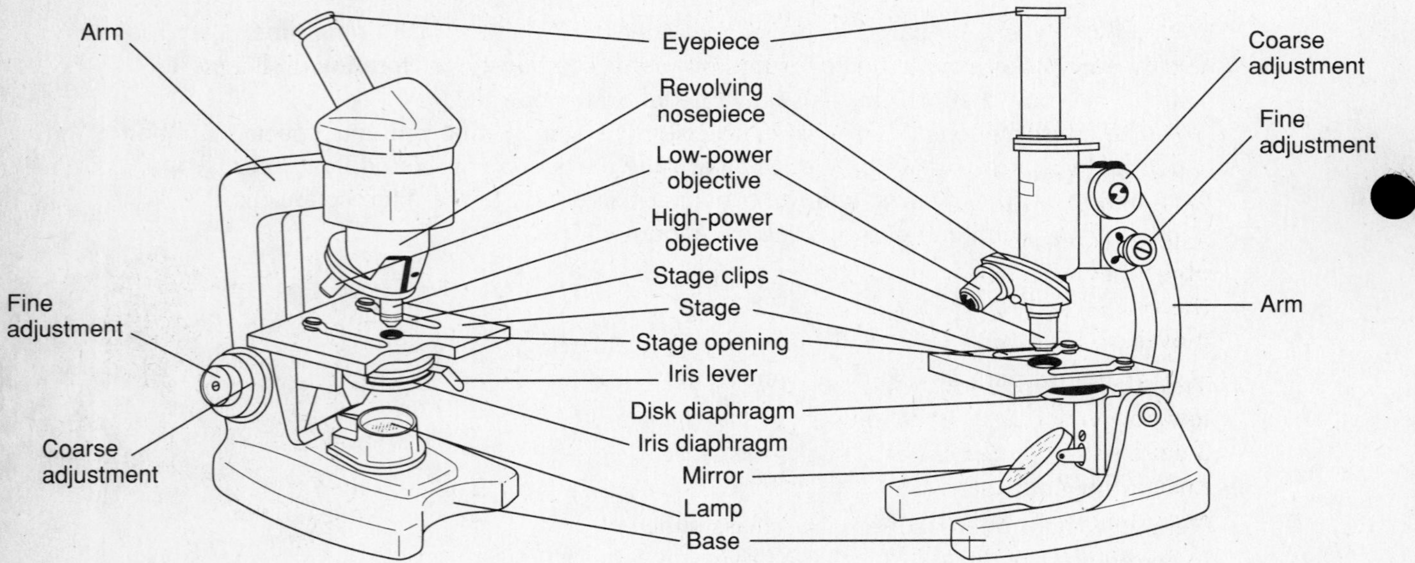

Figure 2

3. Notice the numbers etched on the objectives and on the eyepiece. Each number is followed by an "X" that means "times." For example, the low-power objective may have the number "10X" on its side, as shown in Figure 3. That objective magnifies an object 10 times its normal size. Record the magnifications of your microscope in the Data Table. The total magnification of a microscope is calculated by multiplying the magnification of the objective by the magnification of the eyepiece. For example:

magnification of low-power objective	x	magnification of eyepiece	=	total magnification
10X	x	10X	=	100X

Use the above formula to complete the Data Table.

Name ______________________ Class ____________ Date ____________

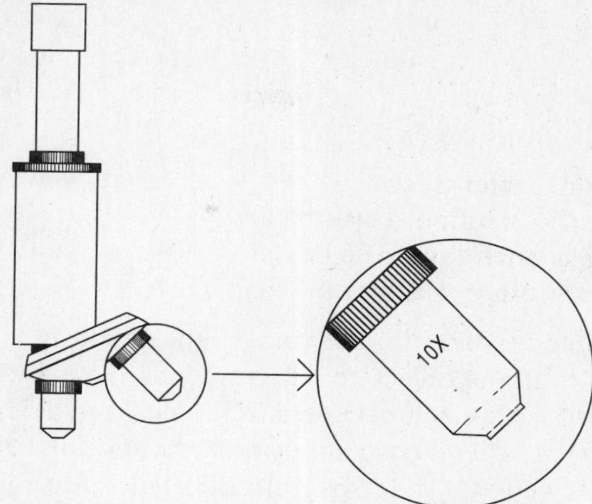

Figure 3

4. Before you use the microscope, clean the lenses of the objectives and eyepiece with lens paper. **Note:** *To avoid scratching the lenses, never clean or wipe them with anything other than lens paper. Use a new piece of lens paper on each lens you clean. Never touch a lens with your finger. The oils on your skin may attract dust or lint that could scratch the lens.*

5. Use a soft cloth or a piece of cheesecloth to wipe the stage and the mirror or light.

Part B. Use of a Compound Light Microscope

1. Look at the microscope from the side as shown in Figure 4. Locate the coarse adjustment knob which moves the objectives up and down. Practice moving the coarse adjustment knob to see how it moves its objectives with each turn.

2. Turn the coarse adjustment so that the low-power objective is positioned about 3 cm from the stage. Locate the revolving nosepiece. Turn the nosepiece until you hear the high-power objective click into position. See Figure 5. When an objective clicks into position, it is in the proper alignment for light to pass from the light source through the objective into the viewer's eye. Now turn the nosepiece until the low-power objective clicks back into position. **Note:** *Always look at the microscope from the side when moving an objective so that the microscope does not hit or damage the slide.*

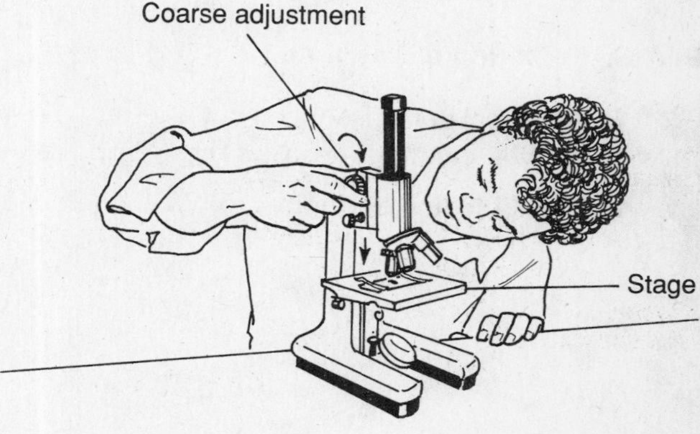

Figure 4

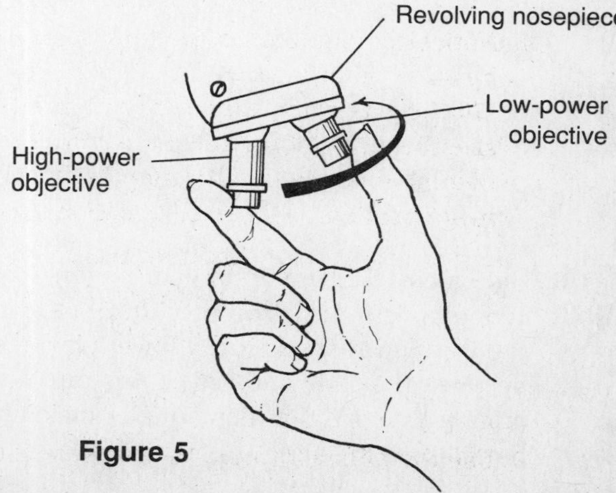

Figure 5

3. If your microscope has an electric light source, plug in the cord and turn on the light. If your microscope has a mirror, turn the mirror toward a light source such as a desk lamp or window. **CAUTION:** *Never use the sun as a direct source of light.* Look through the eyepiece. Adjust the diaphragm to permit sufficient light to enter the microscope. The white circle of light you see is the field of view. If your microscope has a mirror, move the mirror until the field of view is evenly illuminated.

4. Place a prepared slide on the stage so that it is centered over the stage opening. Use the stage clips to hold the slide in position. Turn the low-power objective into place. Look at the microscope from the side and turn the coarse adjustment so that the low-power objective is as close as possible to the stage without touching it.

5. Look through the eyepiece and turn the coarse adjustment to move the low-power objective away from the stage until the object comes into focus. To avoid eyestrain, keep both eyes open while looking through a microscope. **CAUTION:** *To avoid moving the objective into the slide, never lower the objective toward the stage while looking through the eyepiece.*

6. Turn the fine adjustment to bring the object into sharp focus. You may wish to move the diaphragm to adjust the amount of light so that you can see the object more clearly. In the appropriate place in Observations, draw what you see through the microscope. Note the magnification.

7. Look at the microscope from the side and rotate the nosepiece until the high-power objective clicks into position. Look through the eyepiece. Turn the fine adjustment to bring the object on the slide into focus. **CAUTION:** *Never use the coarse adjustment when focusing the high-power objective lens. This could break your slide or damage the lens.* In the appropriate place in Observations, draw what you see through the microscope. Note the magnification.

8. Remove the slide. Move the low-power objective into position.

Part C. Preparing a Wet Mount

1. Use a pair of scissors to cut a letter "e" from a piece of newspaper. Cut out the smallest letter "e" you can find. Position the "e" on the center of a clean glass slide.

2. Use a medicine dropper to place one drop of water on the cut piece of newspaper. See Figure 6.

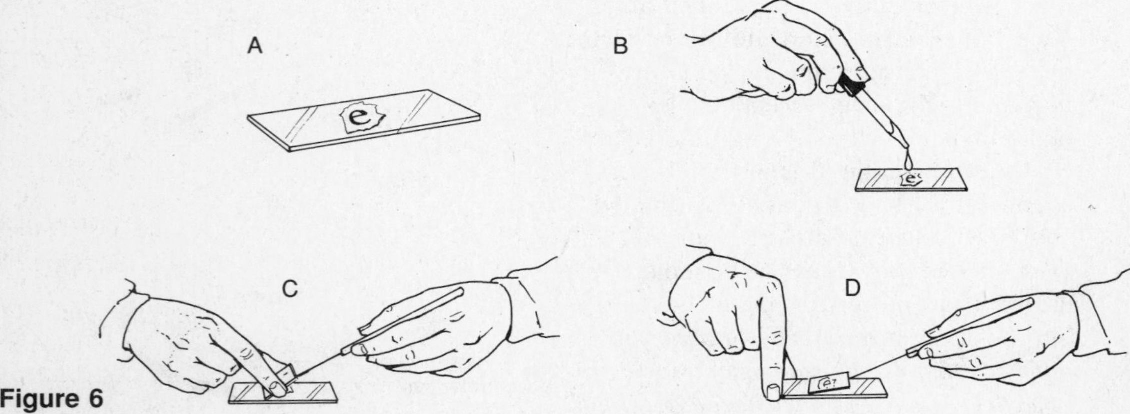

Figure 6

3. Hold a clean coverslip in your fingers as shown in Figure 6. Make sure the bottom edge of the coverslip is in the drop of water. Use a dissecting needle to slowly lower the coverslip onto the wet newspaper. Slowly lowering the coverslip prevents air bubbles from being trapped between the slide and the coverslip. The type of slide you have just made is called a wet mount. Practice the technique of making a wet mount until you can do so without trapping air bubbles on the slide.

Name ______________________ Class __________ Date __________

4. Center the wet mount of the letter "e" on the stage with the "e" in its normal upright position. **Note:** *Make sure the bottom of the slide is dry before you place it on the stage.* Turn the low-power objective into position and bring the letter "e" into focus. In the appropriate place in Observations, draw the letter "e" as seen through the microscope. Note the magnification.

5. Turn the high-power objective into position and bring the letter "e" into focus. In the appropriate place in Observations, draw the letter "e" as seen through the microscope. Note the magnification.

6. Rotate the nosepiece until the low-power objective clicks back into position and bring the letter "e" into focus. While looking through the eyepiece, move the slide to the left. Notice the way the letter seems to move. Now move the slide to the right. Again notice the way the letter seems to move. Move the slide up and observe the direction the letter moves. Move the slide down and observe the direction the letter moves.

7. Take apart the wet mount. Clean the slide and coverslip with soap and water. Carefully dry the slide and coverslip with paper towels and return them to their boxes.

8. Rotate the low-power objective into position and use the coarse adjustment to place it as close to the stage as possible without touching. Carefully pick up the microscope and return it to its storage area.

Observations

1. Label the parts of the microscope shown below.

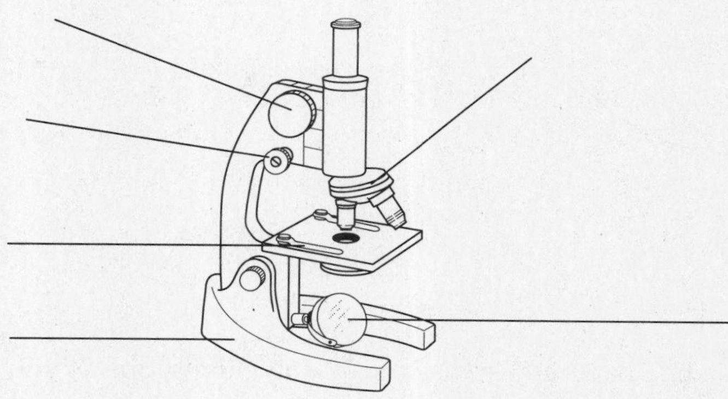

2. Fill in the magnification of each objective and the eyepiece of your microscope. To determine the total magnification, multiply the magnification of each objective by the magnification of the eyepiece.

Data Table

Objective	Magnification of Objective	Magnification of Eyepiece	Total Magnification
Low power			
High power			
Other			

3. Make a detailed drawing of the object on your prepared slide as seen under low power and high power.

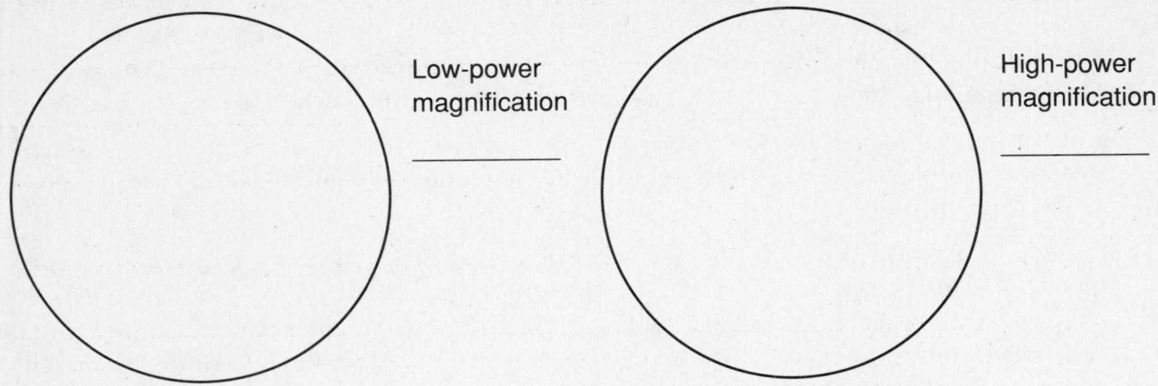

4. Make a detailed drawing of the letter "e" as seen under low power and high power.

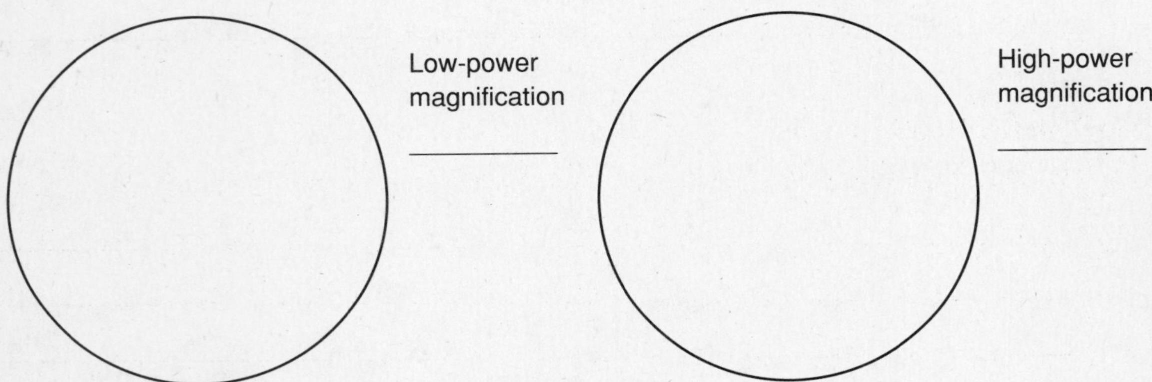

Analysis and Conclusions

1. Why do you place one hand under the base of the microscope as you carry it?

____

____

2. What kind of light source do you have on your microscope? _______________

____

3. How is the image of an object seen through the high-power objective different from the image seen through the low-power objective? _______________

____

____

Name ______________________ Class __________ Date __________

4. How does the letter "e" as seen through the microscope differ from the way an "e" normally appears? ______________________

5. When you move the slide to the left, in what direction does the image appear to move?

6. When you move the slide up, in what direction does the image appear to move?

Critical Thinking and Application

1. Explain why a specimen to be viewed under the microscope must be thin.

2. Why should a glass slide and a coverslip be held by their edges?

3. Why should you use a piece of lens paper only once? ______________________

4. Why is it a good idea to place your microscope at least 10 cm from the edge of the table?

5. Suppose you were observing an organism through the microscope and noticed that it moved toward the bottom of the slide and then it moved to the right. What does this tell you about the actual movement of the organism? ______________________________

____

____

____

Going Further

1. Obtain some common objects, such as a piece of cotton, a piece of nylon, a small piece of a color photograph from a magazine, and so on. View each object under the low-power and high-power objectives of the microscope. Make a drawing for each object. Describe the appearance of the objects when viewed under a microscope. How did each object differ from the way you see it with the unaided eye?
2. Use an appropriate resource to investigate the development of the microscope and the advancements that have been made in microscope technology. Write a report or make an oral presentation as directed by your teacher.

Name ______________________ Class ____________ Date ____________

LABORATORY SKILLS 6

Using the Bunsen Burner

Pre-Lab Discussion

Sometimes a biologist needs to heat materials. In the laboratory, one of the most efficient ways to do this is to use a Bunsen burner. Bunsen burners are made in a variety of designs. In every one, however, a mixture of air and gas is burned. In most Bunsen burners, the amounts of air and gas can be controlled. In some laboratories, electric hot plates or portable liquid-petroleum burners are used instead of Bunsen burners.

In this investigation, you will learn the parts of the Bunsen burner and their functions. You will also learn how to use the Bunsen burner safely in the laboratory.

Problem

How can the Bunsen burner be safely used to heat materials in the laboratory?

Materials *(per group)*

Bunsen burner	Beaker tongs
Ring stand	Iron ring
2 250-mL beakers	100-mL graduated cylinder
Wire gauze	Flint striker or matches
Metric ruler	Clock with second hand
Heat-resistant gloves	

Safety

Put on a laboratory apron if one is available. Put on safety goggles. Handle all glassware carefully. Tie back loose hair and clothing when using the Bunsen burner. Use extreme care when working with heated equipment or materials to avoid burns. Note all safety alert symbols next to the steps in the Procedure and review the meanings of each symbol by referring to the symbol guide on page 10.

Procedure

1. Examine your burner when it is *not* connected to the gas outlet. If your burner is the type that can easily be taken apart, unscrew the barrel from the base and locate the parts shown in Figure 1. As you examine the parts, think about their functions.
    - The *barrel* is the area where the air and gas mix.
    - The *collar* can be turned to adjust the intake of air. If you turn the collar so that the holes are larger, more air will be drawn into the barrel.

- The *air intake openings* are the holes in the collar through which air is drawn in.
- The *base* supports the burner so that it does not tip over.
- The *gas intake tube* brings the supply of gas from the outlet to the burner.
- The *spud* is the small opening through which the gas flows. The small opening causes the gas to enter with great speed.

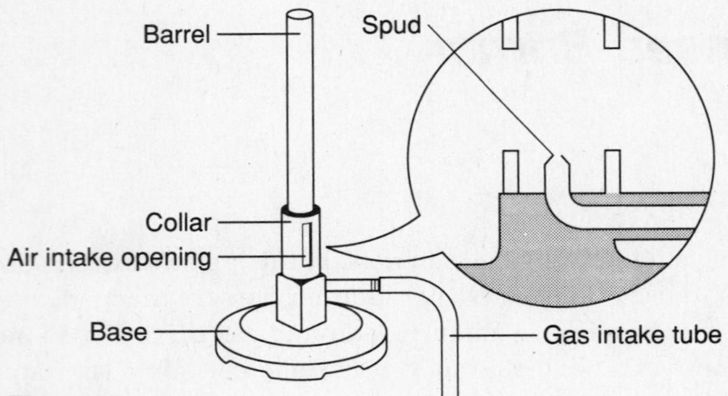

Figure 1

2. Reassemble the Bunsen burner if necessary and connect the gas intake tube to the gas outlet. **CAUTION:** *Put on safety goggles.* Make sure that the burner is placed away from flammable materials.
3. Adjust the collar so that the air intake openings are half open. If you use a match to light the burner, light the match and hold it about 2 cm above and just to the right of the barrel. Hold the match in this position while you open the gas outlet valve slowly until it is fully open. **CAUTION:** *To avoid burns on your hands, always use extreme care when handling lighted matches.* The burner can also be turned off by using the valve. Do not lean over the burner when lighting it.
4. If you use a flint striker to light the burner, hold the striker in the same position you would hold a lighted match. To light the burner with a striker, you must produce a spark at the same time you open the gas valve.
5. Practice lighting the burner several times. Every member of your group should be given the opportunity to light the burner.
6. The most efficient and hottest flame is blue in color and has distinct regions as shown in Figure 2. Adjust the collar so that the flame is blue and a pale blue inner cone is visible.

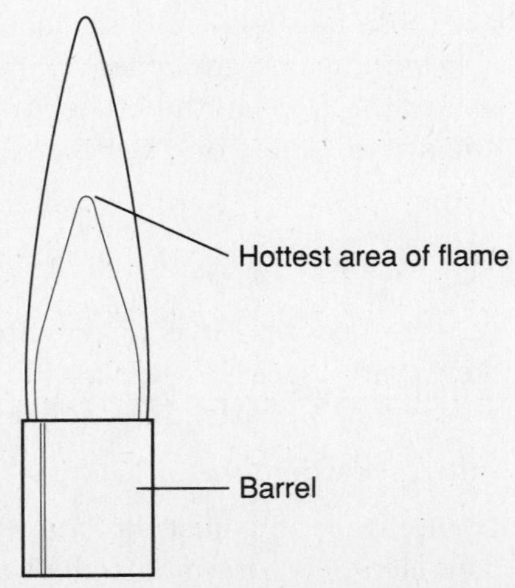

Figure 2

Name ______________________ Class ____________ Date ____________

7. Adjust the flow of gas until the flame is about 6 cm high. Some burners have a valve in the base to regulate the flow of gas, but the flow of gas can always be adjusted at the gas outlet valve. After adjusting the flow of gas, shut off the burner. Leave your safety goggles on for the remainder of the investigation.

8. Arrange the apparatus as pictured in Figure 3.

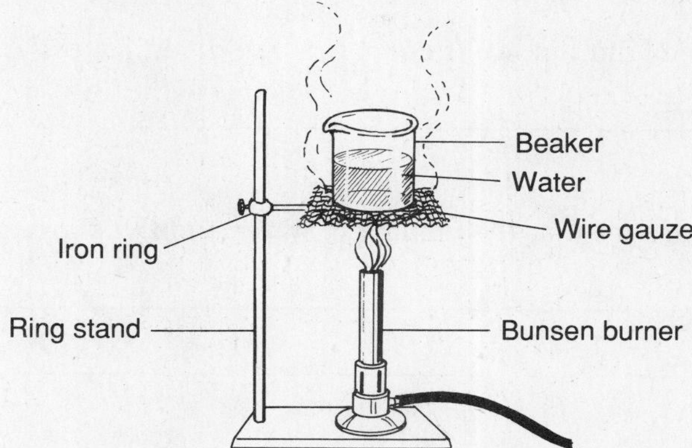

Figure 3

9. Adjust the iron ring so that the bottom of the beaker is about 2 cm above the mouth of the burner barrel. Measure 100 mL of water in the graduated cylinder and pour it into one of the beakers.

10. Light the burner and heat the beaker of water. The bottom of the beaker should just be touching the top of the inner cone of the flame. In the Data Table, record the time it takes for the water to start boiling rapidly. Using the beaker tongs, carefully remove the beaker and pour out the water.

11. Repeat the procedure with the other beaker at a height of about 8 cm above the mouth of the barrel. **CAUTION:** *Be very careful when raising the iron ring. If it is too warm to touch, use heat-resistant gloves.* In the Data Table, record the time it takes for the water to start boiling rapidly at this height. **Note:** *Be sure that the starting temperature of the water is the same in each trial.*

Observations

Data Table

Height Above Burner (cm)	**Time to Boil** (min)
2	
8	

Analysis and Conclusions

1. What would happen if the air intake openings were made very small? ______________________

____

2. If the burner does not light after the gas outlet valve is opened, what might be wrong?

3. Where is the hottest part of the flame of a Bunsen burner?

4. At what height, 2 cm or 8 cm, did the water come to a rapid boil faster?

5. Why is it important to make sure that the volume of water and the starting temperature are the same in each trial?

Critical Thinking and Application

1. Why is it important to wear safety goggles when using a Bunsen burner?

2. Why is it important to tie back loose hair and clothing when using a Bunsen burner?

3. In addition to the items mentioned in questions 1 and 2, what other safety precautions should be followed before lighting a Bunsen burner?

4. State two reasons why a blue flame is preferred over a yellow flame in a Bunsen burner.

Going Further

Test the ability of different kinds of laboratory burners, such as a hot plate, to heat water to boiling. Determine if there is a difference in the efficiency with which different burners are able to heat objects.

Name ______________________ Class __________ Date __________

LABORATORY SKILLS 7

Preparing Laboratory Solutions

Pre-Lab Discussion

A *solution* is a type of mixture in which one substance dissolves in another. In a solution, the substance that is dissolved is called the *solute*. The substance that does the dissolving is called the *solvent*. The most common solvent is water. Most solutions cannot easily be separated by simple physical means such as filtering.

Solutions in which water is the solvent, or *aqueous solutions*, are important to all types of living organisms. Marine microorganisms spend their entire lives in the ocean, an aqueous solution of water, salt, and other substances. Most of the nutrients needed by plants are in aqueous solution in moist soil. Plasma, the liquid part of the blood, is an aqueous solution containing dissolved nutrients and gases.

In this investigation, you will learn some of the techniques used to prepare laboratory solutions. You will also learn some of the proper uses of a triple-beam balance and a filtering apparatus.

Problem

What are some of the different ways in which laboratory solutions can be prepared?

Materials *(per group)*

Sodium chloride
10 mL of red food coloring
100-mL graduated cylinder
Filter paper
Funnel
2 100-mL beakers
Weighing paper
Triple-beam balance
Scoop
10-mL graduated cylinder
Ring stand
Iron ring

Safety

Put on a laboratory apron if one is available. Put on safety goggles. Handle all glassware carefully. Always use special caution when working with laboratory chemicals, as they may irritate the skin or cause staining of the skin or clothing. Never touch or taste any chemical unless instructed to do so. Note all safety alert symbols next to the steps in the Procedure and review the meanings of each symbol by referring to the symbol guide on page 10.

Procedure

Part A. Preparing the Weight/Volume Solution

To prepare a solution of a given percentage, dissolve the number of grams of solid solute equal to the percentage in enough water to make 100 mL of the solution. Use the following steps to prepare a 5% sodium chloride solution.

1. Place a piece of weighing paper on the pan of the triple-beam balance and find its mass.
2. Add exactly 5 g to the value of the mass of the weighing paper and move the riders of the balance to this number.
3. Using the scoop, add a small amount of sodium chloride at a time to the paper on the balance until the pointer rests on zero.
4. Add the 5 grams of sodium chloride to the 100-mL graduated cylinder.
5. Add enough water to bring the volume of the solution to 100 mL. Answer question 1 in Observations.
6. Dispose of this solution according to your teacher's instructions.

Part B. Preparing a Volume/Volume Solution

To prepare a solution of a given percentage, dissolve the number of milliliters of liquid solute equal to the percentage in enough solvent to make 100 mL of the solution. Use the following steps to prepare a 10% colored water solution.

1. Measure 10 mL of red food coloring in the small graduated cylinder and pour it into the large graduated cylinder. **CAUTION:** *Use special caution when working with red food coloring to avoid staining your hands or clothing.*
2. Add enough water to the large graduated cylinder to bring the volume to 100 mL. Answer question 2 in Observations.
3. Keep this solution for use in Part C of this investigation.

Part C. Reducing the Concentration of a Solution

To reduce the concentration of an existing solution, pour the number of milliliters of the existing solution that is equal to the percentage of the new concentration into a graduated cylinder. Add enough solvent to bring the volume in milliliters to an amount equal to the percentage of the original solution. Use the following steps to reduce a 10% colored water solution to a 1% solution.

1. Pour the 10% colored water solution you prepared in Part B into a 100-mL beaker.
2. Measure 1 mL of the 10% solution in the small graduated cylinder.
3. Add enough water to the graduated cylinder to bring the volume to 10 mL. Answer question 3 in Observations.
4. Dispose of the 1% solution in the graduated cylinder according to your teacher's instructions. Keep the 10% solution in the beaker for use in Part D of this investigation.

Part D. Filtering

1. Prepare filter paper as shown in Figure 1. Fold a circle of filter paper across the middle. Fold the resulting half-circle to form a quarter-circle. Open the folded paper into a cone, leaving a triple layer on one side and a single layer on the other.
2. Support a funnel as shown in Figure 2. Place the cone of the filter paper in the funnel and wet the paper so that it adheres smoothly to the walls of the funnel. Set a clean beaker beneath the funnel in such a way that the stem of the funnel touches the side of the beaker.
3. Pour the 10% colored water solution prepared in Part B slowly into the funnel. Do not let the mixture overflow the filter paper. As the mixture filters through the filter paper, record your observations in the Data Table.

Name ______________________ Class __________ Date __________

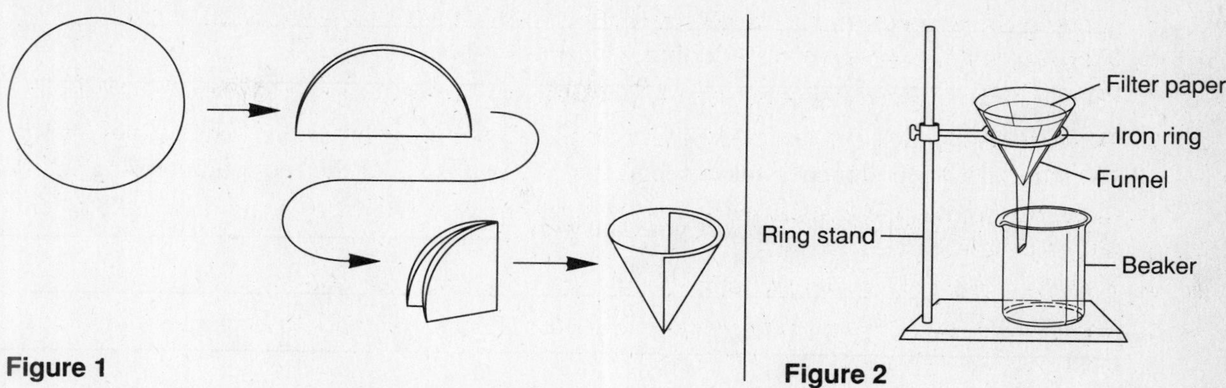

Figure 1

Figure 2

4. After all of the solution has passed through the filter paper into the beaker, observe the appearance of the filter paper. Record your observations in the Data Table.
5. Carefully remove the filter paper from the funnel and dispose of it and the colored water solution according to your teacher's instructions.

Observations

Data Table

Appearance of Liquid Before Filtering	Appearance of Liquid After Filtering	Appearance of Filter Paper After Filtering

1. What happens to the sodium chloride crystals as water is mixed with them?

____

2. What happens to the red food coloring as water is mixed with it? ______________

____

3. What differences do you observe between the 10% and 1% solutions of colored water?

____

Analysis and Conclusions

1. Why is a chemical placed on a piece of weighing paper instead of directly on the pan of the balance when its mass is being measured? ______________

____

____

2. Relate the colors of the 10% and 1% colored water solutions to the number of solute molecules each solution contains. ______

______

______

3. Was the filter paper successful in separating the two parts of the red food coloring solution? Use your observations to support your answer. ______

______

______

Critical Thinking and Application

1. Describe the procedure needed to prepare a 30% sugar solution. ______

______

______

______

______

2. Describe the procedure needed to produce a 20% liquid bleach solution.

______

______

______

3. Describe the procedure needed to reduce an 80% starch solution to a 20% solution.

______

______

______

______

Going Further

Solution concentrations can be expressed in a number of different ways, including molarity (the number of moles of solute per liter of solution) and molality (the number of moles of solute per kilogram of solvent). Using a chemistry reference text, describe the procedures used to prepare 1 molar and 1 molal concentrations.

Name ______________________ Class ____________ Date ____________

LABORATORY SKILLS 8

Using Graphing Skills

Pre-Lab Discussion

Recorded data can be plotted on a graph. A graph is a pictorial representation of information recorded in a data table. It is used to show a relationship between two or more different factors. Two common types of graphs are line graphs and bar graphs.

In this investigation, you will interpret and construct a bar graph and a line graph.

Problem

How do you correctly interpret and construct a line graph and a bar graph?

Materials

No special materials needed

Procedure

Part A. Interpreting Graphs

1. The type of graph that best shows the relationship between two variables is the line graph. A line graph has one or more lines connecting a series of points. See Figure 1. Along the horizontal axis, or x–axis, you will find the most consistent variable in the experiment. Along the vertical axis, or y-axis, you will find the other variable.

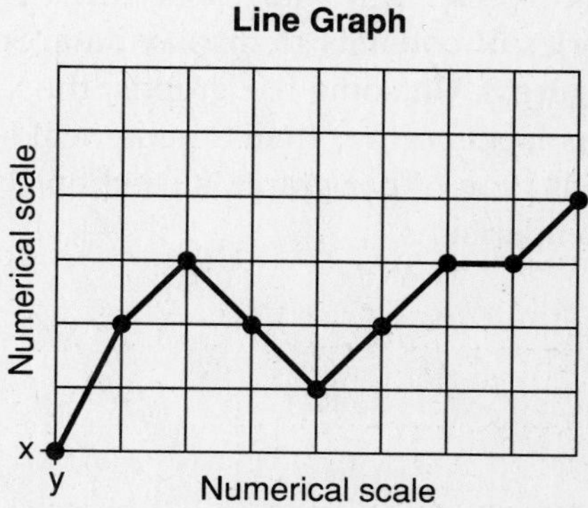

Figure 1

2. Use the line graph in Figure 2 to answer questions 1 through 6 in Observations.

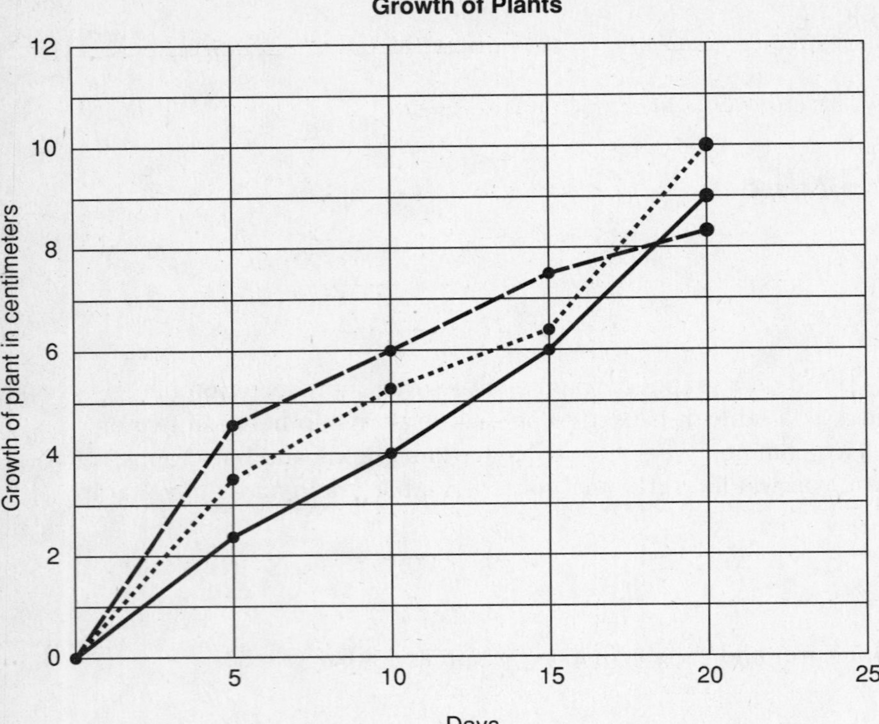

Figure 2

3. A bar graph is another way of showing relationships between variables. A bar graph also contains an x-axis and a y-axis. But instead of points, a bar graph uses a series of columns to display data. See Figure 3. On some bar graphs, the x-axis has labels rather than a numerical scale. This type of bar graph is used only to show comparisons.

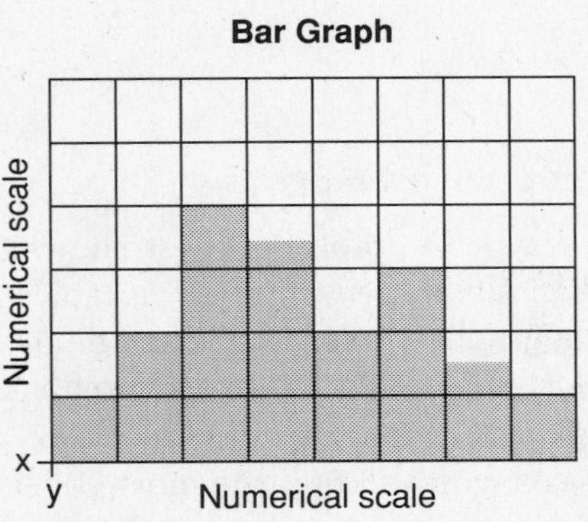

Figure 3

Name ______________________ Class __________ Date __________

4. Use the bar graph in Figure 4 to answer questions 7 through 11 in Observations.

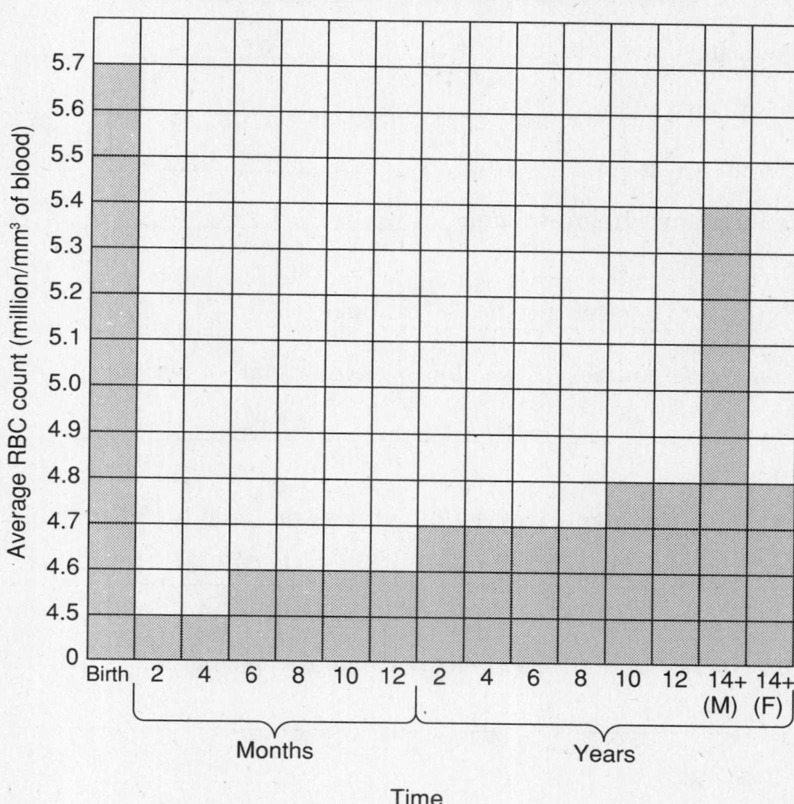

Figure 4

Part B. Constructing Graphs

1. When plotting data on a graph, you must decide which variable to place along the x-axis and which variable to place along the y-axis. Label the axes of your graph accordingly. Then you must decide on the scale of each axis; that is, how much each unit along the axis represents. Scales should be chosen to make the graph as large as possible within the limits of the paper and still include the largest item of data. If the scale unit is too large, your graph will be cramped into a small area and will be hard to read and interpret. If the scale unit is too small, the graph will run off the paper. Scale units should also be selected for ease of locating points on the graph. Multiples of 1, 2, 5, or 10 are easiest to work with.
2. Use the information recorded in Data Table 1 to construct a line graph on the grid provided in number 12 of Observations. You should label each axis, mark an appropriate scale on each axis, plot the data, connect the points, and give your graph a title.
3. Use the information recorded in Data Table 2 to construct a bar graph on the grid provided in number 13 of Observations. You should label each axis, mark an appropriate scale on each axis, plot the data, darken the columns of the graph, and give your graph a title.

Observations

Part A. Interpreting Graphs

Use the line graph in Figure 2 to answer questions 1 through 6.

1. Which plant grew the tallest? ______________________________

2. How many plants grew to be at least 6 cm tall? ______________________________

3. Which plant grew the fastest in the first five days? ______________________________

4. Which line represents plant 2? ______________________________

5. After 10 days, how much had plant 3 grown? ______________________________

6. How long did it take for plant 1 to grow 6 cm? ______________________________

Use the bar graph in Figure 4 to answer questions 7 through 11.

7. At birth, what is the average number of red blood cells per mm^3 of blood?

____

8. What appears to happen to the number of red blood cells between birth and 2 months?

____

9. What happens to the number of red blood cells between the ages of 6 and 8 years?

____

10. Between what ages is a human likely to have 4.6 million red blood cells?

____

11. After 14 years of age, do males or females have a higher red blood cell count?

____

Part B. Constructing Graphs

Data Table 1 Breathing Rate of the Freshwater Sunfish

Temperature (°C)	Breathing Rate (per minute)
10	15
15	25
18	30
20	38
23	60
25	57
27	25

Name ______________________ Class ____________ Date ____________

12. Use the grid below to construct a line graph for the information shown in Data Table 1.

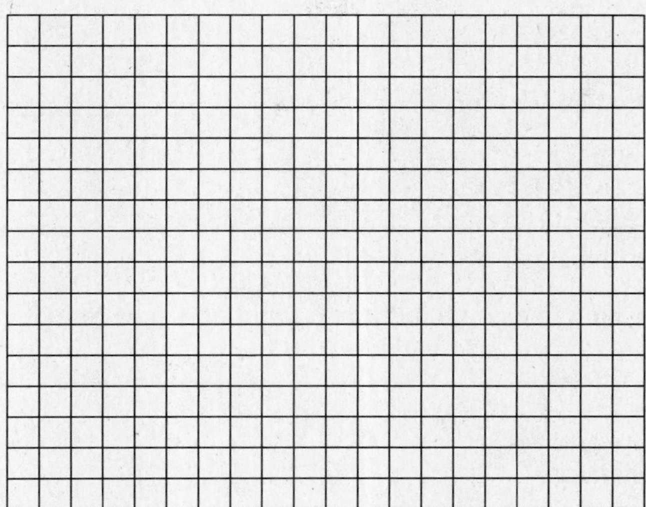

Data Table 2 Average Rainfall in Willamette Valley

Month	Jan.	Feb.	Mar.	April	May	June	July	Aug.	Sept.	Oct.	Nov.	Dec.
Rainfall (mL)	15	21	28	24	16	8	2	1	2	3	5	10

13. Use the grid below to construct a bar graph for the information shown in Data Table 2.

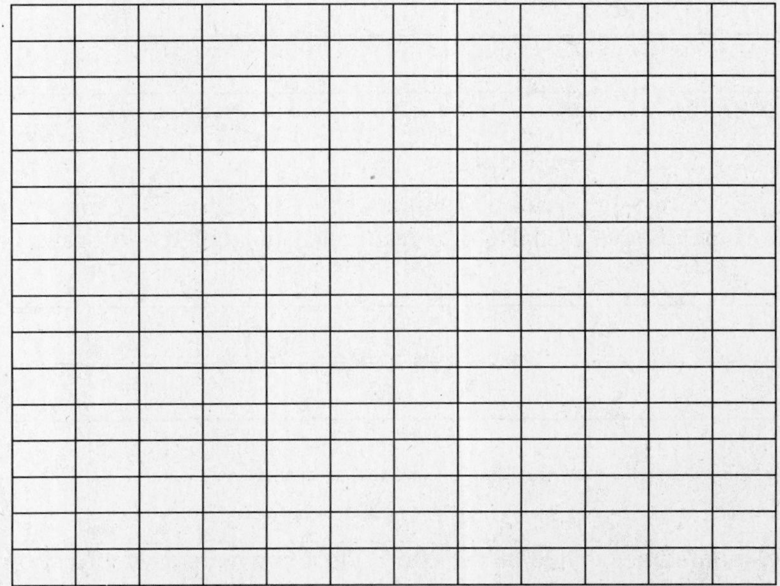

Analysis and Conclusions

1. How is a graph similar to a data table? ______________________________

____

____

2. How is a line graph different from a bar graph? ______________________________

____

____

3. Does a steep curve on a line graph indicate a rapid or a slow rate of change?

____

Critical Thinking and Application

1. You are conducting an experiment to measure the gain in mass of a young mouse over a ten-week period. In constructing a graph to represent your data, which variable should you place along the x-axis and which variable should you place along the y-axis? Explain your answer.

____

____

____

____

2. What is an advantage of using multiple lines on a line graph? (See Figure 2.)

____

____

____

3. Why is it important to have all parts of a graph clearly labeled and drawn?

____

____

____

Going Further

A circle graph (sometimes called a "pie chart") is a convenient way to show the relative sizes of the parts that together form a whole body of data. Look through magazines and newspapers to find examples of circle graphs. Construct a chart listing the similarities and differences between circle graphs, line graphs, and bar graphs.

Name ______________________ Class __________ Date __________

A Metric Scavenger Hunt

Pre-Lab Discussion

In many biology investigations, precise measurements must be made before conclusions can be formed. For everyday measuring we still use English units such as the inch, quart, and pound. But for scientific work, the metric system, or International System of Units (SI), is used. The universal language of science, the metric system allows scientists everywhere to share information.

In this investigation, you will have to "think metric" as you select objects that will correctly match the descriptions on your "metric scavenger hunt."

Problem

How closely can you estimate the metric measurements of everyday items?

Materials *(per group)*

Metric ruler
10-mL graduated cylinder
100-mL graduated cylinder
Triple-beam balance
Celsius thermometer

Safety

Handle all glassware carefully. Note all safety alert symbols next to the steps in the Procedure and review the meanings of each symbol by referring to the symbol guide on page 10.

Procedure

1. Carefully read the description of each object you are to find in the "metric scavenger hunt."

METRIC SCAVENGER HUNT

a. An object that when filled to the top with water has a volume between ______ and ______ milliliters.

b. A metallic object that has a mass less than ______ grams.

c. A coin that has a diameter (the distance from one side of a circle through the center to the other side) between ______ and ______ millimeters.

d. An object with a length between ______ and ______ centimeters.

e. An item that when half filled with water has a volume between ______ and ______ liters.

f. An object with a normal temperature of 37°C. (*Hint:* You brought this object to class!)

g. An object that has a mass between ______ and ______ centigrams.

h. An object that has a volume between ______ and ______ cubic centimeters.

i. An object with a length between ______ and ______ decimeters. (You may not be able to take this object to your lab station.)

j. A substance with a temperature between ______ °C and ______ °C.

k. A person with a mass between ______ and ______ kilograms.

2. Complete Data Table 1 by listing several objects that you think might fit each description. (*Hint:* Look around the room, check your purse and pockets, and do not forget items you have brought to class.)

3. From your list of possible objects, select the one item that you feel will best fit each description. Take the selected item to your lab station.

4. Using the proper measuring tool, make the appropriate measurement and record your answer in Data Table 2. Remember to indicate the correct units.

Observations

Data Table 1

Description	Possible Objects
a	
b	
c	
d	
e	
f	
g	
h	
i	
j	
k	

Name ______________________ Class __________ Date __________

Data Table 2

Description	Object	Measurement	Was Object Within Given Range?
a			
b			
c			
d			
e			
f		Not required	
g			
h			
i			
j			
k			

Analysis and Conclusions

1. Which objects were easiest to find? Hardest? Why? ______________________

2. Was it easy to "think metric"? Why? ______________________

3. Would students in other parts of the world have difficulty "thinking metric"? Why?

4. Why is it important that all scientific measurements be made using metric units?

Critical Thinking and Application

1. The metric system is based on the number 10 and is a decimal system of measurement. What United States system is based on the number 10? ______________________

____

____

2. The United States is one of the few countries in the world that does not use the metric system for everyday measuring. Do you think Congress should pass a law requiring the use of the metric system throughout the country? Explain your answer. ______________________

____

____

3. If the United States did convert to the metric system, how would you implement this change?

____

Going Further

1. In the laboratory investigation, you have worked with SI units for measuring length, mass, temperature, and volume. Other SI base units include the second, ampere, kelvin, mole, and candela. Find out what these SI base units are used to measure and report your findings to your classmates.
2. Conduct a survey to determine the familiarity of different age groups with the metric system. Questions might include the base units for different types of measurements and the prefixes used for making smaller and larger measurements.

Name ______________________ Class ____________ Date ____________

Observing the Uncertainty of Measurements

Pre-Lab Discussion

Biologists use a wide variety of laboratory tools to make scientific measurements. Some basic tools of measurement are the meterstick, triple-beam balance, graduated cylinder, and Celsius thermometer.

The accuracy of a scientific measurement depends on three things: the smallest unit on the measuring scale, the ability of the observer to read the scale properly, and the degree of precision of the measuring instrument and scale.

In this investigation, you will learn about the uncertainty of measurement and how accurately matter can be measured using common laboratory instruments.

Problem

How accurately can matter be measured?

Materials *(per station)*

Station 1: meterstick
Station 2: meterstick
Station 3: metric ruler
regular object
Station 4: 100-mL graduated cylinder
150-mL beaker of colored liquid
Station 5: triple-beam balance
small pebble
Station 6: 100-mL graduated cylinder
150-mL beaker of water
irregular object
Station 7: Celsius thermometer
250-mL beaker of ice and water
paper towel

Safety

Put on a laboratory apron if one is available. Handle all glassware carefully. Note all safety alert symbols next to the steps in the Procedure and review the meanings of each symbol by referring to the symbol guide on page 10.

Procedure

1. Station 1: Use the meterstick to measure the length and width of your science classroom. If the room has an irregular shape, measure the longest width and the longest length. Express your measurements in meters and record them in the Data Table.

2. Station 2: Use the meterstick to measure the length and width of your desk or lab table. If the table has an irregular shape, measure the longest width and the longest length. Express your measurements in centimeters and record them in the Data Table.

3. Station 3: Use the metric ruler to find the volume of the regular object. Volume is found by multiplying the length times the width times the height of the object. Express the volume in cubic centimeters and record it in the Data Table.

4. Station 4: Use the graduated cylinder to find the volume of the colored liquid in the beaker. Remember to always read a graduated cylinder at the bottom curve of the meniscus. Pour the liquid back into the beaker. Express your measurement in milliliters and record it in the Data Table.

5. Station 5: Make certain that the riders on the triple-beam balance are moved all the way to the left and that the pointer rests on zero. Place the pebble on the pan on the triple-beam balance. Move the riders until the pointer is at zero. Express your measurement in grams and record it in the Data Table. Remove the pebble and return all riders to the far left of the balance.

6. Station 6: Fill the graduated cylinder half full with water from the beaker. Find the volume of the irregular object. Express the volume of the object in cubic centimeters and record it in the Data Table. Carefully remove the object from the graduated cylinder. Pour the water back into the beaker.

7. Station 7: Use the Celsius thermometer to find the temperature of the ice water. Express the temperature in degrees Celsius and record it in the Data Table. Remove the thermometer and carefully dry it with a paper towel.

8. Your teacher will construct a large class data table on the chalkboard. Have one member from your group record your data on the class data table.

Observations

Data Table

Station	Measurement
1	
2	
3	
4	
5	
6	
7	

1. What is the smallest unit shown on the meterstick? ____________________

2. What is the smallest unit shown on the metric ruler? ____________________

3. What is the smallest unit shown on the graduated cylinder? ____________________

____

Name ______________________ Class __________ Date __________

4. What is the smallest unit on the triple-beam balance? ______________________

5. What is the smallest unit on the Celsius thermometer? ______________________

____

Analysis and Conclusions

1. Study the class data table. Do each group's measurements have the exact same value for each station? ______________________

____

2. Which stations had measurements that were most nearly alike? Explain why these measurements were so similar. ______________________

____

____

____

3. Which stations had measurements that were most varied? Explain why these measurements were so varied. ______________________

____

____

____

Critical Thinking and Application

1. Why is it important for the riders on a triple-beam balance to always start at the far left of the balance? ______________________

____

____

2. Of the following graduated cylinders—100 mL, 25 mL, or 10 mL—which would you use to most accurately measure 8 mL of a liquid? Explain your answer.

____

____

____

3. Which is more accurate for measuring liquid volume, a beaker or a graduated cylinder? Explain your answer. ____

____

____

4. Describe two variables that might affect the accuracy of your measurement of volume when using a metric ruler to measure the dimensions of an object. ______________________

____

____

____

____

Going Further

1. Calculate the average (mean) of the measurements at each station. To do this, add each measurement to find a total and then divide by the number of measurements. An average of several measurements is usually more precise than a single measurement.
2. Repeat the investigation, this time aiming for the best precision and accuracy possible. Record the measurements as you did before and calculate a class average for each station. Compare the measurements and averages for the two trials. Discuss factors that led to more precise measurements.

Name ______________________ Class __________ Date __________

Characteristics of Life

Pre-Lab Discussion

Dogs, cats, people, and trees are living things. Rocks, water, and air are nonliving things. Distinguishing between living and nonliving things seems simple, yet the differences have been a source of argument among scientists for many years. But scientists agree that all living things are made up of one or more cells, reproduce, grow and develop, obtain and use energy, and respond to the environment.

To help you observe some of the characteristics of life, you will work with bromthymol blue, an indicator that is used to determine the presence of carbon dioxide. Carbon dioxide is produced when many living organisms burn their food to produce the energy that allows them to carry out the processes of life.

In this investigation, you will observe characteristics that indicate yeast are living organisms.

Problem

What characteristics of life can you observe that indicate yeast are living organisms?

Materials *(per group)*

Bromthymol blue in dropper bottle
Straw
4 large test tubes
3 small test tubes
3 one-hole rubber stoppers
with glass tube inserts
to fit large test tubes
Glass-marking pencil
3 pieces of plastic hose, 30–40 cm long
Test tube rack
Graduated cylinder
2 glass slides
2 coverslips
2 medicine droppers
Microscope
Yeast and molasses mixture
Yeast and water mixture
Water and molasses mixture

Safety

Put on a laboratory apron if one is available. Put on safety goggles. Handle all glassware carefully. Always use special caution when working with chemicals, as they may irritate the skin or cause staining of the skin or clothing. Never touch or taste any chemical unless instructed to do so. Bromthymol blue is a dye that can stain your clothing and your skin. Note all safety alert symbols next to the steps in the Procedure and review the meanings of each symbol by referring to the symbol guide on page 10.

Procedure

Part A

1. Add 30 drops of bromthymol blue to a test tube. **CAUTION:** *Bromthymol blue is a dye. Use care to prevent spills.*
2. Place a straw in the test tube, making sure that the straw extends into the bromthymol blue.
3. Gently blow bubbles through the straw into the bromthymol blue. **CAUTION:** *Do not inhale through the straw.* See Figure 1. Record your observations.

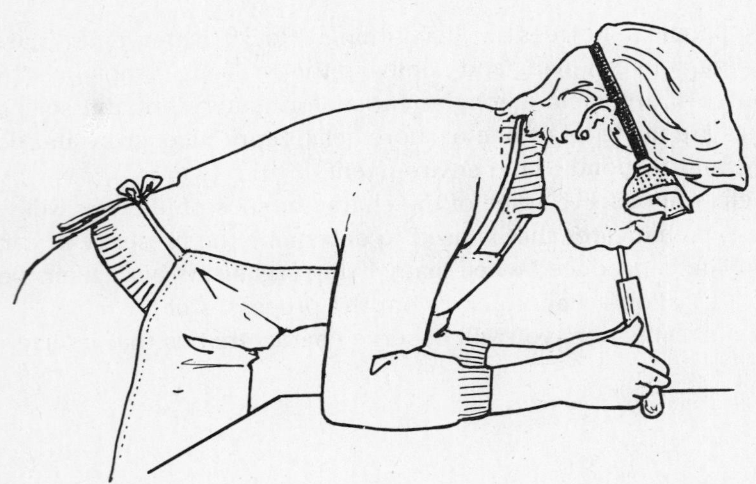

Figure 1

Part B

1. Using a glass-marking pencil, label three large test tubes A, B, and C.
2. Put 5 mL of bromthymol blue in each small test tube and place the test tubes in the test tube rack. **CAUTION:** *Bromthymol blue is a dye. Use care to prevent spills.*
3. Put 30 mL of the yeast and molasses mixture in test tube A.
4. Place a rubber stopper with a glass tube insert firmly into the mouth of test tube A.
5. Attach a plastic hose to the glass tube in the stopper. Place test tube A in the test tube rack next to one of the small test tubes containing bromthymol blue.
6. Put 30 mL of the yeast and water mixture in test tube B.
7. Place a rubber stopper with a glass tube insert firmly into the mouth of test tube B.
8. Attach a plastic hose to the glass tube in the stopper. Place test tube B in the test tube rack next to one of the small test tubes containing bromthymol blue.
9. Put 30 mL of the water and molasses mixture in test tube C.
10. Place a rubber stopper with a glass tube insert firmly in the mouth of test tube C.
11. Attach a plastic hose to the glass tube in the stopper. Place test tube C in the test tube rack next to the third small test tube containing bromthymol blue.

Name ______________________ Class __________ Date __________

12. Place the free end of the plastic hose attached to test tube A in the small test tube containing bromthymol blue, making sure it almost touches the bottom. Do the same thing for test tube B and its small test tube and again for test tube C and its small test tube. See Figure 2.

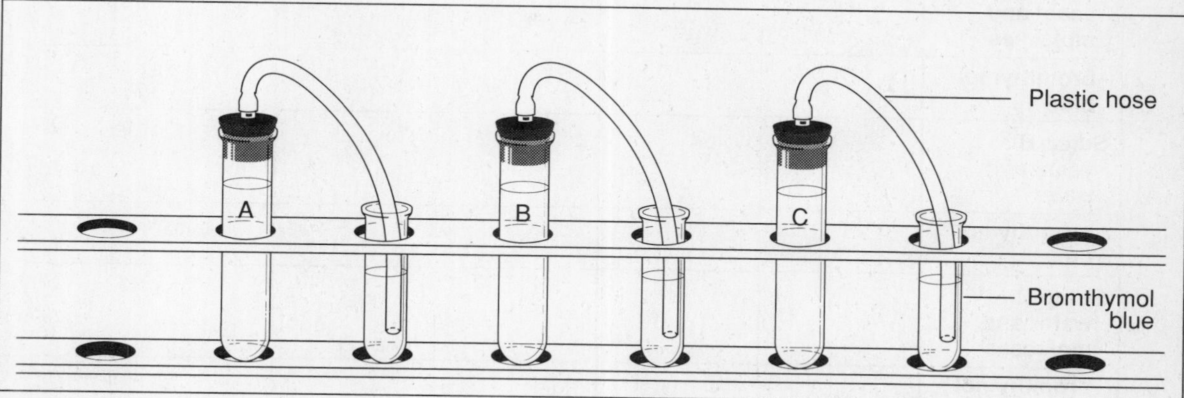

Figure 2

13. Record your initial observations in the Data Table.
14. After 30 minutes, again record your observations in the Data Table.

Part C

1. With a glass-marking pencil, label one glass slide A and a second glass slide B.
2. Using a clean medicine dropper, make a wet mount of the yeast and molasses mixture.
3. Using a clean medicine dropper, make a wet mount of the yeast and water mixture.
4. Examine each slide under the high-power objective of a microscope. You may need to adjust the diaphragm to see the yeast clearly. Be sure to note the presence of any *buds*—small yeast cells still attached to larger yeast cells. Record your observations.

Observations

Part A

1. What color was the bromthymol blue before you exhaled into the test tube?

____

2. What color was the bromthymol blue after you exhaled into the test tube?

____

Part B

Data Table

	Initial Observations			After 30 Minutes		
	Color	Presence of bubbles	Presence of CO_2	Color	Presence of bubbles	Presence of CO_2
Setup A yeast and molasses						
bromthymol blue						
Setup B yeast and water						
bromthymol blue						
Setup C water and molasses						
bromthymol blue						

In which test tube(s) did you observe a chemical reaction in which carbon dioxide was produced? How did you know a chemical reaction was occurring?

____

____

Part C

1. In the spaces below, draw what you observed on slides A and B under the microscope.

Magnification ____________ Magnification ____________

2. In which mixture were more buds present? ____

Name ______________________ Class __________ Date __________

Analysis and Conclusions

Part A

Why did the bromthymol blue change color when you exhaled into the test tube?

____

____

Part B

1. What does the production of carbon dioxide gas indicate about yeast?

____

2. How do you know that the carbon dioxide gas was produced by the yeast?

____

____

Part C

1. What does the presence of buds indicate about the yeast? ______________

____

2. Why were more buds present in one of the mixtures? ______________

____

____

3. What have you observed in this laboratory investigation (all three parts) that allows you to classify yeast as living organisms? ______________

____

____

Critical Thinking and Application

The baking industry uses yeast as an ingredient in bread dough to make the dough "rise." Using the recipe below, answer the questions that follow.

Biscuits

5 cups flour
$\frac{1}{4}$ cup sugar
3 teaspoons baking powder
1 teaspoon baking soda
2 cups buttermilk

1 cup shortening
1 package yeast
2 tablespoons water
1 teaspoon salt

Dissolve yeast in 2 tablespoons of warm, not hot, water. Let yeast and water mixture stand for 5 minutes or until bubbly. Mix dry ingredients. Cut in shortening. Add buttermilk and yeast mixture to the dry ingredients. Stir and then knead a few times. Roll out to $\frac{1}{2}$ inch thickness. Cut with a biscuit cutter. Bake at 400° for 12 to 15 minutes.

1. How does yeast make biscuit dough rise? ______________________________

____

____

2. The recipe instructs the cook to dissolve the yeast in "warm, not hot" water. Why?

____

3. Why does the recipe instruct the cook to let the yeast and water "stand for 5 minutes or until bubbly"? ______________________________

____

____

Going Further

1. Design an experiment that would test the effect of different temperatures on the rate of carbon dioxide production by yeast.
2. Write a report on how the alcohol industry uses yeast.

Name ______________________ Class ____________ Date ____________

4

Measuring with a Microscope

Pre-Lab Discussion

The microscope, developed more than three hundred years ago, is the basic tool of the biologist. The microscope enables biologists to investigate living things and objects that are too small to be seen with the unaided eye. The microscope is able to magnify these tiny specimens by means of lenses located in the eyepiece and objectives. The light microscope is also capable of revealing fine detail. This ability to reveal fine detail is known as resolving power. The type of microscope that you will be using throughout your study of biology is the compound light microscope.

Although it is interesting and informative to observe specimens under the microscope, it is often difficult to know the actual size of the object being observed. Magnification causes us to lose the idea of actual size. You cannot hold up a ruler to a paramecium or a plant cell while it is under the microscope. Therefore size must be measured indirectly—that is, it must be compared with the size of something you already know. The diameter of the microscope field seen through the eyepiece is a convenient standard to use. To measure objects under the microscope, a unit called the micrometer (μm) is used. One micrometer equals 0.001 millimeter.

In this investigation, you will develop skill in using the compound light microscope. You will also learn how to estimate the sizes of objects under the microscope.

Problem

How is the compound microscope used to make measurements of microscopic specimens?

Materials *(per group)*

Microscope	Transparent metric ruler
Lens paper	Prepared slides

Safety

Always handle the microscope with extreme care. You are responsible for its proper care and use. Use caution when handling glass slides as they can break easily and cut you. Note all safety alert symbols next to the steps in the Procedure and review the meanings of each symbol by referring to the symbol guide on page 10.

Procedure

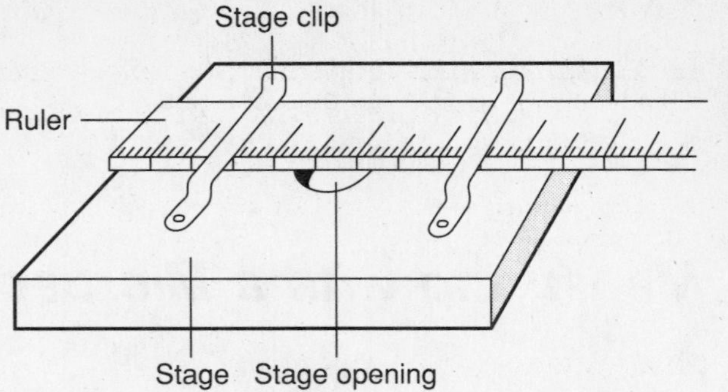

Figure 1

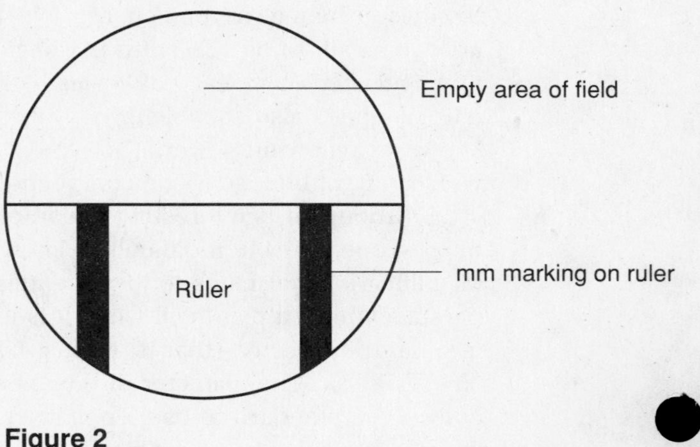

Figure 2

1. Take a microscope from the storage area and place it about 10 centimeters from the edge of the laboratory table.
2. Carefully clean the eyepiece and objective lenses with lens paper.
3. Examine the markings on a metric ruler. Decide which marks indicate millimeter lengths. Place the ruler on the stage so that it covers half of the stage opening, as shown in Figure 1.
4. Prepare your microscope for low-power observation of the ruler.
5. Look through the eyepiece. Focus on the edge of the ruler using the coarse adjustment. Adjust the position of the ruler so that the view in the low-power field is similar to Figure 2.
6. Place the center of one mark at the left side of the field of view. Make sure that the edge of the ruler is exactly across the center of the field. If the ruler sticks to your fingers, use the eraser end of a pencil to arrange it.
7. Note that 1 millimeter is the distance from the middle of one mark to the middle of the next mark. The diameter of the low-power field measures 1 millimeter plus a fraction of another. In Observations, record the measurement of the low-power field diameter in millimeters, expressing the length to the nearest tenth of a millimeter.
8. In Observations, record the measurement of the low-power field diameter in micrometers.
9. You cannot measure the diameter of the high-power field using the process you have just completed. Viewing a ruler under high power presents problems with light and focusing. Also, the high-power field diameter is less than 1 millimeter. But you can obtain the high-power field diameter indirectly. You know the low-power field diameter and the magnifying power of both objectives. Since the magnification of the objectives is inversely proportional to the field size, you can use this formula:

$$\text{high-power field diameter} = \frac{\text{low-power field diameter} \times \text{low-power magnification}}{\text{high-power magnification}}$$

   In Observations, record the high-power field diameter in micrometers. Show your calculations.

10. Now that you know the diameter of your field size under both low and high power, you can estimate the sizes of the objects you view under the microscope by comparing them with the diameter of the field of vision. For example, if a tiny organism takes up approximately one-half of a field of view that is 1000 micrometers in diameter, then its size is about one-half of 1000 micrometers, or 500 micrometers.

Name ______________________ Class __________ Date __________

11. Obtain prepared slides of various organisms and practice estimating their lengths. Write the name of the organism or part you examine and its estimated size in micrometers in the Data Table.

12. When you have finished examining the organisms in step 11, return your microscope to the storage area.

Observations

1. Measurement of the low-power field diameter = ______________________

2. Measurement of the low-power field diameter = __________ micrometers.

3. Low-power magnification = ______________________

4. High-power magnification = ______________________

5. Use the formula shown in step 9 of Procedure to calculate the high-power field diameter. Show your calculations.

Data Table

Name of Object	Measurement of Object (µm)

Analysis and Conclusions

1. How many micrometers are in 1 millimeter? ______________________

2. How many micrometers are in 1 meter? ______________________

3. What happens to the field of view when you change from low-power magnification to high-power magnification? ______________________

____

____

4. How many times is the magnification increased when you change from low-power to high-power magnification? ______

______

______

5. How many times is the diameter of a field decreased when you change from low-power to high-power magnification? ______

______

______

Critical Thinking and Application

1. Approximately 500 of a certain type of bacteria can fit across your low-power field of vision. What is the approximate size of 1 bacterium? ______

______

2. Approximately 7 of a certain type of protist can fit across your high-power field of vision. What is the approximate size of 1 protist? ______

______

3. If a microscope has a low-power magnification of 100X, a high-power magnification of 600X, and a low-power field diameter of 1800 micrometers, what is the high-power field diameter in micrometers? ______

______

4. If 20 objects fit across a low-power field of view whose field diameter is 3000 micrometers, what is the approximate size of each object? ______

______

Going Further

Make a temporary wet-mount slide of a culture of mixed protists. Choose one protist and observe it under low and high power. Estimate its length in micrometers.

Name ______________________ Class __________ Date __________

Making Predictions Using Indirect Evidence

Pre-Lab Discussion

How do scientists know what the inside of an atom looks like? After all, it is impossible to see inside an atom. Scientists have used observations of how atoms behave to develop a model of atomic structure. These observations are one type of indirect evidence for the structure of an atom.

Another type of indirect evidence comes from making predictions. Based on the model they have developed, scientists predict how atoms will behave in certain circumstances. They design experiments to test these predictions. If the predictions are shown to be accurate, they are taken as additional indirect evidence that the model is correct.

In this investigation, you will use indirect evidence to determine the properties of objects you cannot see.

Problem

How can you determine the characteristics of something you cannot observe directly?

Materials *(per pair of students)*

Shoe box, wrapped or with cover, that contains small objects
Triple-beam balance
Bar magnet

Procedure

1. Obtain a sealed shoe box from your teacher.
2. Without opening the box, perform several tests on the box to determine the characteristics of the contents. Tests might include tipping it, shaking it, sliding it, checking for magnetic attraction, and finding its mass. In each case, record your observations in the Data Table. Observations might include sound or the way the contents behave (roll, slide, etc.). You might also wish to find the mass of an identical box and lid that is empty. The difference in mass would be the mass of the object or objects inside.
3. On the basis of your observations, sketch and label the box and its contents in Figure 1 in Analysis and Conclusions.

Observations

Describe some of the ways in which you tested the unknown objects to determine their characteristics. Record your descriptions in the Data Table.

Data Table

Test	Observations

Analysis and Conclusions

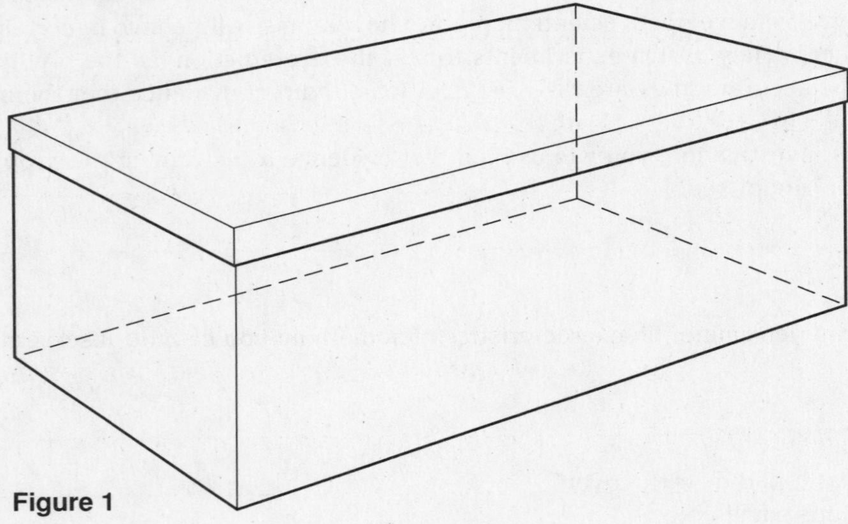

Figure 1

1. What senses did you use to determine a model of the object or objects inside the box?

____

____

2. How does this investigation compare with the way in which scientists have learned about the atom? ____

____

____

Name ______________________ Class ____________ Date ____________

3. Do you think that all the boxes used in this investigation contain the same number and types of objects? What evidence would you use to support your answer?

Critical Thinking and Application

1. In what ways could you test the accuracy of your model of the contents of the box without opening the box?

2. What kinds of special instruments could also be used to make indirect observations of the contents of the box?

3. If you used some of these special instruments and found that the results did not agree with what you originally thought to be contained within the box, what would you need to do?

4. If your teacher allows you to do so, open the box and directly observe the object(s) inside or look at a list of the objects contained within the box. How accurate is your indirect determination?

What does this tell you about the current model of the atom? ______________________

____

____

____

____

Going Further

1. Glue an object inside a shoe box. Trade boxes with another student who has done the same. Using knitting needles, determine the location and shape of the object by sticking the needles through the box.
2. How would you determine what kind of liquid was inside of a completely filled, closed, opaque bottle? How would it be helpful if you had an identical empty bottle?

Physical and Chemical Changes

Pre-Lab Discussion

Matter is constantly changing. The two kinds of changes that occur in matter are physical and chemical changes. In a physical change, no new substances are formed. However, physical properties such as size, shape, color, or phase may change. Dissolving, melting, evaporating, and grinding are physical changes.

As a result of a chemical change, one or more new substances with new and different properties are formed. The new substances are different from the original substance. Chemical changes, which are often called chemical reactions, are taking place around you and even inside you all the time. Respiration and digestion are chemical changes you could not live without. Photosynthesis, or the food-making process in green plants, is a chemical change. Burning, the rusting of iron, and the changing colors of leaves in autumn are other examples of chemical changes.

In this investigation, you will observe physical and chemical changes and learn to recognize each type of change when it occurs.

Problem

What are the differences between physical and chemical changes?

Materials *(per group)*

Birthday candle
Aluminum foil (10 cm x 10 cm)
Modeling clay
Test tube rack
3 test tubes
Matches
Scoop
Celsius thermometer
10-mL graduated cylinder
Antacid tablet
Magnesium ribbon (2 cm long)
10 mL 1 M hydrochloric acid
Table salt
Dropper bottle of 0.1 M silver nitrate

Safety

Put on a laboratory apron if one is available. Put on safety goggles. Handle all glassware carefully. Always use special caution when working with laboratory chemicals, as they may irritate the skin or cause staining of the skin or clothing. Never touch or taste any chemical unless instructed to do so. Note all safety alert symbols next to the steps in the Procedure and review the meanings of each symbol by referring to the symbol guide on page 10.

Procedure

1. Take a small piece of modeling clay and place it on the square of aluminum foil. Firmly place a candle in the clay so that it is well supported. Light the candle and allow it to burn while you continue the rest of the investigation. Record your observations of the burning candle.
2. Add a small scoop of table salt to a test tube that has been half-filled with tap water. Place your thumb over the top of the test tube and shake to dissolve the salt. Record your observations. Using the dropper, add 5 drops of silver nitrate to the salt water. Record your observations.
3. Place 10 mL of hydrochloric acid in a test tube. Place the thermometer in the solution and record the temperature. Cut off a small piece of magnesium ribbon and place it in the test tube. Record your observations of the reaction. Record the temperature of the hydrochloric acid at the end of the reaction.
4. Place 10 mL of water in a test tube. Place the thermometer in the water and record the temperature. Break the antacid tablet into small pieces and add them to the test tube of water. Record your observations of the reaction. Record the temperature of the water at the end of the reaction.

Observations

1. What did you observe as the candle burned? ____________________

____

____

What was left after the candle burned? ____________________

____

2. What did you observe when you added the salt to the water in the test tube and shook it?

____

What did you observe when the silver nitrate was added to the salt water?

____

____

3. Original temperature of the hydrochloric acid = ____________

Temperature of the hydrochloric acid after magnesium is added = ____________

What did you observe when the magnesium was added to the hydrochloric acid?

____

____

____

Name ______________________ Class ____________ Date ____________

4. Original temperature of the water = ____________

Temperature of the water after antacid is added = ____________

What did you observe when the antacid tablet was added to the water?

____

____

____

Analysis and Conclusions

1. Identify each of the following as either a physical change or a chemical change. Give a reason for your answer.

a. Melting candle wax ______________________

b. Burning a candle ______________________

c. Dissolving salt in water ______________________

d. Adding silver nitrate to salt water ______________________

e. Cutting a piece of magnesium ribbon ______________________

f. Adding magnesium metal to hydrochloric acid ______________________

g. Breaking an antacid tablet into small pieces ______________________

h. Adding an antacid tablet to water ______________________

2. What happened to the temperature of the hydrochloric acid when you added the magnesium to it? ______________________

3. What happened to the temperature of the water when you added the antacid tablet to it?

____

4. Describe two observations you might make when a physical change occurs.

____

____

5. Describe two observations you might make when a chemical change occurs.

____

____

Critical Thinking and Application

1. How could you prove that dissolving the salt in water resulted in only a physical change?

____

____

2. How could you prove that adding magnesium to hydrochloric acid resulted in a chemical change? ____

____

____

3. The following changes often indicate that a chemical change has occurred. But they can also indicate the occurrence of a physical change. Explain how each change might result from a physical, not a chemical, change.

   a. Change of color ____

   ____

   b. Loss of mass ____

   ____

   c. The substance seems to "disappear." ____

   ____

Going Further

Write out a recipe that involves cooking or baking. Identify each step in the recipe as resulting in either a physical change or a chemical change in the ingredients.

Using Acid-Base Indicators to Test Unknown Substances

Pre-Lab Discussion

The science laboratory is not the only place where acids and bases are found. Many items commonly found at home are acids or bases. For example, many of the foods you eat contain acids. Many commonly used cleaning products owe their effectiveness to the fact that they are alkaline, or contain bases.

Indicators are special chemicals that can show whether a given substance is an acid, a base, or neither. Indicators usually react with an acid or a base to form a slightly different chemical with a different color. Two examples of indicators are litmus paper (blue and red) and pH paper. Blue litmus paper turns red in an acid and stays blue in a base. Red litmus paper turns blue in a base and stays red in an acid. One type of pH paper turns a different color at each of several pH values ranging from 2 to 10.

In this investigation, you will test a number of substances using litmus paper and pH paper.

Problem

How can you determine whether a substance is an acid or a base?

Materials *(per group)*

10 mL of each test liquid
10 small test tubes
Test tube rack
10-mL graduated cylinder
Glass stirring rod
Vials of red and blue litmus paper
Vial of pH paper
Distilled water

Safety

Put on a laboratory apron if one is available. Put on safety goggles. Handle all glassware carefully. Always use special caution when working with laboratory chemicals, as they may irritate the skin or cause staining of the skin or clothing. Never touch or taste any chemical unless instructed to do so. Note all safety alert symbols next to the steps in the Procedure and review the meanings of each symbol by referring to the symbol guide on page 10.

Procedure

1. Use the graduated cylinder to measure 10 mL of distilled water into one of the test tubes.
2. Dip the glass stirring rod into the test tube containing the distilled water and then touch it to a piece of red litmus paper. Get a bit more on the stirring rod and touch it to a piece of blue litmus paper. Get a bit more on the rod and touch it to a piece of pH paper. Record your observations in the Data Table.
3. Based on the color change you observe in the pH paper, approximate the pH of the distilled water. Record this value in the Data Table.
4. Taking care to rinse and dry the stirring rod and graduated cylinder thoroughly, repeat steps 1 through 3 with the other substances provided by your teacher. Be sure to write the names of the substances in the Data Table.

Observations

Data Table

Substance	Color of Indicator			Approximate pH of Substance
	Blue Litmus	Red Litmus	pH Paper	
Distilled Water				

Analysis and Conclusions

1. Which substances are acids? ______________________________

____

2. Which substances are bases? ______________________________

____

3. Which substances are neutral? ____

4. Which substance is probably the strongest acid? Explain your conclusion.

5. Which substance is probably the strongest base? Explain your conclusion.

6. Is litmus paper useful in determining the exact pH of a substance? Explain your answer.

Critical Thinking and Application

1. Describe three situations in which acid-base indicators might be useful in everyday life.

2. Suppose you are manufacturing a certain type of cosmetic. You know that it can be slightly acidic, but it should not be strongly acidic. Which of the indicators you have just studied will help you determine the degree of acidity? Explain your answer.

3. You may have found that the results you obtained were different from those of other groups.

What variables might have affected your results? ______________________________

____

____

____

4. Design an experiment in which you could investigate which of two antacids is more effective in

neutralizing stomach acid. ______________________________

____

____

____

____

Going Further

1. Test the pH of the substances used in this investigation with different indicators: phenolphthalein, phenol red, methyl orange, bromthymol blue, red cabbage juice, and grape juice, for example.
2. Using indicators, test the acidity of various soils near your home or school. Find out which kinds of plants grow well in acidic soil.

Name ______________________ Class __________ Date __________

Identifying Organic Compounds

Pre-Lab Discussion

The most common organic compounds found in living organisms are lipids, carbohydrates, proteins, and nucleic acids. Common foods, which often consist of plant materials or substances derived from animals, are also combinations of these organic compounds. Simple chemical tests with substances called indicators can be conducted to determine the presence of organic compounds. A color change of an indicator is usually a positive test for the presence of an organic compound.

In this investigation, you will use several indicators to test for the presence of lipids, carbohydrates, and proteins in particular foods.

Problem

How are indicators used to test for the presence of organic compounds?

Materials *(per group)*

10 test tubes
Test tube rack
Test tube holder
Masking tape
Bunsen burner or hot plate
Iodine solution
20 mL honey solution
20 mL egg white and water mixture
20 mL corn oil
20 mL lettuce and water mixture
20 mL gelatin and water solution
20 mL melted butter
20 mL potato and water mixture
20 mL apple juice and water mixture
20 mL distilled water
20 mL unknown substance
Paper towels
600-mL beaker
Brown paper
Sudan III stain
Biuret reagent
Benedict's solution

Safety

Put on a laboratory apron if one is available. Put on safety goggles. Handle all glassware carefully. Always use special caution when using any laboratory chemicals, as they may irritate the skin or cause staining of the skin or clothing. Never touch or taste any chemical unless instructed to do so. Use extreme care when working with heated equipment or materials to avoid burns. Note all safety alert symbols next to the steps in the Procedure and review the meanings of each symbol by referring to the symbol guide on page 10.

Procedure

Part A. Testing for Lipids

1. Obtain 9 test tubes and place them in a test tube rack. Use masking tape to make labels for each test tube. As shown in Figure 1, write the name of a different food sample (listed in Materials) on each masking-tape label. Label the ninth test tube "distilled water."
2. Fill each test tube with 5 mL of the substance indicated on the masking-tape label. Add 5 drops of Sudan III stain to each test tube. Sudan III stain will turn red in the presence of lipids.
3. Gently shake the contents of each test tube. **CAUTION:** *Use extreme care when handling Sudan III to avoid staining hands or clothing.* In the Data Table record any color changes and place a check mark next to those substances testing positively for lipids.
4. Wash the test tubes thoroughly.
5. For another test for lipids, divide a piece of brown paper into 9 equal sections. In each section, write the name of one test substance. See Figure 2.

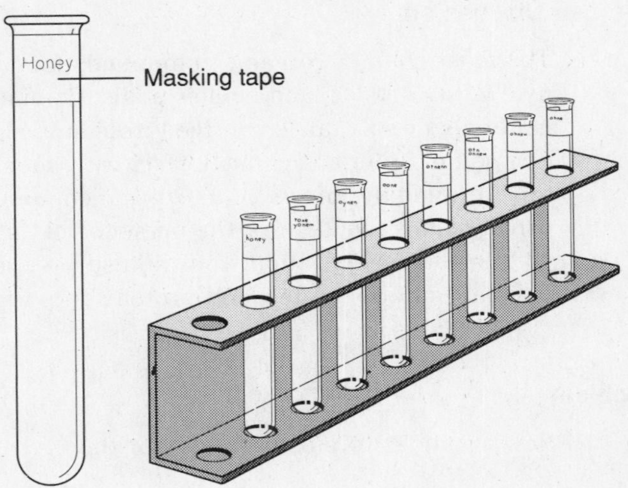

Figure 1

Honey	Egg white	Corn oil	Lettuce	Gelatin
Butter	Potato	Apple juice	Distilled water	

Figure 2

6. In each section, rub a small amount of the identified food onto the brown paper. Rub the food until a "wet" spot appears on the paper. With a paper towel, rub off any excess pieces of food that may stick to the paper. Set the paper aside until the spots appear dry—about 10 to 15 minutes.
7. Hold the piece of brown paper up to a bright light or window. You will notice that some foods leave a translucent spot on the brown paper. The translucent spot indicates the presence of lipids.

Part B. Testing for Carbohydrates

1. Sugars and starches are two common types of carbohydrates. To test for starch, refill each cleaned test tube with 5 mL of the substance indicated on the masking-tape label. Add 5 drops of iodine solution to each test tube. Iodine will change color from yellow-brown to blue-black in the presence of starch.

2. Gently shake the contents of each test tube. **CAUTION:** *Use extreme caution when using iodine as it is poisonous and can also stain hands and clothing.* In the Data Table, record any color changes and place a check mark next to those substances testing positive for starch.

3. Wash the test tubes thoroughly.

4. For a sugar test, set up a hot-water bath as shown in Figure 3. Half fill the beaker with tap water. Heat the water to a gentle boil. **CAUTION:** *Use extreme care when working with hot water. Do not let the water splash onto your hands.*

5. While the water bath is heating, fill each cleaned test tube with 5 mL of the substance indicated on the masking-tape label. Add 10 drops of Benedict's solution to each test tube. When heated, Benedict's solution will change color from blue to green, yellow, orange, or red in the presence of a single sugar, or monosaccharide.

6. Gently shake the contents of each test tube. **CAUTION:** *Use extreme caution when using Benedict's solution to avoid staining hands or clothing.*

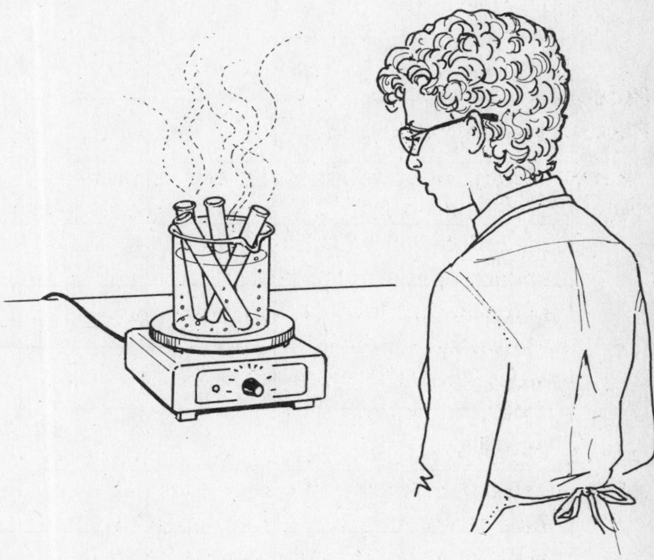

Figure 3

7. Place the test tubes in the hot-water bath. Heat the test tubes for 3 to 5 minutes. With the test tube holder, remove the test tubes from the hot-water bath and place them back in the test tube rack. **CAUTION:** *Never touch hot test tubes with your bare hands. Always use a test tube holder to handle hot test tubes.* In the Data Table, record any color changes and place a check mark next to any substances that test positive for a single sugar.

8. After they have cooled, wash the test tubes thoroughly.

Part C. Testing for Proteins

1. Put 5 mL of the appropriate substance in each labeled test tube. Add 5 drops of Biuret reagent to each test tube. **CAUTION:** *Biuret reagent contains sodium hydroxide, a strong base. If you splash any reagent on yourself, wash it off immediately with water. Call your teacher for assistance.*

2. Gently shake the contents of each test tube. Biuret reagent changes color from yellow to blue-violet in the presence of protein. In the Data Table, record any changes in color and place a check mark next to any substances that test positively for protein.

3. Wash test tubes thoroughly.

Part D. Testing an Unknown Substance for Organic Compounds

1. Obtain a sample of an unknown substance from your teacher and pour it into the remaining test tube. Repeat the tests described in Parts A, B, and C of the Procedure to determine the main organic compounds in your sample. Record your results in the Data Table.

2. Wash the test tube thoroughly.

Observations

Data Table

Substance	Lipid Test		Carbohydrate Test				Protein Test	
	Sudan color	Lipids Present (✓)	Iodine color	Starches present (✓)	Benedict color	Sugar present (✓)	Biuret color	Proteins present (✓)
Honey								
Egg white								
Corn oil								
Lettuce								
Gelatin								
Butter								
Potato								
Apple juice								
Distilled water								
Unkown								

Analysis and Conclusions

1. Which test substances contain lipids? ______________________________

____

2. Which test substances contain starch? ______________________________

____

3. Which test substances contain single sugar? ______________________________

____

4. Which test substances contain protein? ______________________________

____

5. Which test substances did not test positive for any of the organic compounds?

____

Name ______________________ Class ____________ Date ____________

6. What is the purpose of using distilled water as one of your test substances?

7. What do all of the indicators you have used have in common?

Critical Thinking and Application

1. Your brown lunch bag has a large, translucent spot on the bottom. What explanation could you give for this occurrence?

2. What conclusion could you make if a positive test for any of the organic compounds occurred in the test tube containing only distilled water?

3. A very thin slice is removed from a peanut and treated with Sudan III stain. Then a drop of Biuret reagent is added to the peanut slice. When you examine the peanut slice under a microscope, patches of red and blue-violet are visible. What conclusions can you draw from your examination?

Going Further

1. Test each food from a school lunch for the presence of lipids, starch, single sugars, and proteins. Construct a data table to summarize your findings.
2. Make three-dimensional models of the molecules of the organic compounds described in this investigation. Use polystyrene spheres, gum drops, or pieces of clay to represent the atoms of each molecule. Each type of atom should have its own distinct color. Use toothpicks to represent the bonds between atoms.

Name ______________________ Class ____________ Date ____________

9

Characteristics of Prokaryotic and Eukaryotic Cells

Pre-Lab Discussion

Cells are the basic units of structure and function of all living things. There are two major divisions into which all cells fall—prokaryotic and eukaryotic.

Prokaryotic cells are cells that lack a nucleus and membrane-bound organelles. Bacteria and related microorganisms are prokaryotes. *Eukaryotic cells* are cells that contain a nucleus and membrane-bound organelles. Organisms such as animals, plants, fungi, and protists are all eukaryotes.

In this investigation, you will observe several prepared slides to examine the differences between prokaryotic and eukaryotic cells. You will also use these differences to classify an unknown specimen.

Problem

What are the differences between prokaryotic and eukaryotic cells?

Materials *(per group)*

Microscope
Lens paper
Prepared slides of prokaryotic and eukaryotic cells

Safety

Always handle the microscope with extreme care. You are responsible for its proper care and use. Use caution when handling glass slides as they can break easily and cut you. Note all safety alert symbols next to the steps in the Procedure and review the meanings of each symbol by referring to the symbol guide on page 10.

Procedure

1. Take a microscope from the storage area and place it about 10 centimeters from the edge of the laboratory table.
2. Carefully clean the eyepiece and objective lens with lens paper.
3. Place your first prepared slide on the microscope stage so that it is centered over the stage opening. Hold the slide in position with the stage clips.
4. Using the low-power objective lens, locate the cell(s) under the microscope. Turn the coarse adjustment knob until the cell comes into focus.

5. In the Data Table, write the name of the type of cell that you examined. Describe the general shape of the cell in the space provided. Estimate the length of the cell and record this figure. Refer to Laboratory Investigation 4 if you need to review how to estimate the size of objects under the microscope. Put a check mark next to the cell structures you are able to observe under low power.

6. Switch to the high-power objective lens. **CAUTION:** *When turning to the high-power objective lens, you should always look at the objective from the side of your microscope so that the objective lens does not hit or damage the slide.* Look for cell structures unobservable under low power. Put a check mark next to these structures in the Data Table. Based on your observations, decide if the cell is prokaryotic or eukaryotic and record this in the Data Table.

7. In the appropriate place in Observations, draw and label what you see using the high-power objective lens. Record the magnification of the microscope.

8. Repeat steps 1 through 7 using other prepared slides provided by your teacher.

9. Repeat steps 1 through 7 using an unidentified prepared slide provided by your teacher.

10. When you have finished examining all of the prepared slides, return the microscope to the storage area.

Observations

Data Table

Cell Type	Shape	Size (μm)	Cell Structures							Prokaryotic or Eukaryotic
			Cell wall	Cell membrane	Nucleus	Nuclear envelope	Cytoplasm	Vacuoles	Plastids	
Unknown										

Prepared Slide 1 High-power objective

Magnification ______________

Prepared Slide 2 High-power objective

Magnification ______________

Name ______________________ Class __________ Date __________

Prepared Slide 3 High-power objective

Magnification __________

____

Prepared Slide 4 High-power objective

Magnification __________

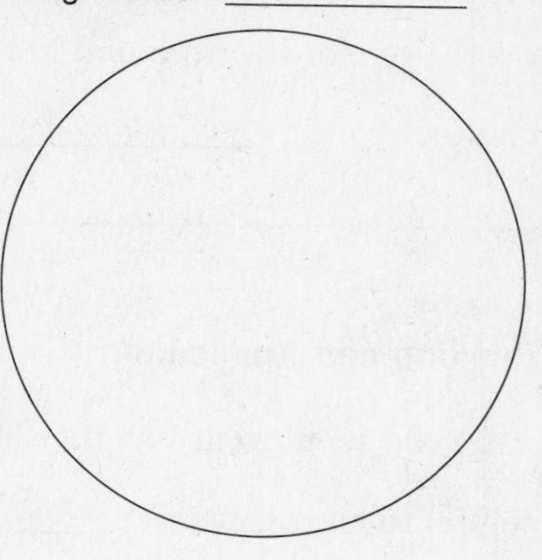

____

Prepared Slide 5 High-power objective

Magnification __________

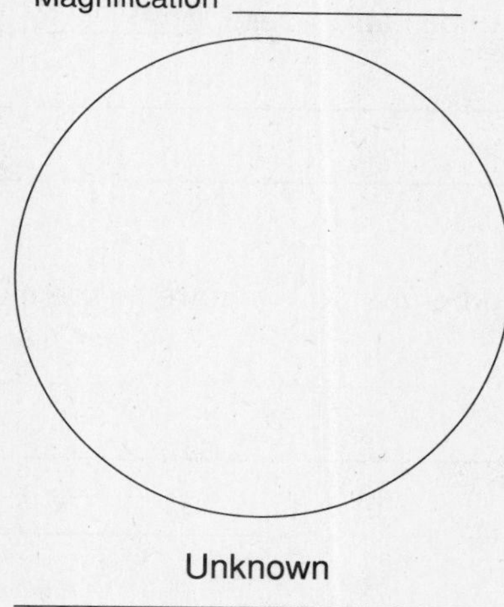

Unknown

____

Analysis and Conclusions

1. Based on your observations, do all cells have the same shape? Support your answer.

____

____

2. Based on your observations, do all cells have the same size? Support your answer.

____

____

3. What cell structures are common to all cells? ______________________

____

4. What cell structures are found only in eukaryotic cells? ______________________

____

5. Are the nuclei always found in the same place within different types of cells? Support your answer. ______________________

____

Critical Thinking and Application

1. Skin cells seem to fit together like pieces of a jigsaw puzzle. How is this arrangement of cells helpful to an organism? ______________________

____

____

2. Why do cells have different shapes and sizes? ______________________

____

____

3. What cell structure might you be able to compare to the main (principal's) office in your school? Explain your answer. ______________________

____

____

Going Further

1. Observe characteristics of living cells by making wet-mount slides of plant and animal tissues or protist cultures. Construct a data table to record the shapes and sizes of the cells and the structures they contain.
2. Think about the cell structures that you were unable to see with a compound light microscope. Use resources from your library to locate electron micrographs of these structures.
3. Research the use of some of the stains used in the preparation of wet-mount slides. Some of the stains that might be included in your report are methylene blue, neutral red, acetocarmine, Congo red, Janus green B, and Sudan III. What cell structures do each of these stains make more visible?

Name ______________________ Class __________ Date __________

Comparing Plant and Animal Cells

Pre-Lab Discussion

Ever since the first microscope was used, biologists have been interested in studying the cellular organization of all living things. After hundreds of years of observations by many biologists, the cell theory was developed. The cell theory states that the cell is the structural and functional unit of living things. Cells contain structures called organelles that carry out life processes. Cells can be classified by the types of organelles they contain. In plant and animal cells, similarities and differences exist because of varied life functions.

In this investigation, you will compare the structures of a typical plant cell *(Elodea)* and a typical animal cell (human).

Problem

How are plant and animal cells alike? How are they different?

Materials *(per group)*

Forceps
Medicine dropper
Elodea leaf
Water
Microscope
Glass slide
Coverslip
Toothpicks
Methylene blue stain
Paper towel
Lens paper

Safety

Put on a laboratory apron if one is available. Put on safety goggles. Always handle the microscope with extreme care. You are responsible for its proper care and use. Use caution when handling glass slides as they can break easily and cut you. Always use special caution when working with laboratory chemicals, as they may irritate the skin or cause staining of the skin or clothing. Never touch or taste any chemical unless instructed to do so. Note all safety alert symbols next to the steps in the Procedure and review the meanings of each symbol by referring to the symbol guide on page 10.

Procedure

Part A. Examining Plant Cells

1. Take a microscope from the storage area and place it about 10 centimeters from the edge of the laboratory table.
2. Carefully clean the eyepiece and objective lenses with lens paper.
3. Place a drop of water in the center of a clean glass slide.
4. With the forceps, remove a leaf from the *Elodea* plant and place it on the drop of water on the slide. Make sure that the leaf is flat. If it is folded, straighten it with the forceps.
5. Carefully place a coverslip over the drop of water and *Elodea* leaf.
6. Place the slide on the stage of the microscope with the leaf directly over the opening in the stage.
7. Using the low-power objective lens, locate the leaf under the microscope. Turn the coarse adjustment knob until the leaf comes into focus.
8. Switch to the high-power objective lens. **CAUTION:** *When turning to the high-power objective lens, you should always look at the objective from the side of your microscope so that the objective lens does not hit or damage the slide.*
9. Observe the cells of the *Elodea* leaf. Draw and label what you see in the appropriate place in Observations. Record the magnification of the microscope.
10. Carefully clean and dry your slide and coverslip.

Part B. Examining Animal Cells

1. Place a drop of water in the center of a clean glass slide.
2. Using the flat end of a toothpick, gently scrape the inside of your cheek. See Figure 1. **CAUTION:** *Do not use force when scraping the inside of your cheek. Only a few cells are needed.* The end of the toothpick will have several cheek cells stuck to it even though you may see nothing but a drop of saliva.
3. Stir the water on the slide with the end of the toothpick to mix the cheek cells with the water. See Figure 2. Dispose of the toothpick as instructed by your teacher.

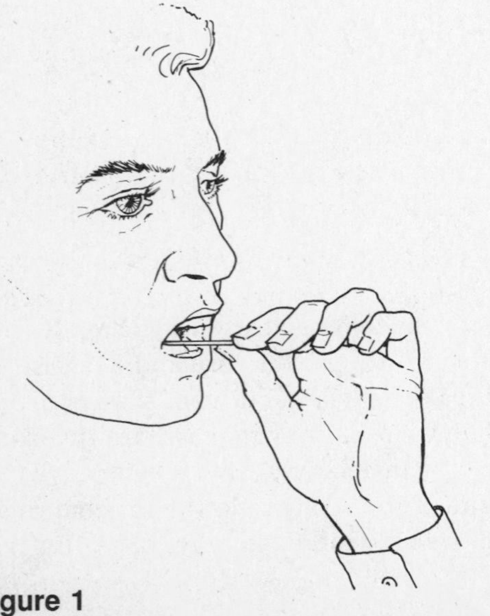

Figure 1

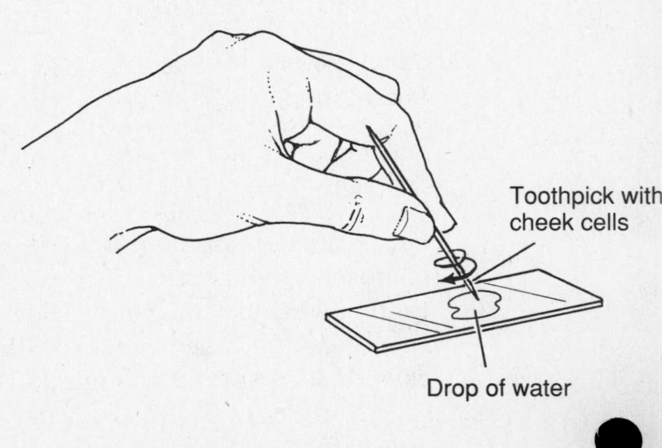

Figure 2

Name ______________________ Class __________ Date __________

4. Put one drop of methylene blue stain on top of the drop of water containing the cheek cells. **CAUTION:** *Use care when working with methylene blue to avoid staining hands and clothing.*

5. Wait one minute, then carefully place a coverslip over the stained cheek cells.

6. To remove the stain from under the coverslip and replace it with clear water, place a piece of paper towel at the edge of one side of the coverslip. Then place a drop of water at the edge of the coverslip on the opposite side. See Figure 3. The stained water under the coverslip will be absorbed by the paper towel. As the stain is removed, the clear water next to the coverslip on the opposite side will be drawn under the coverslip. Discard the paper towel after it has absorbed the stained water.

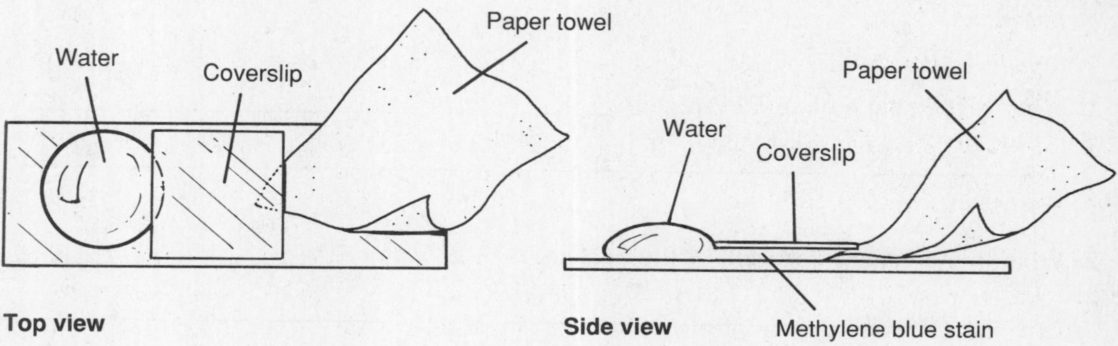

Figure 3

7. Place the slide on the stage of the microscope with the center of the coverslip directly over the opening in the stage.

8. Using the low-power objective lens, locate a few cheek cells under the microscope. **Note:** *You will need to reduce the amount of light coming through the slide in order to see the cells more clearly. Adjust the diaphragm as necessary.*

9. Switch to the high-power objective lens. **CAUTION:** *When turning to the high-power objective lens, you should always look at the objective from the side of your microscope so that the objective lens does not hit or damage the slide.*

10. Observe some cheek cells. Draw and label what you see in the appropriate place in Observations. Record the magnification of the microscope.

11. Carefully clean and dry your slide and coverslip.

12. Return your microscope to the storage area.

Magnification ______________

Magnification ______________

***Elodea* Cells**
(Plant cells)

Cheek Cells
(Animal cells)

Observations

1. What is the shape of an *Elodea* cell? ______________________________

____

2. What is the general location of the nucleus in an *Elodea* cell? ______________________________

____

3. What is the shape of the cheek cell? ______________________________

____

4. What is the general location of the nucleus in a cheek cell? ______________________________

____

Analysis and Conclusions

1. How are plant and animal cells similar in structure? ______________________________

____

____

2. How are plant and animal cells different in structure? ______________________________

____

____

Name ____________________ Class __________ Date __________

3. Why are stains such as methylene blue used when observing cells under the microscope?

Critical Thinking and Application

1. What is an advantage of using a wet-mount preparation instead of a dry-mount preparation in the study of living cells?

2. Explain why you could not use an oak leaf in this investigation.

3. Why is it possible to easily collect cells by gently scraping the inside of your cheek?

4. In general, the surface of a tree has a harder "feel" than does the surface of a dog. What cell characteristic of each organism can be used to explain this difference?

5. If you were given a slide containing living cells of an unknown organism, how would you identify the cells as either plant or animal? ______________________________

____

____

____

Going Further

1. Remove the skin from fruits and vegetables such as tomatoes, onions, and leeks. Prepare wet-mount slides for each skin and observe the cells under the low-power and high-power objectives of your microscope. Sketch and label what you see. What cell structures do these cells have in common? How do these cells compare with animal cells?
2. Make a wet-mount slide of an *Elodea* leaf using a 10% salt solution (10 g of table salt dissolved in 100 mL of water) instead of water. Examine the leaf under low power and high power. Describe any differences in characteristics between these *Elodea* cells and the cells observed in water.

Name ______________________ Class __________ Date __________

Plant Pigments

Pre-Lab Discussion

Photosynthesis begins when light is absorbed by pigments in the plant cell. One technique for separating and identifying these pigments is paper chromatography. In paper chromatography, solvent moves up the paper carrying with it dissolved substances—in this case plant pigments. The pigments are carried along at different rates because they are not equally soluble in the solvent and are attracted in different degrees to the paper.

Many green leaves contain pigment colors that are not seen until autumn because they are hidden by the chlorophyll. A few plants have leaves that are red, orange, or yellow all year long.

In this investigation, you will use paper chromatography to determine what differences exist in the plant pigments of various colors of leaves. You will also determine which leaves or which parts of leaves contain the chlorophyll necessary to carry out photosynthesis.

Problem

What plant pigments can be found in different colored leaves?

Materials *(per group)*

2 pieces of filter paper
2 150-mL beakers
2 glass plates (covers for beakers)
Coin
70% isopropyl alcohol (rubbing alcohol)
Fresh spinach leaf
Red leaf such as a *Coleus* leaf
Scissors
Stapler
Metric ruler
Pencil

Safety

Put on a laboratory apron if one is available. Put on safety goggles. Handle all glassware carefully. Alcohol is flammable. Do not expose it to heat or flames. Do this laboratory investigation in a well-ventilated room. Be careful when handling sharp instruments. Note all safety alert symbols next to the steps in the Procedure and review the meanings of each symbol by referring to the symbol guide on page 10.

Procedure

1. Make two filter-paper rectangles that are each approximately 12 cm by 7 cm. Using a pencil, draw a base line 1.5 cm from the bottom of the long side of each rectangle. See Figure 1.

Figure 1

2. Place a spinach leaf over the pencil line on one of the rectangles. Roll the coin over the leaf so that a horizontal green line is transferred to the pencil line. Repeat this step with the red leaf and the second filter-paper rectangle. See Figure 2.

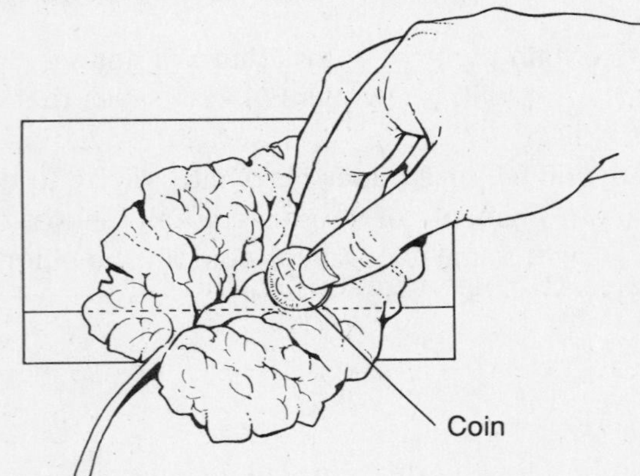

Figure 2

3. Add just enough isopropyl alcohol to each of the two beakers to cover the bottom. Do not add more than 1 cm to ensure that the pigment line will not be submerged when the paper is lowered into the beaker. **CAUTION:** *Avoid inhaling the alcohol.*

4. Make a cylinder of each piece of filter paper by stapling it end to end. Lower each paper cylinder into a beaker containing alcohol. See Figure 3. The solvent will begin to move up the paper and cause the pigments to move as well.

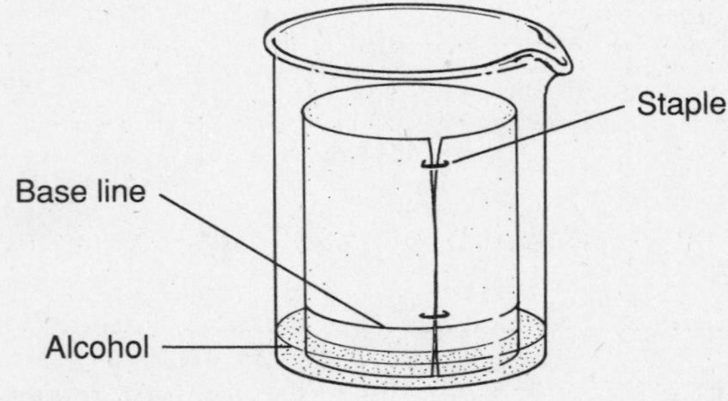

Figure 3

Name ______________________ Class ____________ Date ____________

5. Cover each beaker with a glass plate. Do not disturb the beakers for approximately 15 minutes, or until the solvent is about 1 cm from the top of the paper.

6. When the solvent is about 1 cm from the top of the paper, remove the paper and mark the farthest point of the solvent's progress (front line) with your pencil before this line evaporates.

7. Allow the filter-paper cylinders to dry, and then make a sketch of the chromatogram. Some possible colors and the pigments they represent are: faint yellow–carotenes; yellow–xanthophyll; bright green–chlorophyll *a;* yellow-green–chlorophyll *b;* red–anthocyanin

Observations

Data Table

	Line	Color Observed	Probable Pigment
Spinach	1 (base line)		
	2		
	3		
	4		
	5 (front line)		
Red Leaf	1 (base line)		
	2		
	3		
	4		
	5 (front line)		

Analysis and Conclusions

1. How many pigments were separated in each type of leaf? ______________________

____

____

2. How did the pigments in the spinach leaf compare with the pigments in the red leaf?

____

____

3. Which of these leaves can carry on photosynthesis? Explain your answer.

Critical Thinking and Application

1. Photosynthesis requires the green pigment chlorophyll. Explain how a Japanese maple tree, having only red leaves, can carry on photosynthesis.

2. Many trees have leaves that are green in the summer and red, yellow, or orange in autumn. Where were these colors during the summer? How can they suddenly appear in autumn?

3. In addition to separating plant pigments, what are some other possible applications for paper chromatography?

Going Further

Perform a similar experiment to compare the light pink portion of the *Coleus* leaf to the dark purple portion, or to compare leaves from plants grown in the dark to leaves from plants grown in the light.

Name ______________________ Class __________ Date __________

What Plants Do With Sunlight

Pre-Lab Discussion

The energy used by living things ultimately comes from the sun. Plants produce food by changing radiant energy to chemical energy. Plants do this in a process called photosynthesis. Photosynthesis captures light energy and stores it in glucose molecules. The glucose molecules are then almost immediately bound together into chainlike molecules of starch. You can test for the presence of starch by using iodine solution, which turns blue or blue-black in the presence of starch.

In this investigation, you will observe what plants do with sunlight. You will use the starch test to compare the contents of the leaves of plants that have grown with and without the presence of light. The covered leaves will be the experimental leaves and the uncovered one the control leaf.

Problem

What do plants do with sunlight?

Materials *(per group)*

Geranium or other green plant that has been in the dark for 2 days
Aluminum foil
Dark construction paper
Scissors
1000-mL beaker
250-mL beaker
150 mL 95% ethyl alcohol
200 mL tap water
Hot plate
Medicine dropper
Forceps
Iodine solution
Paper clips
3 petri dishes

Safety

Put on a laboratory apron if one is available. Put on safety goggles. Alcohol is flammable. Do not expose it to heat or flames. Do this laboratory investigation in a well-ventilated room. Always use special caution when working with chemicals, as they may irritate the skin or cause staining of the skin or clothing. Never touch or taste any chemical unless instructed to do so. Note all safety alert symbols next to the steps in the Procedure and review the meanings of each symbol by referring to the symbol guide on page 10.

Procedure

Part A. Preparing the Plant

1. Completely cover one of the leaves on the plant with aluminum foil. You may need to secure the foil with a paper clip. On a second leaf, paper clip a piece of construction paper that has been cut in the shape of the letter X. See Figure 1.
2. Identify each leaf you are using with a masking-tape label that gives your group number or name and your class period. Wrap the masking tape loosely around the petiole of the leaf. See Figure 1.
3. Place the plant in bright light for 2 to 3 days.

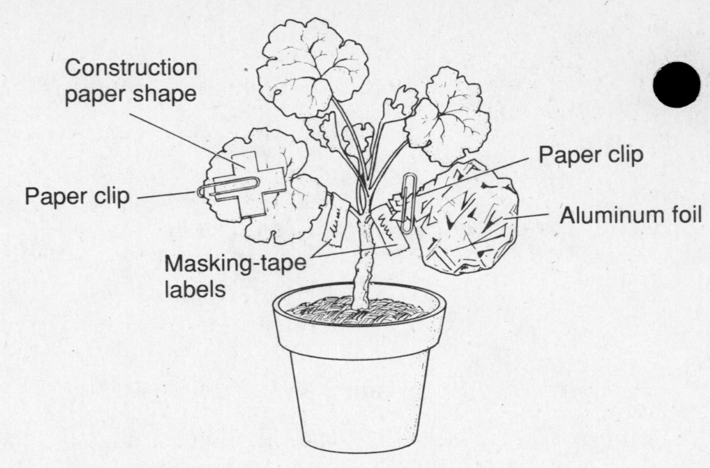

Figure 1

Part B. Testing the Leaves

1. After 2 to 3 days, remove 3 leaves from the plant: the 2 experimental leaves (1 covered with aluminum foil and 1 with paper shape attached) and 1 control leaf (leaf that has had no treatment). Remove the masking-tape labels, but in order to identify the leaves, cut their petioles different lengths according to the following:

   no petiole—control leaf (no treatment)
   half petiole—leaf with paper shape attached
   full petiole—leaf covered with aluminum foil

2. Your teacher has set up a water bath in which the alcohol is heating. Remove the aluminum foil and the paper shape from the 2 leaves. Then drop all 3 leaves into the boiling water in the larger beaker of the water bath. Allow the leaves to remain in the boiling water for 1 minute. **CAUTION:** *Do not remove the beaker containing alcohol from the water while you use the water around it.* See Figure 2.
3. Remove the leaves from the boiling water with forceps and drop them into the hot alcohol in the inner beaker. Keep them in the hot alcohol until they have lost most of their color.
4. When the leaves have lost most of their color, remove them from the alcohol with forceps and dip them into the hot water again for a few seconds.
5. Place the leaves into three different petri dishes.
6. Use the medicine dropper to cover each of the leaves with iodine solution. After one minute, pour off the excess iodine solution and record your observations in the Data Table.

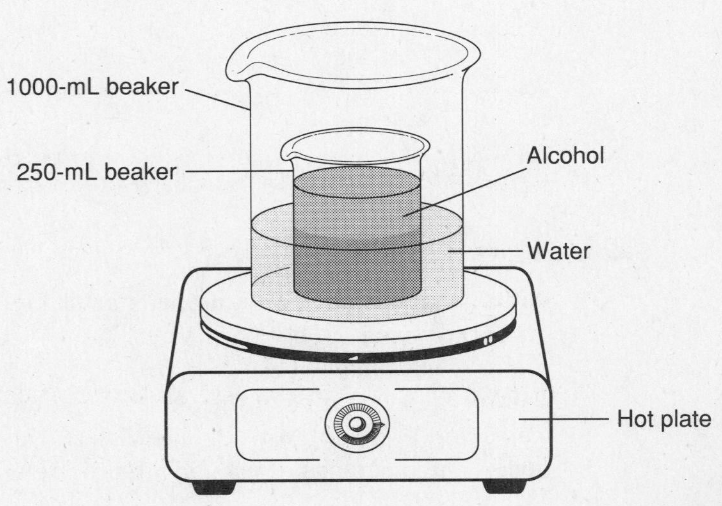

Figure 2

Name ______________________ Class __________ Date __________

Observations

Data Table

Leaf	Treatment	Effect of Iodine
1	covered with aluminum foil	
2	covered with paper shape	
3	no covering—full light	

Analysis and Conclusions

1. What does a positive iodine test (blue or blue-black) indicate is present in the leaf?

____

2. Which leaves showed a positive iodine test? ______________________

____

____

3. What environmental factor is necessary in order to get a positive iodine test?

____

____

4. Can a plant store energy in its leaves in the absence of light? ______________

____

____

5. In those leaves in which you observed a negative iodine test, what can you say about the process of photosynthesis? ______________________

____

____

6. What do plants do with sunlight? ____

____

____

____

Critical Thinking and Application

1. Photosynthesis produces glucose. What is the relationship between glucose and starch?

____

____

____

2. Explain the importance of plants to the survival of animals. ______________________

____

____

____

3. Does photosynthesis occur at night? Explain. ______________________________

____

____

4. Some paleontologists hypothesize that dinosaurs became extinct after the impact of a huge meteorite darkened the atmosphere of the Earth with dust. Using knowledge gained in this laboratory investigation, explain how this darkening could affect animals.

____

____

____

____

Going Further

Investigate starch production in plants by varying the color or intensity of the light used on the plants, rather than the presence or absence of light.

Name ______________________ Class ____________ Date ____________

Assembling a Protein Molecule

Pre-Lab Discussion

DNA and RNA, the two types of nucleic acids found in cells, determine which protein molecules a cell makes, or synthesizes. Protein molecules, formed by sequencing twenty different amino acids in various combinations, are important to living things because they control biological pathways, direct the synthesis of organic molecules, and are responsible for cell structure and movement. DNA carries the information for the synthesis of all the proteins in code form. The three different types of RNA carry out the DNA instructions to synthesize proteins.

During *transcription,* the DNA code is transcribed by mRNA into the language of protein synthesis. Three base *codons* of mRNA carry this information to the ribosomes, where *translation* occurs. During translation, mRNA codons are translated into a protein molecule when tRNA *anticodons* bring the correct amino acids to the ribosome. The amino acids brought side-by-side by tRNA form *peptide bonds* and become protein molecules.

In this investigation you will model the process of protein synthesis carried out by the cells of your body.

Problem

How can you and your classmates carry out the different processes of protein synthesis and assemble a protein molecule?

Materials *(per class)*

Large index cards
String, 70 cm long

Procedure

1. Obtain a large index card from your teacher. On this card, you will find your assignment for this investigation. You may have the role of an amino acid, a DNA triplet, an mRNA codon, or a tRNA anticodon. Put the card around your neck and make sure you understand your role in protein synthesis.
2. Report to the area of the cell where you will carry out your role.

3. When your teacher identifies the amino acids that make up the protein to be synthesized, use Figure 1 to determine the DNA triplets that code for that protein. Complete the Data Table for the protein to be synthesized.

4. The DNA nucleotides should form a double stranded DNA molecule in which the DNA triplets will code for the announced protein.

5. The DNA molecule unzips to allow the mRNA codons to form. Once the mRNA codons form and leave the nucleus, the DNA molecule reforms.

6. The mRNA codons move to the ribosomes and line up in the correct sequence.

7. On the ribosome, tRNA anticodons with the proper amino acids pair up with the correct mRNA codons.

8. As the mRNA moves along the ribosome, peptide bonds form between the amino acids. When the protein molecule has been made, mRNA, tRNA, and the protein leave the ribosome and return to the cytoplasm.

9. Repeat steps 1 through 8 to form another protein.

Amino Acid	DNA Triplet
Alanine	CGT
Glutamine	GTT
Glutamic Acid	CTT
Leucine	GAT
Lysine	TTT
Phenylalanine	AAA
Proline	GGC
Serine	AGC
Tyrosine	ATG
Valine	CAA

Figure 1

Observations

1. Using Figure 1, complete the Data Table for each protein molecule.

Data Table

Protein	Amino Acids	DNA Triplet	mRNA Codon	tRNA Anticodon
1	Proline Glutamic Acid Alanine			
2	Lysine Glutamine Valine			
3	Leucine Proline Tyrosine			
4	Phenylalanine Glutamine Proline			
5	Lysine Serine Leucine			

Name ______________________ Class __________ Date __________

2. In what ways do DNA and RNA molecules differ? ______________________

3. How was mRNA formed? ______________________

4. How do mRNA and tRNA differ? ______________________

5. Where in the cell are protein molecules formed? ______________________

Analysis and Conclusions

1. What is the role of DNA in protein synthesis? ______________________

2. What is the role of mRNA in protein synthesis? ______________________

3. Which step in the procedure represents transcription? Explain your answer.

4. Which step in the procedure represents translation? Explain your answer.

5. What would happen to Protein 1 if the first DNA triplet was TTT instead of GGC?

Critical Thinking and Application

1. How do the processes of transcription and translation differ?

2. How do DNA replication and DNA transcription differ?

3. If an incorrect nucleotide is in a DNA molecule due to a mutation, will protein synthesis be affected? Explain your answer.

Going Further

Design a system of communication using four different colors to represent each letter of the alphabet. (*Hint:* Use the genetic triplet code as a model.)

Simulating Protein Synthesis

Pre-Lab Discussion

Genes are the units that determine inherited characteristics, such as hair color and blood type. Genes are lengths of DNA molecules that determine the structure of polypeptides (the building blocks of proteins) that our cells make. The sequence of nucleotides in DNA determines the sequence of amino acids in polypeptides, and thus the structure of proteins.

In a process called *transcription,* which takes place in the nucleus of the cell, messenger RNA (mRNA) reads and copies the DNA's nucleotide sequences in the form of a complementary RNA molecule. Then the mRNA carries this information in the form of a code to the ribosomes, where protein synthesis takes place. The code, in DNA or mRNA, specifies the order in which the amino acids are joined together to form a polypeptide. The code words in mRNA, however, are not directly recognized by the corresponding amino acids. Another type of RNA called transfer RNA (tRNA) is needed to bring the mRNA and amino acids together. As the code carried by mRNA is "read" on a ribosome, the proper tRNAs arrive in turn and give up the amino acids they carry to the growing polypeptide chain. The process by which the information from DNA is transferred into the language of proteins is known as *translation.*

In this investigation, you will simulate the mechanism of protein synthesis and thereby determine the traits inherited by fictitious organisms called CHNOPS. CHNOPS, whose cells contain only one chromosome, are members of the kingdom Animalia. A CHNOPS chromosome is made up of six genes (A, B, C, D, E, and F), each of which is responsible for a certain trait.

Problem

How can the traits on a particular chromosome be determined? How can these traits determine the characteristics of an organism?

Materials *(per student)*

Blue pencil
Orange pencil

Procedure

1. To determine the trait for Gene A of your CHNOPS, fill in the information in the box labeled Gene A in the Data Table. Notice the sequence of nucleotides in DNA. On the line provided, write the sequence of nucleotides of mRNA that are complementary to DNA. Then, on the line provided, write the sequence of nucleotides of tRNA that are complementary to mRNA.
2. In order to determine the sequence of amino acids, match each tRNA triplet with the specific amino acid in Figure 1. Using a - (hyphen) to separate each amino acid number, record this information in the appropriate place in the Data Table.
3. Using Figure 2, find the trait that matches the amino acid sequence. Record this information in the appropriate place in the Data Table.

4. Repeat steps 1 through 3 for the remaining genes (B through F).

5. Using all the inherited traits, sketch your CHNOPS in the space provided.

tRNA Triplet	Amino Acid Number
ACC	20
AGC	16
CGA	2
AAC	4
CGC	3
GGG	5
AGG	7
AAA	8
UUU	9
GGU	12
UAU	13
CCC	1
AUC	6
CUA	10
GGA	11

Figure 1

Amino Acid Sequence	Trait
20-11-13	hairless
20-12-13	hairy
20-21-21	plump
13-14-15	skinny
16-2	four-legged
12-7-8-1	long nose
5-7-8-1	short nose
9-8	no freckles
9-4	freckles
11-3-2	blue skin
11-3-3	orange skin
6-6-10	male
6-6-14	female

Figure 2

Observations

Data Table

Gene A	Gene B	Gene C
DNA ACC GGT TAT	**DNA** AGC CGA	**DNA** TTT AAC
mRNA ________	**mRNA** ________	**mRNA** ________
tRNA ________	**tRNA** ________	**tRNA** ________
Amino acid sequence ________	**Amino acid sequence** ________	**Amino acid sequence** ________
Trait ________	**Trait** ________	**Trait** ________
Gene D	**Gene E**	**Gene F**
DNA GGA CGC CGA	**DNA** GGG AGG AAA CCC	**DNA** ATC ATC CTA
mRNA ________	**mRNA** ________	**mRNA** ________
tRNA ________	**tRNA** ________	**tRNA** ________
Amino acid sequence ________	**Amino acid sequence** ________	**Amino acid sequence** ________
Trait ________	**Trait** ________	**Trait** ________

Name ______________________ Class ____________ Date ____________

Analysis and Conclusions

1. Distinguish between translation and transcription.

2. What is the specific site for transcription and translation in the cell?

3. How many tRNA nucleotides form an anticodon that will attach to the mRNA codon?

Critical Thinking and Application

1. Suppose you knew the makeup of specific proteins in a cell. How would you determine the particular DNA code that coded for them? ____________________

2. How could one change in a DNA nucleotide alter the formation of the translated protein? (An example would be the difference between normal and sickle-cell hemoglobin.)

Going Further

Create two additional traits for your CHNOPS and give their initial DNA sequence, mRNA codon, and tRNA anticodon. Include the resulting amino acid sequence.

Name ______________________ Class __________ Date __________

Investigating the Limits of Cell Growth

Pre-Lab Discussion

In multicellular organisms, growth is accomplished by the production of more cells by cell division. Cell division will occur only when the cells have reached a size large enough to ensure that the resulting daughter cells will have all of the necessary materials and structures for survival.

The rate of exchange of materials entering (food, oxygen, and water) and leaving (waste products) the cell through the cell membrane is determined by the cell's surface area. And the rate at which these materials are used within the cell depends upon the cell's volume, or the amount of space within the cell. A number of factors, such as the reduction of food and changes in temperature, can cause cells to stop growing and dividing.

In this investigation, you will observe one factor that limits cell growth.

Problem

What is one factor that limits cell growth?

Materials *(per group)*

Plastic spoon
3 agar-phenolphthalein blocks
(1 cm, 2 cm, and 3 cm)
250-mL beaker
100 mL of 0.4% sodium hydroxide solution
Paper towel
Scalpel
Metric ruler

Safety

Put on a laboratory apron if one is available. Put on safety goggles. Handle all glassware carefully. Always use special caution when working with laboratory chemicals, as they may irritate the skin or cause staining of the skin or clothing. Never touch or taste any chemical unless instructed to do so. Be careful when handling sharp instruments. Note all safety alert symbols next to the steps in the Procedure and review the meanings of each symbol by referring to the symbol guide on page 10.

Procedure

1. Using the plastic spoon, place the 3 agar-phenolphthalein blocks in the beaker. Pour 100 mL of the 0.4% sodium hydroxide solution over the blocks. **CAUTION:** *Use extreme care when working with chemicals. Do not allow them to come into contact with the skin or clothing.* Allow the blocks to remain in the solution undisturbed for 10 minutes.

2. While the blocks are soaking, use the formulas given below to determine the surface area, volume, and surface-area-to-volume ratio for each block. Record this information in Data Table 1.

$$\text{surface area} = \text{number of surfaces} \times \text{length} \times \text{width}$$

$$\text{volume} = \text{length} \times \text{width} \times \text{height}$$

$$\text{surface-area-to-volume ratio} = \frac{\text{surface area}}{\text{volume}}$$

3. After 10 minutes, remove the blocks with the plastic spoon. Place them on the paper towel and blot them dry.

4. Cut each block in half with the scalpel. **CAUTION:** *Be careful when handling sharp instruments.* Using the metric ruler, measure the distance in millimeters that the pink color has diffused into each block. Then measure the distance from the end of the pink color to the center of the cube. Record these measurements in the appropriate places in Data Table 2.

Observations

Data Table 1

Cube Size	Surface Area (cm^2)	Volume (cm^3)	Surface-Area-to-Volume Ratio
3 cm			
2 cm			
1 cm			

Data Table 2

Cube Size	Distance Pink Color Has Diffused into Block (mm)	Distance from End of Pink Color to Block's Center (mm)
3 cm		
2 cm		
1 cm		

Name ______________________ Class __________ Date __________

Analysis and Conclusions

1. Which block has the greatest surface area? ______________________

____

2. Which block has the greatest surface-area-to-volume ratio? ______________________

____

3. In which block did the pink color diffuse the most? Explain your answer.

____

____

____

4. If the blocks were actual cells, which would be the most efficient in terms of permitting materials to enter and leave the cell? ______________________

____

____

____

____

5. Based on this investigation, what one factor may limit the growth of an individual cell?

____

____

____

____

Critical Thinking and Application

1. What happens to the surface-area-to-volume ratio of a cell as it grows?

____

____

2. Propose a hypothesis that explains why the growth of a cell decreases as its size increases.

____

____

____

____

Going Further

Calculate the surface-area-to-volume ratios for cells that are 0.1 cm and 0.01 cm on a side. Which has the greater surface area in proportion to its volume?

Name ________________ Class ________ Date ________

16

Determining the Time Needed for Mitosis

Pre-Lab Discussion

Mitosis is the process by which the cell nucleus is divided into two nuclei. Mitosis takes place in four phases: prophase, metaphase, anaphase, and telophase. The period between one mitosis and the next is called interphase. Chromosome replication occurs during interphase.

Organisms such as the common intestinal bacteria *E. coli* can complete mitosis in 30 minutes. Other cells require days. In some cells, such as human muscle cells, mitosis never occurs.

In this investigation, you will determine the time required for plant and animal cells to go through each phase of mitosis.

Problem

How can the time needed for each phase of mitosis be determined in a plant and an animal cell?

Materials *(per group)*

Textbook
Prepared slide of plant mitosis (onion root tip)
Prepared slide of animal mitosis (whitefish blastula)
Microscope

Safety

Handle all glassware carefully. Always handle the microscope with extreme care. You are responsible for its proper care and use. Use caution when handling glass slides as they can break easily and cut you. Note all safety alert symbols next to the steps in the Procedure and review the meanings of each symbol by referring to the symbol guide on page 10.

Procedure

1. Review the phases of plant and animal cell mitosis by studying the illustrations and photographs on pages 164 through 171 in your textbook.
2. Place the prepared onion root tip slide on the stage of the microscope. Using the low-power objective, focus on the cells just above the tip of the root. Switch to high power and count the total number of cells in the field of view. Record this information in the appropriate place in Data Table 1.
3. Without changing the field of view, count the number of cells in each phase of mitosis: prophase, metaphase, anaphase, and telophase. Record this information in the appropriate place in Data Table 1.
4. To determine the approximate proportion of time a cell spends in each phase of mitosis, divide the number of cells in each phase by the total number of cells in the field of view. To convert each decimal to a percent, multiply by 100. Record this information in the appropriate place in Data Table 1.
5. Repeat steps 2 through 4 using the prepared animal mitosis slide. Record all of the information in the appropriate places in Data Table 2.

Observations

Data Table 1

	Plant Cell Mitosis		
Phase	**Number of Cells in Phase**	$\frac{\textbf{Number of Cells in Phase}}{\textbf{Total Number of Cells}}$	**Percentage of Time Spent in Phase**
Prophase			
Metaphase			
Anaphase			
Telophase			
Total number of cells in field of view			

Data Table 2

	Animal Cell Mitosis		
Phase	**Number of Cells in Phase**	$\frac{\textbf{Number of Cells in Phase}}{\textbf{Total Number of Cells}}$	**Percentage of Time Spent in Phase**
Prophase			
Metaphase			
Anaphase			
Telophase			
Total number of cells in field of view			

Name ______________________ Class __________ Date __________

Analysis and Conclusions

1. In which phase of plant cell mitosis is the most time spent? In which phase of animal cell mitosis? ______________________

____

2. In which phase of plant cell mitosis is the least time spent? In which phase of animal cell mitosis? ______________________

____

3. Based on this investigation, what is the total percentage of time the plant and animal cells spend undergoing mitosis? ______________________

____

4. What percentage of the time are the plant and animal cells not undergoing mitosis?

____

____

5. What are the plant and animal cells doing when they are not undergoing mitosis?

____

Critical Thinking and Application

1. Determine the percentages of time spent in each phase of mitosis for the onion root tip using the total number of cells undergoing mitosis instead of the total number of cells in the field of view.

$$\text{percentage} = \frac{\text{Number of cells in phase}}{\text{Number of cells undergoing mitosis}}$$

prophase ______________

metaphase ______________

anaphase ______________

telophase ______________

2. In the Pre-Lab Discussion you were told that *E. coli* is capable of undergoing mitosis in 30 minutes. Using the information from the previous problem, determine how long *E.coli* spends in each phase of mitosis.

prophase ____________________

metaphase ____________________

anaphase ____________________

telophase ____________________

Going Further

Using reference sources, investigate the relationship between mitosis and cancer. What phases of mitosis might you expect to observe most frequently?

Name ______________________ Class __________ Date __________

17

The Principles of Genetics

Pre-Lab Discussion

Through his careful and detailed studies with garden pea plants, the Austrian monk and scientist Gregor Mendel determined that heredity is controlled by factors that are passed from generation to generation with predictable results. Mendel's experiments and conclusions led to the formulation of several basic genetic principles.

The principle of *dominance* states that some factors (alleles) are dominant, whereas others are recessive. The effects of a dominant allele are seen even if it is present with a contrasting recessive allele. The principle of *segregation* states that during gamete (sperm and egg) formation, the alleles for a trait separate, so that each gamete has only one of the alleles for that trait. The principle of *independent assortment* states that as gametes form, the genes for various traits separate independently of one another.

In this investigation, you will construct Punnett squares to observe the principles of genetics that were based on Mendel's work. You will also simulate a two-factor cross to learn the important role probability plays in the study of genetics.

Problem

How can the characteristics of offspring be predicted?

Materials *(per group)*

Textbook
Pencil
Masking tape
Scissors
4 disks
Paper cup

Safety

Be careful when handling sharp instruments. Note all safety alert symbols next to the steps in the Procedure and review the meanings of each symbol by referring to the symbol guide on page 10.

Procedure

Part A. Constructing Punnett Squares for One-Factor and Two-Factor Crosses

1. Examine Figure 1. The capital letter T is used to represent the dominant allele for tallness in the stem length for pea plants. The lowercase letter t is used to represent the recessive allele for shortness in pea plants.
2. Notice the genotypes, or gene combinations, that result from the parental (P_1) generation cross between a female purebred tall (TT) pea plant and a male recessive short (tt) pea plant. All of the offspring that make up the first filial (F_1) generation are tall (Tt). Their phenotypes, or visible characteristics, are all tall.
3. Figure 2 shows the results of a one-factor cross between the offspring from the F_1 generation. Observe the genotypic and phenotypic ratios for the second filial (F_2) generation. When writing such ratios, the number(s) for the dominant genotype(s) or phenotype(s) comes first. In this example, 1/4 of the offspring are TT, 1/2 (or 2/4) are Tt, and 1/4 are tt. The genotypic ratio is therefore 1/4:2/4:1/4, or 1:2:1. Three fourths of the offspring have tall stems and 1/4 have short stems. The phenotypic ratio is therefore 3/4:1/4 or 3:1.
4. In soybeans, purple flower color is dominant and white flower color is recessive. Let P represent the dominant allele (purple), and p the recessive allele (white).
5. Predict the probable genotypes of the first filial (F_1) generation of offspring by completing the Punnett square in Data Table 1. Indicate the genotypic and phenotypic ratios of the F_1 generation.

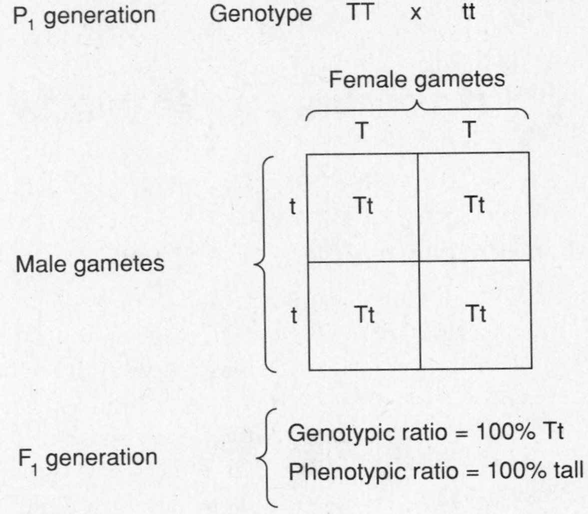

Figure 1

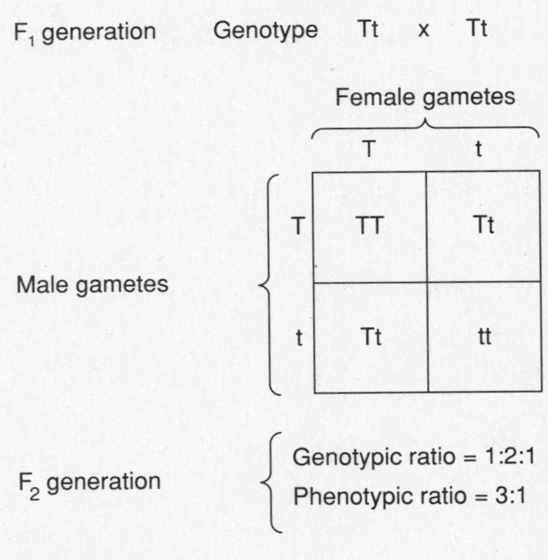

Figure 2

6. In Data Table 2, cross two plants from the F_1 generation. Write the appropriate genotypes for this cross on the lines provided. Predict the probable genotypes of this cross by completing the Punnett square in Data Table 2.
7. Indicate the genotypic and phenotypic ratio of the F_2 generation in Data Table 2.
8. In corn plants, rough seed shape (R) is dominant over smooth seed shape (r) and yellow seeds (Y) are dominant over white seeds (y). Determine the probable seed shape and color of the F_1 generation of offspring whose parents are heterozygous for the traits by completing the Punnett square in Data Table 3.
9. After you have completed the Punnett square in Data Table 3, fill in Data Table 4.

Name ______________________________ Class ____________ Date ______________

Part B. Simulating a Two-Factor Cross

1. To simulate the two-factor cross discussed in step 8 of Part A, cut 8 pieces of masking tape just large enough to fit on the disks. Place the tape on both sides of all 4 disks. On one side of 2 of the disks write R. On the other side of these disks, write r. On one side of the remaining 2 disks write Y. On the other side of these disks, write y. Each labeled disk represents the alleles in the heterozygous plant. The tossing together of the labeled disks represents the crossing of the heterozygous plants.
2. Place the 4 labeled disks into the paper cup. Holding one hand over the mouth of the cup, shake it so that the disks are tossed. Then empty the cup onto your desk or laboratory table. Record the results of each toss by making a tally mark in the appropriate place in Data Table 5.
3. Toss the 4 disks a total of 48 times, recording the results of each toss in Data Table 5. Count the tally marks for each phenotypic combination and record these totals in the appropriate place in Data Table 5.

Observations

Data Table 1

Purebred purple female x White male

P_1 generation PP x pp

Female gametes

Male gametes

F_1 generation
- Genotypic ratio = ______________
- Phenotypic ratio = ______________

Data Table 2

F_1 generation ______ x ______

Female gametes

Male gametes

F_2 generation
- Genotypic ratio = ______________
- Phenotypic ratio = ______________

Data Table 3

P_1 generation RrYy x RrYy

		Female gametes			
		RY	Ry	rY	ry
Male gametes	RY				
	Ry				
	rY				
	ry				

Data Table 4

Genotype	Genotypic Ratio	Phenotype	Phenotypic Ratio
RRYY			
RrYy			
RrYY			
RRYy			
RRyy			
Rryy			
rrYY			
rrYy			
rryy			

Name ______________________ Class ____________ Date ____________

Data Table 5

Phenotypes	Genotypes	Number Expected for 16 Offspring	Number Expected for 48 Offspring	Toss Results	Total Number Observed
Rough, yellow seeds	RRYY RrYy RRYy RrYY				
Rough, white seeds	RRyy Rryy				
Smooth, yellow seeds	rrYY rrYy				
Smooth, white seeds	rryy				

Analysis and Conclusions

1. When two homozygous plants with contrasting traits are crossed, what are the expected genotypes for the offspring? ____________________

____

2. What is the expected genotypic ratio for a one-factor cross of two hybrid organisms?

____

3. What is the expected phenotypic ratio for a one-factor cross of two hybrid organisms?

____

4. In your simulated two-factor cross, why might your actual experimental values have been different from the values you expected? ____________________

____

____

5. Why is it helpful to conduct a large number of trials when simulating genetic crosses?

____

____

____

Critical Thinking and Application

1. How is an understanding of genetics useful to an animal or plant breeder?

2. Is it possible for two organisms to have different phenotypes but the same genotype? Explain your answer.

3. Is it possible for two organisms to have different genotypes but the same phenotype? Explain your answer.

4. How could a guinea pig breeder determine whether a rough-coated guinea pig is homozygous or heterozygous for this trait?

5. In dogs, wire hair is due to a dominant gene W. Two wire-hair dogs were mated and produced a puppy with smooth hair. What were the genotypes of the two parent dogs?

Going Further

Repeat Part B of this investigation using a cross between genotypes RrYy and Rryy. Construct a Punnett square to represent your expected results and calculate expected genotypic and phenotypic ratios. Tally the results of your 48 tosses and compare your experimental values with the expected values.

Name ______________________ Class __________ Date __________

18

Biochemical Genetics

Pre-Lab Discussion

Modern genetics is concerned with the role of genes and their control over enzyme action. *Enzymes* regulate metabolism and control biochemical reactions that may alter the phenotype, or the physical appearance, of an organism. Put another way, the phenotype of an individual is a reflection of its biochemical composition.

In green plants, glucose manufactured by the process of photosynthesis is converted to starch by the enzyme phosphorylase. By investigating this conversion reaction, differences in biochemical activity between smooth and wrinkled peas can be determined.

In this investigation, you will demonstrate the biochemical difference between smooth and wrinkled peas in their ability to convert glucose to starch.

Problem

How can biochemical genetics be studied?

Materials *(per group)*

Water-soaked smooth pea
Water-soaked wrinkled pea
5 dry smooth peas
5 dry wrinkled peas
Microscope
2 10-mL graduated cylinders
Mortar and pestle
Glass-marking pencil
Paper towels
Glass slide
Coverslip
Forceps
Toothpicks
3 medicine droppers
Lugol solution
Petri dish with glucose agar

Safety

Put on a laboratory apron if one is available. Put on safety goggles. Always use special caution when working with laboratory chemicals, as they may irritate the skin or cause staining of the skin or clothing. Never touch or taste any chemical unless instructed to do so. Always handle the microscope with extreme care. You are responsible for its proper care and use. Use caution when handling glass slides as they can break easily and cut you. Note all safety alert symbols next to the steps in the Procedure and review the meanings of each symbol by referring to the symbol guide on page 10.

Procedure

Part A. Microscopic Differences Between Smooth and Wrinkled Peas

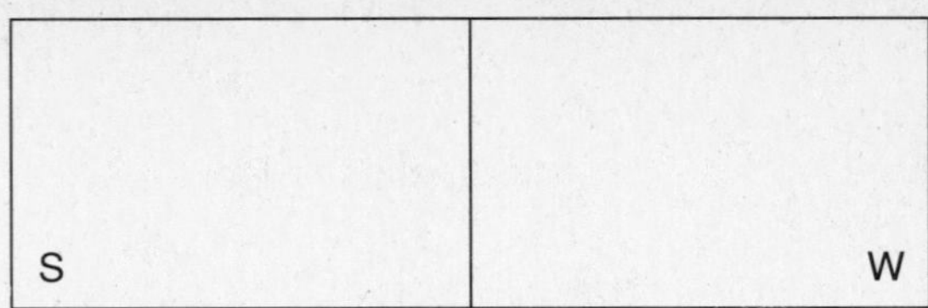

Figure 1

1. With a glass-marking pencil, divide a clean glass slide in half. Label the slide as shown in Figure 1.
2. Obtain a water-soaked smooth pea from your teacher.
3. With forceps, remove the seed coat from the smooth pea. This action exposes the cotyledon.
4. Using a clean toothpick, scrape a small amount of the cotyledon onto the half of the glass slide marked S.
5. Using a medicine dropper, place one drop of Lugol solution on the cotyledon. **CAUTION:** *Lugol solution is a chemical that can stain the skin and clothing.* With a toothpick, mix the Lugol solution and cotyledon scraping together. Discard the toothpick.
6. Repeat steps 3 through 5 using a water-soaked wrinkled pea. Be sure to use a clean toothpick to scrape the cotyledon.
7. Place a coverslip over each cotyledon scraping. Examine the cotyledon scraping from the smooth and wrinkled peas under the low-power objective of the microscope. The Lugol solution stains the starch in the cotyledon scrapings dark blue or black. Sketch what you see in the appropriate place in Observations. Record the magnification of the microscrope.
8. Switch to the high-power objective. **CAUTION:** *When switching to the high-power objective, always look at the objective from the side of your microscope so that the objective does not hit or damage the slide.* Sketch what you see in the appropriate place in Observations. Record the magnification of the microscope.

Part B. Biochemical Differences Between Smooth and Wrinkled Peas

1. With a glass-marking pencil, mark one 10-mL graduated cylinder S for smooth peas. Mark the other 10-mL graduated cylinder W for wrinkled peas.
2. Place 5 dry smooth peas into the mortar. Grind the peas into a fine powder with the pestle.
3. Place 1 mL of powdered smooth peas into the 10-mL graduated cylinder marked S. Add 3 mL of water. Set the mixture aside for 5 minutes or until the solids settle to the bottom of the graduated cylinder. **Note:** *Use a centrifuge if available to separate the powdered peas and the water.*
4. Repeat steps 2 and 3 using the 5 dry wrinkled peas.
5. With a glass-marking pencil, divide the bottom of the petri dish containing glucose agar in half. As shown in Figure 2, mark one half S for smooth peas and the other half W for wrinkled peas.
6. With another medicine dropper, place two drops of the clear liquid from the graduated cylinder marked S on the side of the glucose agar marked S.
7. Using the remaining medicine dropper, place two drops of the clear liquid from the graduated cylinder marked W on the side of the glucose agar marked W.
8. Set the petri dish aside for about 20 minutes to allow the pea extracts to work on the glucose in the agar. You may wish to begin cleaning your laboratory station while you are waiting.

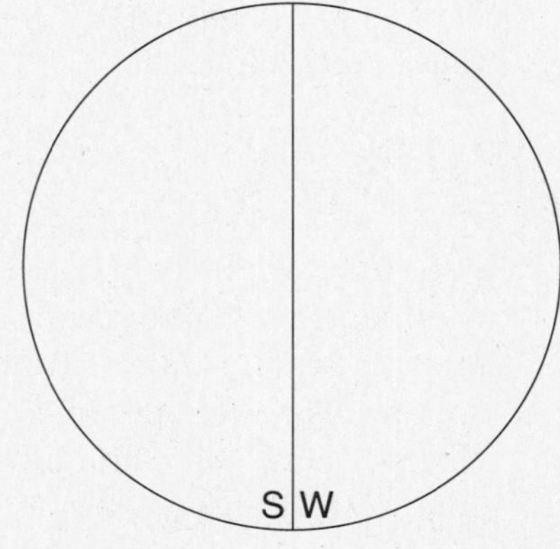

Figure 2

Name ______________________ Class ____________ Date ____________

9. After 20 minutes, use a paper towel to carefully blot any excess extract from either side of the agar. Add a drop of Lugol solution to each half of the petri dish. **CAUTION:** *Lugol solution is a chemical that can stain the skin and clothing.* Look for a dark blue or black color, which indicates the presence of starch. The presence of starch indicates the conversion of glucose to starch by the enzyme phosphorylase. Record your observations in the Data Table.

Observations

Low-power objective

Magnification ____________ Magnification ____________

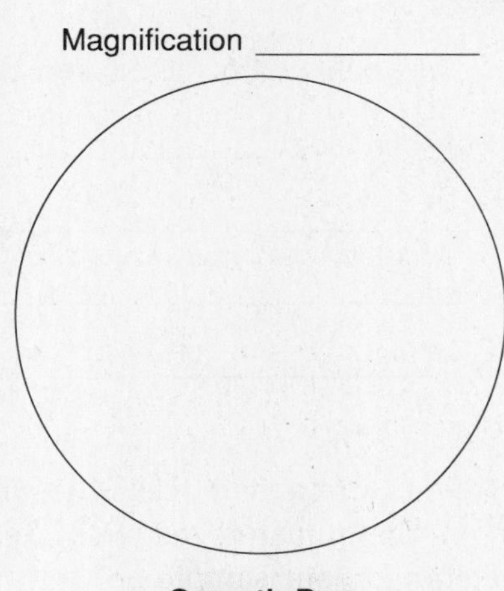

Smooth Pea **Wrinkled Pea**

High-power objective

Magnification ____________ Magnification ____________

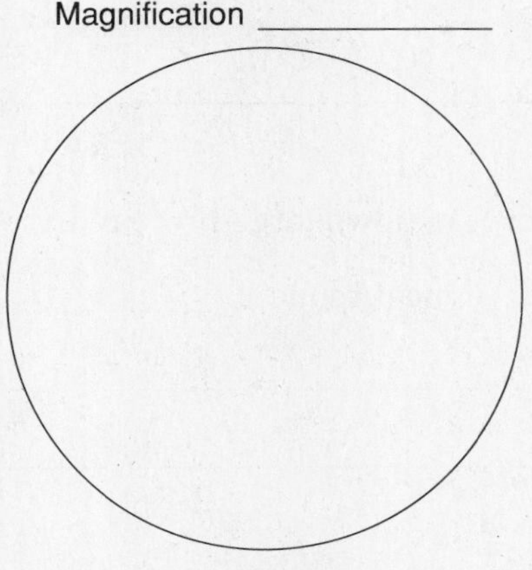

Smooth Pea **Wrinkled Pea**

Data Table

Type of Pea	Starch Present?
Smooth	
Wrinkled	

Analysis and Conclusions

1. How does the shape of the starch grains from the smooth peas compare with those from the wrinkled peas? ____

____

2. Which type of pea contains the enzyme phosphorylase? ______________________________

____

3. Which type of pea does not contain the enzyme phosphorylase? ______________________

____

Critical Thinking and Application

1. All of the differences observed in this investigation are believed to be caused by a single gene.

   What characteristics are produced by each allele of this gene? ______________________

   ____

   ____

   ____

2. Humans vary in their levels of various enzymes. If you were a geneticist who wanted to determine the normal level of a particular enzyme in the human digestive system, what kinds of people should your sample include? How large should your sample be?

   ____

   ____

   ____

3. The enzyme papain, found in the papaya fruit, breaks down large protein molecules and is

   used as a meat tenderizer. How might papain make meat tender? ______________________

   ____

   ____

4. Cancer treatments such as chemotherapy or radiation therapy may cure one person of a certain type of cancer but leave another person with the same type of cancer unaffected. What

   is the most likely explanation of this difference in response? ______________________

   ____

   ____

   ____

Going Further

1. Repeat this investigation using other examples such as yellow and green peas or peas of different varieties.
2. Design experiments to determine the effects of light and/or heat on enzyme activity in seeds.

Name ______________________ Class ____________ Date ____________

19

Human Sex Chromosomes

Pre-Lab Discussion

The body cells of humans contain 23 pairs of chromosomes. Twenty-two pairs of these chromosomes are the same in both males and females and are called *autosomes,* or body chromosomes. The other pair of chromosomes, called *sex chromosomes,* is not the same in males and females. Human females have two matching sex chromosomes called *X chromosomes.* Human males have two dissimilar sex chromosomes—one X chromosome and one *Y chromosome.* The sex chromosomes are responsible for determining the sex and other traits in human offspring.

In 1960, the Canadian physician Murray Barr noticed darkly staining bodies of chromatin in some of the nuclei of the cells of female mammals. These bodies are absent from the nuclei of cells of males. Although females have two X chromosomes, one of them is condensed into a *Barr body,* as these chromatin bodies have come to be called. Many biologists believe that the formation of a Barr body deactivates parts of the second X chromosome in a female. This deactivation prevents the chromosome segments from producing double the necessary amount of enzymes and proteins, which would cause a chemical imbalance in the female.

In this investigation, you will examine human epithelial cells from the inside of the cheek for the presence or absence of Barr bodies.

Problem

How do male and female epithelial cells differ?

Materials *(per group)*

Microscope
Glass slide
Coverslip
Toothpicks
Medicine dropper
Paper towels
Methylene blue
Glass-marking pencil

Safety

Put on a laboratory apron if one is available. Put on safety goggles. Always handle the microscope with extreme care. You are responsible for its proper care and use. Use caution when handling glass slides as they can break easily and cut you. Always use special care when working with laboratory chemicals, as they may irritate the skin or cause staining of the skin or clothing. Never touch or taste any chemical unless instructed to do so. Note all safety alert symbols next to the steps in the Procedure and review the meanings of each symbol by referring to the symbol guide on page 10.

Procedure

1. On the edge of a clean glass slide, use the glass-marking pencil to write an F if you are a female and an M if your are a male. Then place a drop of water in the center of the slide.
2. Using the flat end of the toothpick, gently scrape the inside of your cheek. **CAUTION:** *Do not use force when scraping the inside of your cheek. Only a few cells are needed.* The end of the toothpick will have several cheek cells stuck to it even though you may see nothing but a drop of saliva.
3. Stir the water on the slide with the end of the toothpick in order to mix the cheek cells with the water. Immediately discard the toothpick in a place provided by your teacher.
4. Place one drop of methylene blue on top of the drop of water containing the cheek cells. **CAUTION:** *Use care when working with methylene blue as it may stain the skin and clothing.*
5. Wait one minute, then carefully place a coverslip over the cheek cells.
6. After waiting another minute, place a piece of paper towel at the edge of one side of the coverslip.
7. Place a drop of water at the edge of the coverslip on the side opposite the paper towel. See Figure 1. As methylene blue and water are absorbed by the paper towel, the drop of water on the opposite side of the coverslip will be drawn under the coverslip. This action will help to remove most of the methylene blue stain. Discard the paper towel.

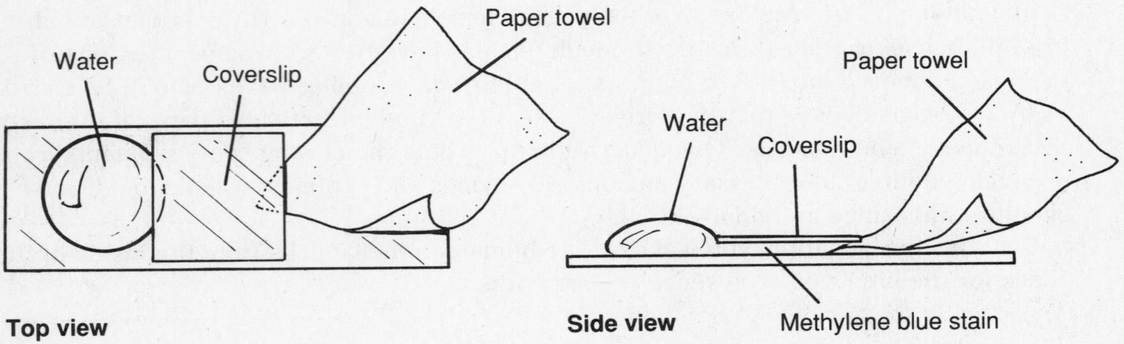

Figure 1

8. Place the slide on the microscope stage and use the low-power objective to locate a few cheek cells. **Note:** *In order to see the cells more clearly, you will need to adjust the diaphragm to reduce the amount of light entering the microscope.*
9. Switch to the high-power objective. **CAUTION:** *When switching to the high-power objective, always look at the objective from the side of the microscope so that the objective does not hit or damage the slide.* Observe your cheek cells under high power. In cheek cells, the Barr body will appear as a darkly staining body inside the nuclear membrane. Females will have one Barr body in many of the cells, whereas males have no Barr bodies.
10. Using the appropriate place in Observations, sketch and label what you see under the microscope. Label the following cell parts: cell membrane, nucleus, and Barr body (if present). Record the magnification of the microscope.
11. Exchange slides with a student of the opposite sex. Repeat steps 8 through 10.
12. Return slides to the appropriate class members. Carefully clean and dry your own slide and coverslip.

Name ______________________ Class __________ Date __________

Observations

High-power objective

Magnification __________

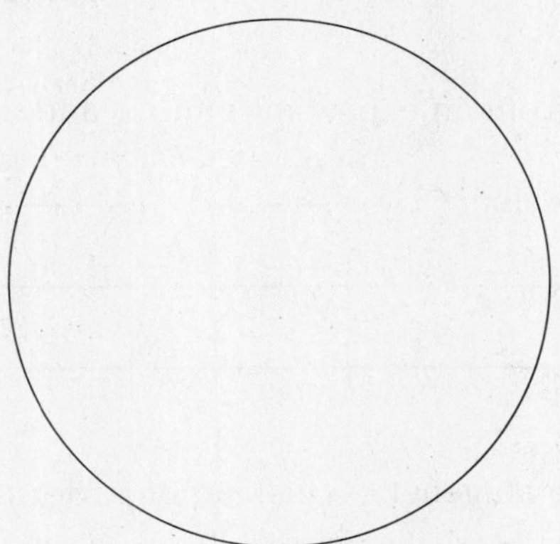

Female Cheek Cell

Magnification __________

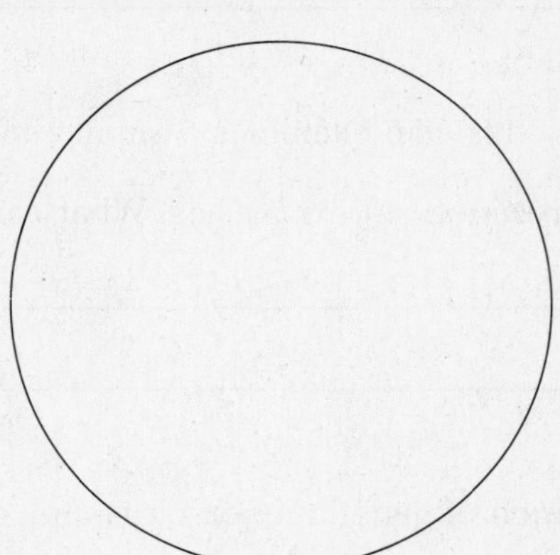

Male Cheek Cell

Analysis and Conclusions

1. What was the purpose of using methylene blue in this investigation?

2. How are male and female cheek cells similar in structure?

3. How are male and female cheek cells different in structure?

Critical Thinking and Application

1. Which parent is responsible for determining the sex of the offspring? Explain your answer.

2. A certain couple has three daughters. The mother is expecting a fourth child. What are the chances that the child will be a boy? Explain your answer. ______________________________

____

____

____

3. A physician examines a small amount of amniotic fluid from an expectant mother and finds the presence of Barr bodies. What can the physician tell the mother? ______________________________

____

____

4. Much of genetic research is aimed at understanding human genetics. Instead of performing genetic experiments on humans, researchers commonly use fruit flies. Explain why.

____

____

Going Further

Interview an obstetrician about the advantages and disadvantages of amniocentesis, which is the procedure that involves the removal of some amniotic fluid from a pregnant woman for examination and analysis.

Name ______________________ Class ____________ Date ____________

20

Karyotypes

Pre-Lab Discussion

Several human genetic disorders are caused by additional, missing, or damaged chromosomes. One way of studying genetic disorders is to observe the chromosomes themselves. In order to do this, cells from a person are grown in a laboratory. After the cells have reproduced a few times, they are treated with a chemical that stops cell division at the metaphase stage. During metaphase, the chromosomes are at the best length for identification.

The cells are treated further, stained, and then placed on glass slides. The chromosomes are observed under the microscope, where they are counted, checked for abnormalities, and photographed. The photograph is then enlarged, and the chromosomes are individually cut out. The chromosomes are identified and arranged in *homologous* pairs. Homologous chromosomes are identical, or matching, chromosomes. The arrangement of homologous pairs is called a *karyotype.*

In this investigation, you will use a sketch of chromosomes to make a karyotype. You will also examine the karyotype to determine the presence of any genetic defects.

Problem

How can chromosomes be observed?

Materials *(per student)*

Scissors
Glue or transparent tape

Safety

Be careful when handling sharp instruments. Note all safety alert symbols next to the steps in the Procedure and review the meanings of each symbol by referring to the symbol guide on page 10.

Procedure

Part A. Analyzing a Karyotype

1. Observe the karyotype in Figure 1. Notice that the two sex chromosomes, pair number 23, do not look alike. They are different because this karyotype is of a male, and a male has an X and a Y chromosome.

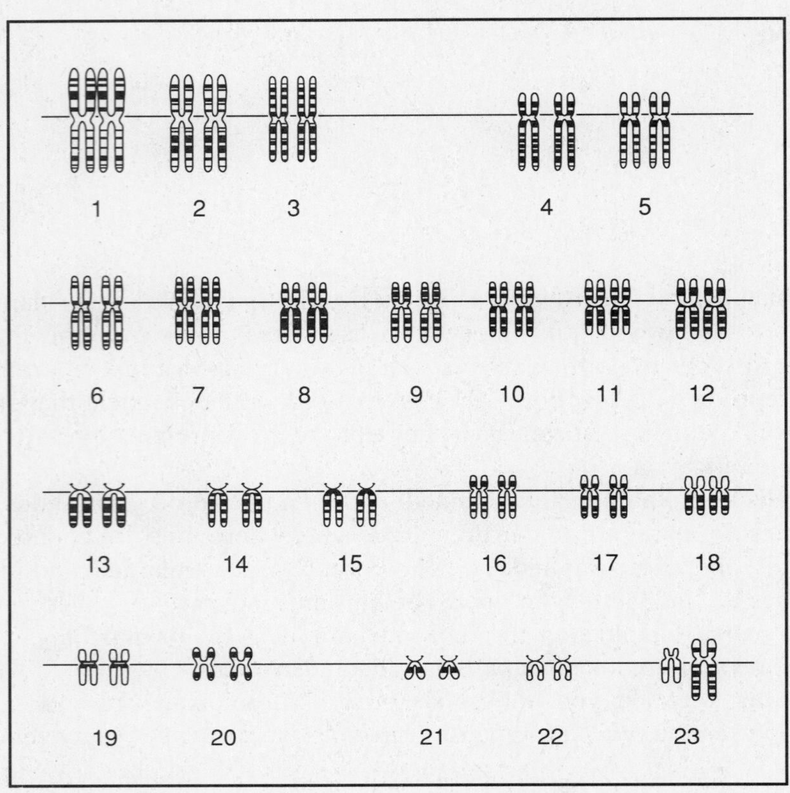

Figure 1

2. Identify the centromere in each pair of chromosomes. The centromere is the area where each chromosome narrows.

Part B. Using a Karyotype to Identify a Genetic Disorder

1. Study the human karyotype in Figure 2. Notice that 23 chromosomes are numbered 1 through 23.
2. To match the homologous chromosomes, look carefully at the unnumbered chromosomes. Note their overall size, the position of the centromere, and the pattern of the light and dark bands. Next to the unnumbered chromosome that is most similar to chromosome 1, write 1.
3. Repeat step 2 for chromosomes 2 through 23. **Note:** *Many genetic disorders involve missing or extra chromosomes.*

Name ______________________ Class ____________ Date ____________

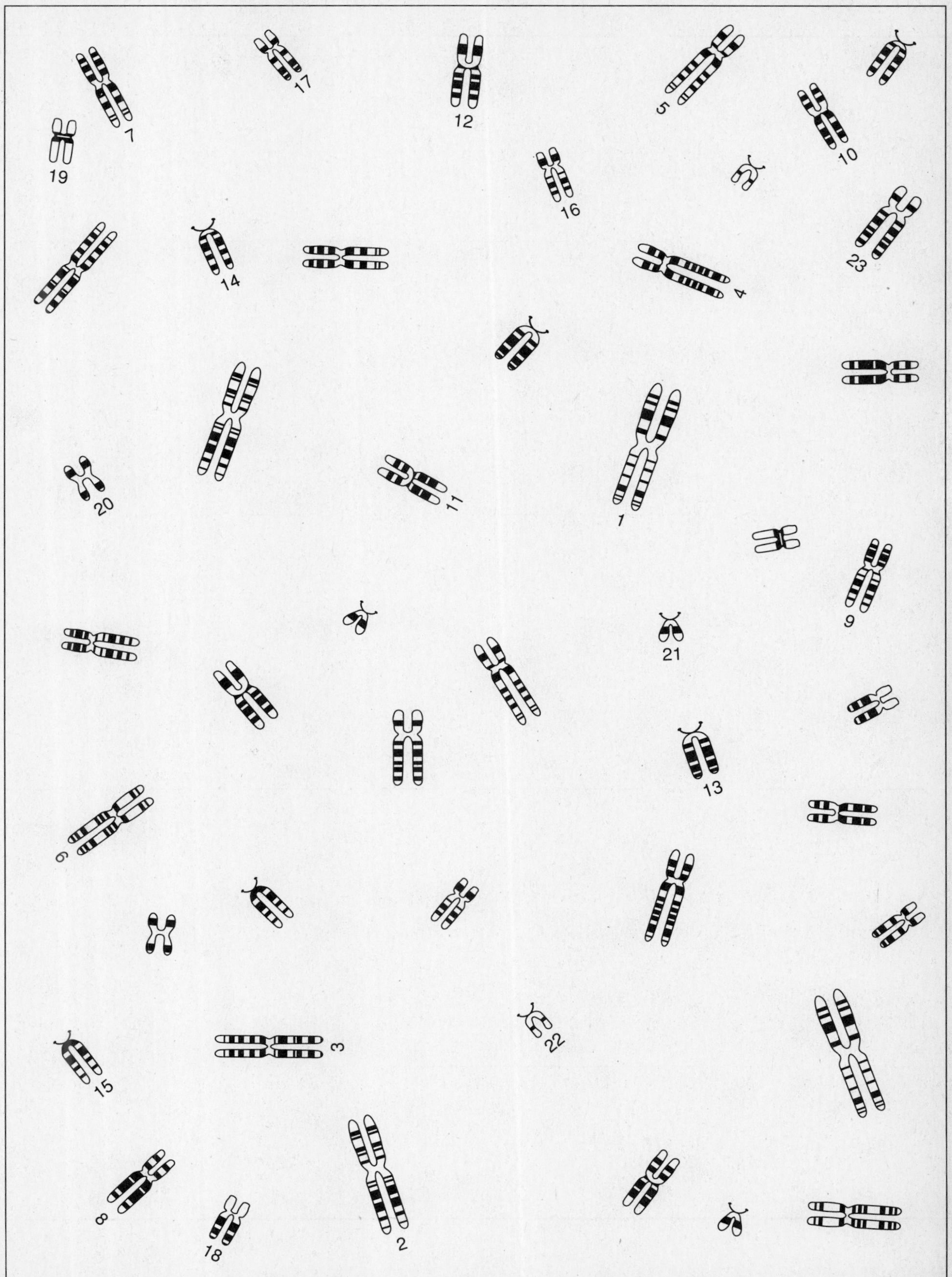

Figure 2

4. After all the chromosomes have been identified, use scissors to cut them out. Arrange the chromosomes in their appropriate place in Figure 3. Note the presence of any genetic defects.

1	2	3	4	5		
6	7	8	9	10	11	12
13	14	15	16	17	18	
19	20	21	22	23		

Figure 3

5. Observe the karyotypes in Figures 4 and 5. Note the presence of any genetic defects.

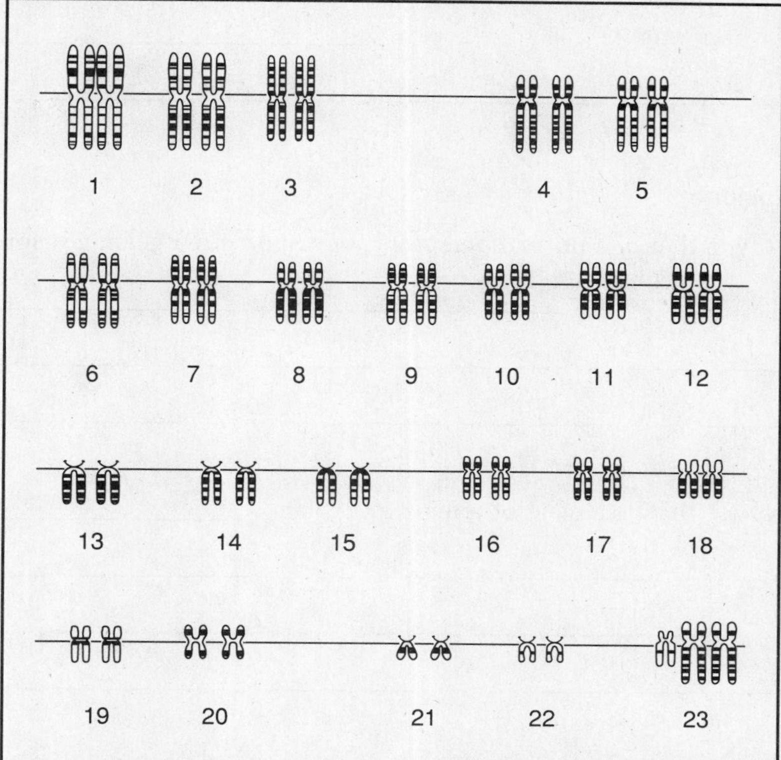

Figure 4

Figure 5

Observations

1. How many autosomes are present in your karyotype? ______________________

2. How many sex chromosomes are present in your karyotype? ______________________

3. Are there any abnormalities? If so, where? ______________________

____

Analysis and Conclusions

1. Is your karyotype that of a normal person or a person with a genetic disorder? If it is the latter, identify the disorder. ______________________

____

____

2. Is your karyotype that of a male or a female? Explain. ______________________

____

____

____

3. How does the karyotype in Figure 4 differ from the karyotype in Figure 1?

____

____

____

____

4. How does the karyotype in Figure 5 differ from the karyotype in Figure 1?

____

____

____

____

5. Do the karyotypes in Figures 4 and 5 exhibit any genetic disorder? If so, identify the disorder.

____

____

____

____

Name ______________________ Class ____________ Date ____________

Critical Thinking and Application

1. What happens during meiosis that ultimately results in a defect characterized by the addition of chromosomes? By the deletion of chromosomes? ______________________

2. The human male determines the sex of his offspring. Explain this statement.

Going Further

Using reference material, research the following human genetic disorders: Patau syndrome and Edwards syndrome. For each disorder find the cause, type of mutation, and characteristics. Construct a karyotype of each.

Name ______________________ Class ____________ Date ____________

Constructing a Human Pedigree

Pre-Lab Discussion

Human traits are often difficult to study for several reasons. Unlike some organisms, which produce large numbers of offspring very quickly, humans reproduce slowly and produce few offspring at one time. Thus human traits must be studied through population sampling and pedigree analysis. A *pedigree* is a diagram that shows the phenotype of a particular genetic trait in a family from one generation to the next. Genotypes for individuals in a pedigree often can be determined with an understanding of inheritance and probability.

In this investigation, you will use both population sampling and pedigree analysis to observe human traits.

Problem

How can pedigree analysis help in the study of human traits?

Materials *(per student)*

No special materials are needed.

Procedure

Part A. Interpreting a Pedigree Chart

1. Figure 1 is a pedigree, or a diagram of a family's pattern of inheritance for a specific trait.

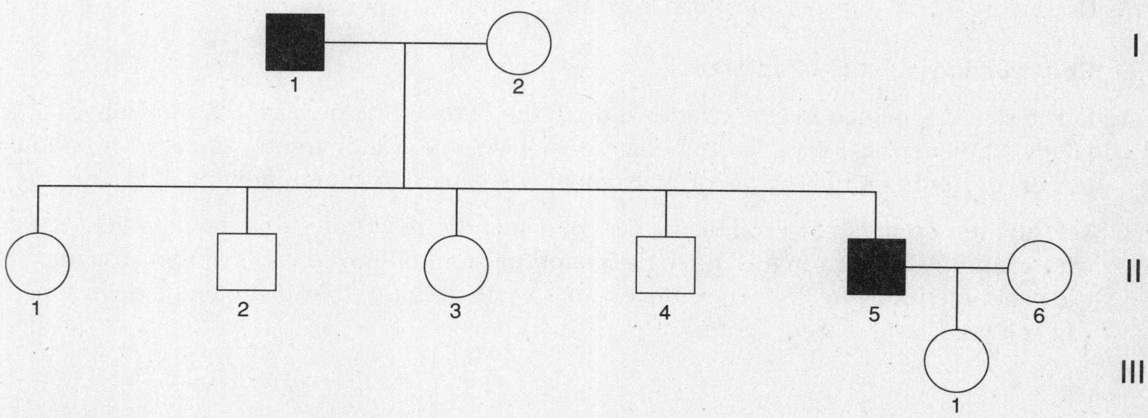

Figure 1

2. Notice that in a pedigree, each person is represented by an Arabic number and each generation is represented by a Roman numeral. In this way, each person can be identified by a generation numeral and an individual number. Males are represented by squares and females by circles. Unshaded symbols (squares or circles) indicate people who are homozygous or heterozygous for the dominant trait. Shaded symbols indicate people who are homozygous for the recessive trait.

3. In Figure 1, I-1 and I-2 are the parents. The horizontal line that connects them is called a marriage line. The vertical line that extends down from the marriage line connects the children to the parents. Children are listed in order of their births from left to right. In other words, the oldest child is always placed on the extreme left. In this pedigree, persons II-1, II-2, II-3, II-4, and II-5 are the children of persons I-1 and I-2.

4. The trait being analyzed in Figure 1 is ear-lobe shape. There are two general ear-lobe shapes, free lobes and attached lobes. See Figure 2. The gene responsible for free ear lobes, represented by the capital letter E, is dominant over the gene for attached ear lobes, represented by the lowercase letter e. People with attached ear lobes are homozygous for the recessive trait and are represented as ee. In Figure 1, I-1 and II-5 are homozygous recessive (ee) and have attached ear lobes. The people represented by the unshaded symbols have two possible genotypes: EE or Ee.

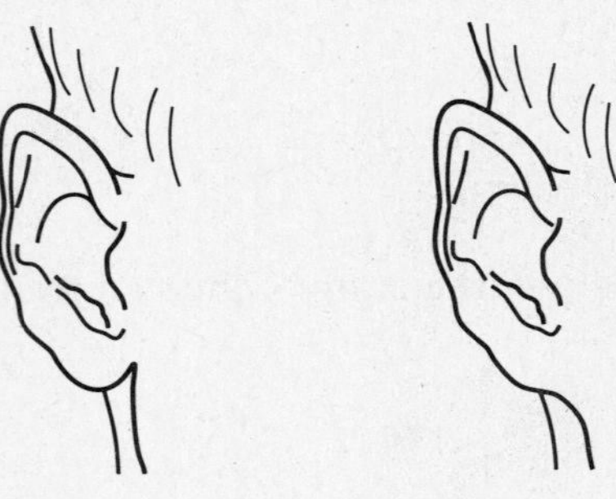

Figure 2

5. Use Figure 1 to complete questions 1 through 8 in Observations.

Part B. Constructing a Family Pedigree

1. In the space provided in Observations, draw the outline of a pedigree of your family or a family who lives near you. In the pedigree, include any grandparents, aunts, uncles, cousins, nieces, or nephews that live nearby. Number the generations and individuals.

2. The ability of a person to roll his or her tongue is the result of the dominant allele R. People who cannot roll their tongues have the genotype rr. People who can roll their tongues have the genotype RR or Rr. If you are developing a pedigree for your own family, determine if you can roll your tongue. See Figure 3.

Name ______________________ Class __________ Date __________

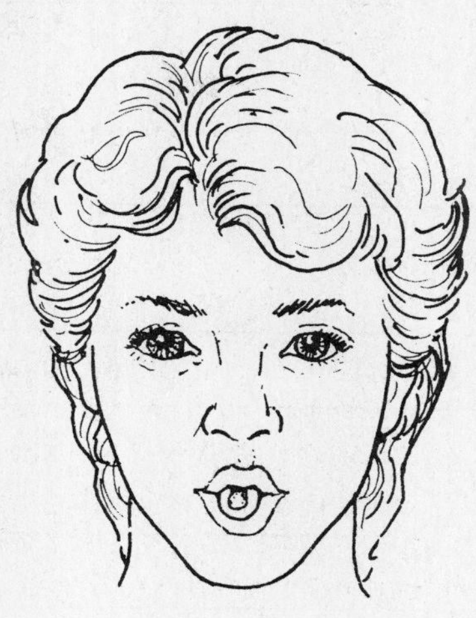

Tongue roller

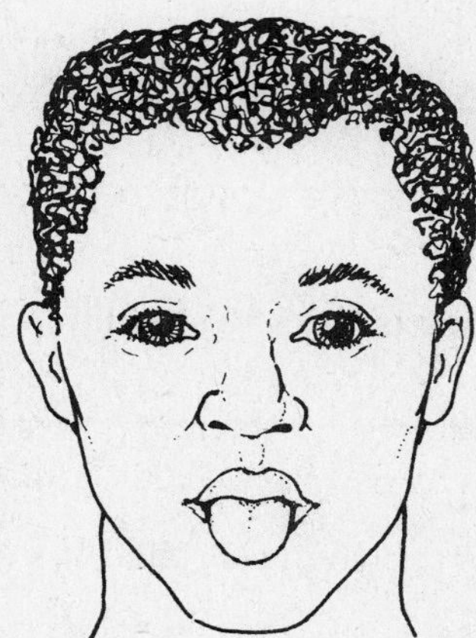

Non-tongue roller

Figure 3

3. If you cannot roll your tongue, enter the genotype rr in the space below your symbol on the pedigree.
4. If you can roll your tongue, enter the genotype R _ for the presence of the dominant gene in the space below your symbol on the pedigree.
5. If you are constructing a pedigree for your own family, survey additional members of your family for their ability to roll their tongue. If you are constructing a pedigree for another family, survey as many members of that family as possible.
6. Record the results of family members you tested in your pedigree.

Observations

1. What is the genotype of I-2? Explain your answer. ______________________

2. What are the genotypes of II-1, II-2, II-3, and II-4? Explain your answer.

3. What are the possible genotypes for II-6? Explain your answer. ____________________

____

____

4. If II-6 is EE, what is the genotype of her child with II-5? ____________________

____

5. What sex is the oldest child in generation II? ____________________

____

6. Who is the youngest child in generation II? ____________________

____

7. Who is the daughter-in-law in this family? ____________________

____

8. How many generations are represented in this pedigree? ____________________

____

Use this space to construct the pedigree for the family you have chosen to study. Correctly number each generation and person. Shade in the symbols for those people who are homozygous recessive. Below the symbol for each person, write as much of the person's genotype as possible.

Name ______________________ Class ____________ Date ____________

Analysis and Conclusions

1. Would you expect the other students in your class to have tongue-rolling pedigrees that are identical to yours? Explain your answer. ______________________

2. Explain why you are not always able to determine the exact genotype for a trait of a person when you construct a pedigree. ______________________

3. If two parents are unable to roll their tongues, is it likely that they will have children who will be able to roll their tongues? Explain your answer. ______________________

Critical Thinking and Application

1. Can the actual traits of an offspring be determined by knowing the traits of the parents? Explain your answer. ______________________

2. A woman received the genes aBcD from her mother and AbCd from her father. Which of the following gene combinations could be present in her gametes: ABCD, abcd, ABCDD, aBccD, ABcd, AaBb? Explain your answer. ______________________

3. If a man who has long eyelashes (LL) marries a woman who has long eyelashes (Ll), what are the possible genotypes and phenotypes of their children? ______________________

4. Complete the pedigree in the figure below. In the spaces below each symbol, write as much of the genotype of each individual as can be determined from the information provided. Assume the shaded symbols represent the homozygous recessive genotype rr.

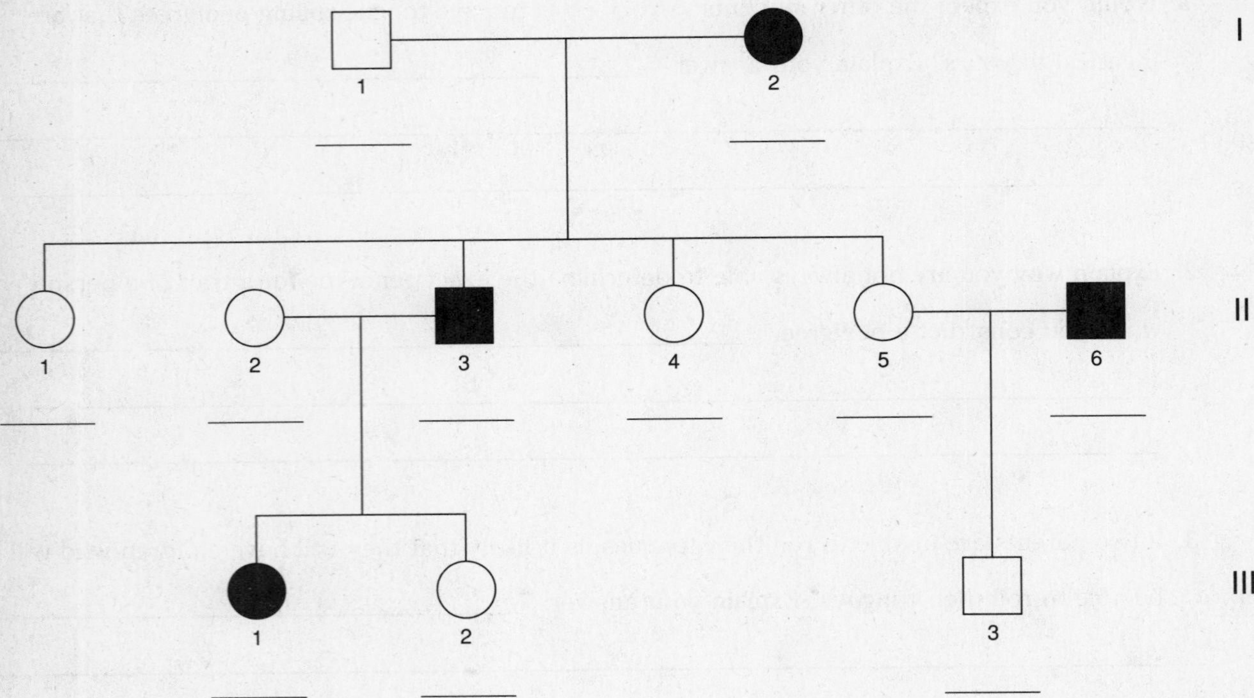

Going Further

Design another pedigree for the family you used in Part B showing any of the following traits: hair on middle joint of fingers, straight little finger, cleft in chin, direction of hair whorl. Determine if each of these traits is dominant or recessive. Apply all appropriate symbols and labels with the possible genotypes of each person. Compare the different pedigrees you have constructed. What are the similarities? Differences?

Investigating Inherited Human Traits

Pre-Lab Discussion

Heredity is the passing on of traits, or characteristics, from parent to offspring. The units of heredity are called genes. Genes are found on the chromosomes in a cell. The combinations of genes for each trait occur by chance.

When one gene in a pair is stronger than the other gene, the trait of the weaker gene is masked, or hidden. The stronger gene is the dominant gene, and the gene that is masked is the recessive gene. Dominant genes are written as capital letters and recessive genes are written as lowercase letters. If both genes in a gene pair are the same, the trait is said to be *homozygous,* or pure. If the genes are not similar, the trait is said to be *heterozygous,* or hybrid. Sometimes genes are neither dominant nor recessive. The result of such a situation is a blending of traits.

The genetic makeup of an individual is known as its genotype. The observable physical characteristics of an individual that are the result of its genotype are known as its phenotype. In humans, the sex of an individual is determined by the particular combination of the two sex chromosomes. Individuals that have two X chromosomes (XX) are females, whereas those with an X and a Y chromosome (XY) are males.

In this investigation, you will observe how the results of different gene combinations produced certain traits.

Problem

How are traits inherited?

Materials *(per pair of students)*

3 textbooks
2 coins
Pencil

Procedure

1. Place the textbooks on the laboratory table so that they form a triangular well in which to toss the coins.
2. Determine which partner will toss for the female and which will toss for the male. Remember that there are two genes per trait.
3. Have the partner who is representing the male flip a coin into the well to determine the sex of the offspring. If the coin lands heads up, the offspring is a female. If the coin lands tails up, the offspring is a male. Record the sex of the offspring in Observations.
4. For all the coin tosses you will now make, heads will represent the dominant gene and tails will represent the recessive gene.

5. You and your partner should now flip your coins into the well at the same time. **Note:** *The coins should be flipped only once for each trait.*

6. Continue to flip the coins for each trait listed in the table in Figure 1. After each flip, record the trait of your offspring by placing a check in the appropriate box in the table.

7. Using the recorded traits, draw the facial features for your offspring in the space provided in Observations.

Traits	Dominant (both heads)	Hybrid (one head, one tail)	Recessive (both tails)
Shape of face	 round (RR)	 round (Rr)	 square (rr)
Cleft in chin	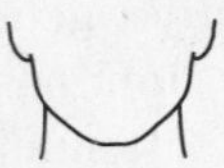 absent (CC)	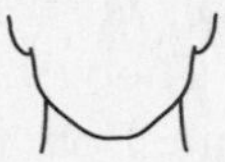 absent (Cc)	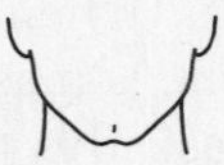 present (cc)
Hair	 curly (HH)	 wavy (Hh)	 straight (hh)
Widow's peak	 present (WW)	 present (Ww)	 absent (ww)
Spacing of eyes	 close together (EE)	 normal distance (Ee)	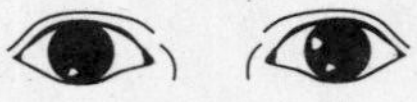 far apart (ee)
Shape of eyes	 almond (AA)	 almond (Aa)	 round (aa)
Position of eyes	 straight (SS)	 straight (Ss)	 slant upward (ss)
Size of eyes	 large (LL)	 medium (Ll)	small (ll)

Figure 1

Name ______________________ Class __________ Date __________

Traits	Dominant (both heads)	Hybrid (one head, one tail)	Recessive (both tails)
Length of eyelashes	long (LL)	long (Ll)	short (ll)
Shape of eyebrows	bushy (BB)	bushy (Bb)	fine (bb)
Position of eyebrows	not connected (NN)	not connected (Nn)	connected (nn)
Size of nose	large (LL)	medium (Ll)	small (ll)
Shape of lips	thick (TT)	normal (Tt)	thin (tt)
Size of ears	large (LL)	normal (Ll)	small (ll)
Size of mouth	large (LL)	medium (Ll)	small (ll)
Freckles	present (FF)	present (Ff)	absent (ff)
Dimples	present (DD)	present (Dd)	absent (dd)

Figure 1 *(continued)*

Observations

Sex of offspring ____________________

Drawing of Offspring

Name ______________________ Class ____________ Date ____________

Analysis and Conclusions

1. What percent chance did you and your partner have of "producing" a male offspring? A female offspring? Explain your answer. ______________________

2. Would you expect the other pairs of students in your class to have an offspring similar to yours? Explain your answer. ______________________

3. If a woman who is homozygous for almond-shaped eyes (AA) marries a man who is heterozygous for almond-shaped eyes (Aa), what are the possible genotypes and phenotypes of their children? ______________________

4. What are the possible genotypes of the parents of a child who has wavy hair (Hh)? ______________________

5. Which traits in this investigation showed a blending of genes? ______________________

Critical Thinking and Application

1. Did you think that anyone in your class has all the same genetic traits that you have? Explain your answer. ______________________

2. How might it be possible for you to show a trait when neither of your parents shows it?

____

____

____

3. Do you think you would have some genetic traits similar to your grandparents? Explain your answer. ____

____

____

4. There is a small village in a mountain valley in Spain where a large number of people are polydactyl (have more than five fingers or toes). Why does this trait tend to be passed on from generation to generation? ____

____

____

____

5. There have been cases in history where a king divorced his queen because she produced only daughters. Using your knowledge of genetics, explain why this was an incorrect move.

____

____

____

Going Further

Repeat this investigation with your partner to "produce" your second offspring. After completing all of your tosses, make a drawing of the offspring. What similarities exist between your first and second offspring? What differences? Would you expect a third offspring to resemble either the first or the second offspring?

Name ______________________ Class __________ Date __________

Cloning

Pre-Lab Discussion

In recent years, plant breeders have produced new and improved varieties of plants by producing polyploid plants. A polyploid plant has one or more extra sets of chromosomes. In plants polyploidy is fairly common. Polyploid varieties of many garden plants are in demand commercially because they produce more foliage and larger fruits.

Once plants with desirable characteristics have been produced by inducing polyploidy, it is possible to obtain exact copies of the plants by *cloning* them. Cloning is the production of genetically identical organisms. Cloning is used to grow uniform and commercially desirable varieties of fruits, vegetables, and flowering plants. Cloning is not limited to plants alone, however. Certain animals such as frogs, toads, and salamanders can also be cloned. Animal cloning is particularly helpful to medical research on cancer and other diseases. It is also being used to understand how cells perform different tasks and how they build up immunities to various diseases.

In this investigation, you will clone a duckweed plant.

Problem

How is a plant cloned?

Materials *(per group)*

- Glass-marking pencil
- Distilled water
- Tongs
- Watch glass
- Duckweed plant culture
- 600-mL beaker
- 5 glass petri dishes (60 mm in diameter)
- 10% bleach solution
- Inoculating loop
- Hand lens
- Bunsen burner
- Ring stand
- Iron ring
- Wire gauze
- Plant growth medium
- Paper towels
- Fluorescent light with timer (if available)

Safety

Put on a laboratory apron if one is available. Put on safety goggles. Handle all glassware carefully. Use extreme care when working with heated equipment or materials to avoid burns. Always use special caution when working with laboratory chemicals, as they may irritate the skin or cause staining of the skin or clothing. Never touch or taste any chemical unless instructed to do so. Note all safety alert symbols next to the steps in the Procedure and review the meanings of each symbol by referring to the symbol guide on page 10.

Procedure

1. Place a duckweed plant along with some water from its container into a watch glass. Examine the plant with a hand lens. Note the elongated flattened body, which consists of leaflike stems called fronds. The fronds float on the water's surface while the threadlike rootlets extend down into the water.

2. To set up a water bath, half fill a 600-mL beaker with distilled water. Bring the water to a gentle, rolling boil. **CAUTION:** *Be very careful when working with heated equipment or materials to avoid burns. Wear safety goggles.*

3. Place the lids and bottoms of the 5 glass petri dishes into the boiling water bath for 3 minutes.

4. After 3 minutes, use tongs to remove the lids and bottoms from the water bath. Place them upside down on paper towels to dry. **Note:** *Do not expose the inside of the dishes to the air.*

5. After the lids and bottoms have cooled, assemble the petri dishes and place them side-by-side in a row. With a glass-marking pencil, number the dishes 1 through 5. **Note:** *To keep from contaminating the dishes, remember to open the lids just enough to carry out the following steps without unnecessary exposure to the air.*

6. Half fill petri dish 1 with the 10% bleach solution. Half fill petri dishes 2, 3, and 4 with distilled water. Half fill petri dish 5 with the plant growth medium. **CAUTION:** *Be very careful when using a bleach solution. It is a strong base and may burn the skin or clothing. Wear safety goggles.*

7. Sterilize the inoculating loop by passing it through the flame of a Bunsen burner until its entire length is heated to a red glow. **CAUTION:** *Be very careful when using an open flame. Wear safety goggles.* Allow the inoculating loop to cool for a few seconds.

8. Using the inoculating loop, remove a duckweed plant from the watch glass and transfer it to petri dish 1. Submerge the plant in the 10% bleach solution for 10 minutes. This action will destroy any organisms that might be on the plant.

9. Transfer the plant to petri dish 2 to rinse off the bleach solution. To ensure the removal of the bleach, gently submerge and agitate the plant. Repeat this process in petri dishes 3 and 4. Then transfer the plant to petri dish 5. **Note:** *Sterilize the loop before each transfer and then pause a few seconds to allow the loop to cool before picking up the plant.*

10. With a hand lens, examine the plant in petri dish 5. Count the number of fronds that the plant has. Record this number in the Data Table.

11. Place petri dish 5 under a fluorescent light. The light should be regulated with a timer to provide 16 hours of light followed by 8 hours of darkness. If a fluorescent light is unavailable, place the petri dish on the ledge of a north-facing window.

12. Thoroughly wash and clean petri dishes 1 through 4, using a paper towel to remove the numbers from them.

13. Observe the duckweed plant each week for three weeks for signs of growth (number of fronds). Record this information in the Data Table.

Name ______________________ Class __________ Date __________

Observations

Data Table

Number of Fronds on Duckweed Plant			
First Day	After First Week	After Second Week	After Third Week

Analysis and Conclusions

1. What is a clone? ____

____

____

2. Why is so much care taken to sterilize equipment in this investigation?

____

____

3. How does cloning differ from sexual reproduction? ____

____

____

4. What happened to the old fronds in the duckweed plant as new ones appeared?

____

____

____

Critical Thinking and Application

1. Of what value is cloning to someone working in a greenhouse? ____

____

____

2. How might cloning be useful in redeveloping a forested area that was destroyed by fire?

3. Explain why the new plant that develops from a stem of another plant is a clone.

4. Although the development of hybrid varieties of corn has tripled average corn yields in the United States since the 1930s, some plant breeders say that genetic uniformity (sameness) has become a problem in terms of plant disease. Why might disease be a problem for cloned plants?

5. Why has cloning been limited thus far to simple organisms and plants?

Going Further

1. Research various techniques that are used for cloning plants. Then try to clone a mature fern from a few tissue cells taken from a leaflet of a frond.
2. Vegetative propagation is a cloning technique in which various parts of a plant such as leaves, stems, and roots are used to produce a new plant. Demonstrate the technique of vegetative propagation by cloning new plants from these plant parts.

Name ______________________ Class ____________ Date ____________

24

Interpreting Events from Fossil Evidence

Pre-Lab Discussion

A *fossil* is any evidence of life in the prehistoric past. Fossils can be actual remains of animals, impressions, carbon residues, or tracks left by a living organism. By using fossil records, scientists are able to piece together the story that fossils tell about the history of the Earth.

In this investigation, you will use some present-day knowledge and common sense to interpret fossil evidence.

Problem

How can fossils relate the history of an era?

Materials *(per student)*

Pencil

Procedure

Part A.

1. Observe the diagram of the fossil footprints in Figure 1.

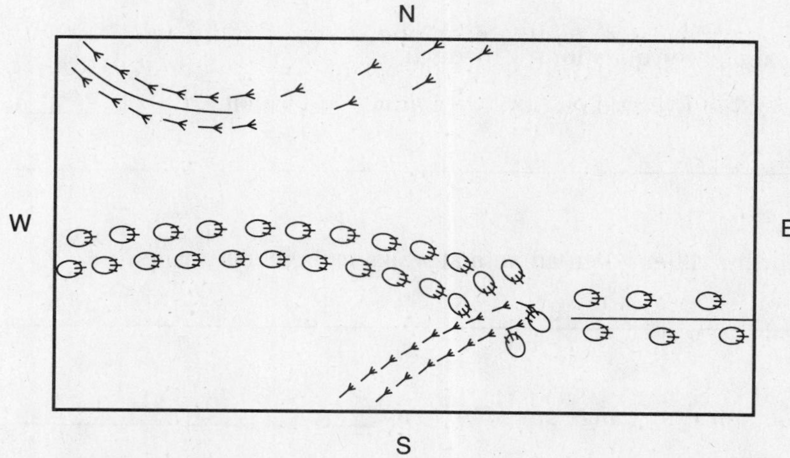

Figure 1

2. Use Figure 1 to answer questions 1 through 6 in Observations.

3. Using the answers to these questions and some common sense, intepret the events that may have occurred in Figure 1. **Note:** *Your interpretation should be consistent with the diagram.*

Part B.

1. Observe the diagram of the fossil footprints in Figure 2.

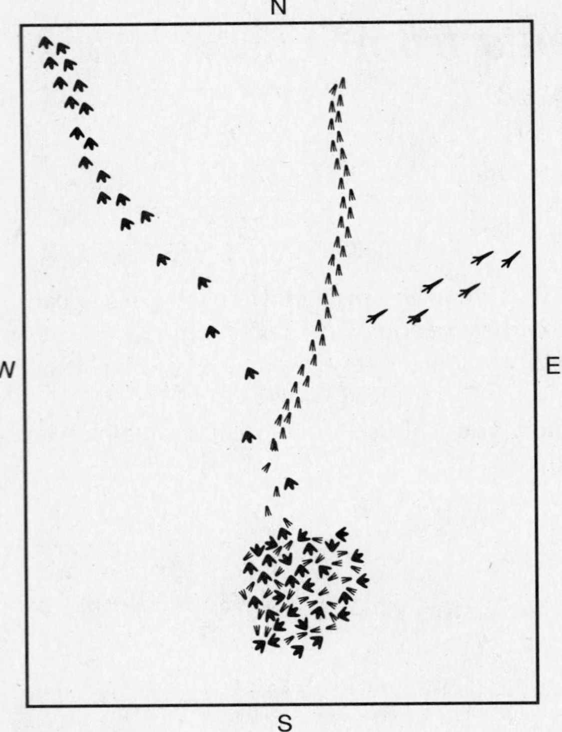

Figure 2

2. Use Figure 2 to answer questions 7 through 9 in Observations.

3. Using the answers to these questions and some common sense, interpet the events that may have occurred in Figure 2. **Note:** *Your interpretation should be consistent with the diagram.*

Observations

Part A. Use Figure 1 to answer questions 1 through 6.

1. How can you tell in which direction the animals are walking? ______________________________

____

2. How do footprints differ when an animal walks from when it runs?

____

3. Did any of the animals change speed? When? ______________________________

____

____

Name ______________________ Class ______________ Date ______________

4. How many different types of animals are represented? ______________________

5. How did the animals interact with each other? ______________________

6. How would footprints be formed and preserved? ______________________

Part B. Use Figure 2 to answer questions 7 through 9.

7. Did any of the animals change speed? When? ______________________

8. How many different types of animals are represented? ______________________

9. How did the animals interact with one another? ______________________

Analysis and Conclusions

Part A. Use Figure 1 to answer questions 1 through 4.

1. In what kind of environment did this all take place? ______________________

2. If all the footprints were made within minutes of one another, which way was the wind blowing? Explain your answer. ______________________

3. What might the lines between the footprints of two of the organisms represent?

4. Write a paragraph interpreting the events in Figure 1. Support your interpretations with evidence from the diagram.

Part B. Use Figure 2 to answer questions 5 through 7.

5. Is it possible that the animals who made the footprints represented in the diagram never actually met each other at the same time? Explain your answer.

6. What might depth of a footprint tell an interpreter?

Name ______________________ Class __________ Date __________

7. Write a paragraph interpreting the events in Figure 2. Support your interpretations with evidence from the diagram.

Critical Thinking and Application

1. In what type of rock would you be most likely to find fossils? Explain your answer.

2. Would you be likely to find the fossil remains of a jellyfish or a worm? Explain your answer.

3. The exposed rock layers of the Grand Canyon are rich with fossil specimens. What does the presence of fossil coral, sponges, shellfish, and trilobites indicate about the past climate of the Grand Canyon area?

4. How could paleontologists reconstruct our present-day environment 25,000 years from now if all written history were to be lost or destroyed? ______________________________

____

____

____

5. Scientists have found fossils of the same kind of organism on different continents. How might this have occurred? ______________________________

____

____

____

Going Further

1. Make a cast of leaves, twigs, footprints, or bones, using plaster of Paris or clay. Relate this process to fossil formation.
2. Find out if there is a relationship between the depth of a footprint and the mass of a living thing. Have different people step in soft clay or soil. Measure the depth of the footprint and determine the relationship between the mass of the person and the depth of the footprint. Organize your information in a chart. Then interpret an "unknown" footprint.

Name ______________________ Class ____________ Date ____________

25

Amino Acid Sequences and Evolutionary Relationships

Pre-Lab Discussion

Homologous structures—those structures believed to have a common origin but not necessarily a common function—provide some of the most significant evidence supporting the theory of evolution. For example, the forelimbs of vertebrates often have different functions and outward appearances, yet the underlying similarity of the bones indicates a common origin. Although homologous structures can be used to demonstrate relationships between similar organisms, they are of little value in determining evolutionary relationships among those structures that are dissimilar.

Another technique used to determine evolutionary relationships is to study the biochemical similarity of organisms. Though molds, aardvarks, and humans appear to have little in common physically, a study of their proteins reveals certain similarities. Biologists have perfected techniques for determining the sequence of amino acids in proteins. By comparing the amino acid sequences in homologous proteins of similar organisms and of diverse organisms, evolutionary relationships that might otherwise go undetected can be determined. Biologists believe that the greater the similarity between the amino acid sequences of two organisms, the closer their relationship. Conversely, the greater the differences, the more distant the relationship. Further, biologists have found that such biochemical evidence compares favorably with other lines of evidence for evolutionary relationships.

In this investigation, you will compare amino acid sequences in proteins of several vertebrates. You will also study amino acid differences and infer evolutionary relationships among some diverse organisms.

Problem

How do amino acid sequences provide evidence for evolution?

Materials *(per student)*

No special materials are needed.

Procedure

Part A. Comparing Amino Acid Sequences

1. Examine Figure 1, which compares corresponding portions of hemoglobin molecules in humans and five other vertebrate animals. Hemoglobin, a protein composed of several long chains of amino acids, is the oxygen-carrying molecule in red blood cells. The sequence shown is only a portion of a chain made up of 146 amino acids. The numbers in Figure 1 indicate the position of a particular amino acid in the chain.

	87	88	89	90	91	92	93	94	95	96	97	98	99	100	101
Human	THR	LEU	SER	GLU	LEU	HIS	CYS	ASP	LYS	LEU	HIS	VAL	ASP	PRO	GLU
Chimpanzee	THR	LEU	SER	GLU	LEU	HIS	CYS	ASP	LYS	LEU	HIS	VAL	ASP	PRO	GLU
Gorilla	THR	LEU	SER	GLU	LEU	HIS	CYS	ASP	LYS	LEU	HIS	VAL	ASP	PRO	GLU
Rhesus monkey	GLN	LEU	SER	GLU	LEU	HIS	CYS	ASP	LYS	LEU	HIS	VAL	ASP	PRO	GLU
Horse	ALA	LEU	SER	GLU	LEU	HIS	CYS	ASP	LYS	LEU	HIS	VAL	ASP	PRO	GLU
Kangaroo	LYS	LEU	SER	GLU	LEU	HIS	CYS	ASP	LYS	LEU	HIS	VAL	ASP	PRO	GLU

	102	103	104	105	106	107	108	109	110	111	112	113	114	115	116
Human	ASN	PHE	ARG	LEU	LEU	GLY	ASN	VAL	LEU	VAL	CYS	VAL	LEU	ALA	HIS
Chimpanzee	ASN	PHE	ARG	LEU	LEU	GLY	ASN	VAL	LEU	VAL	CYS	VAL	LEU	ALA	HIS
Gorilla	ASN	PHE	LYS	LEU	LEU	GLY	ASN	VAL	LEU	VAL	CYS	VAL	LEU	ALA	HIS
Rhesus monkey	ASN	PHE	LYS	LEU	LEU	GLY	ASN	VAL	LEU	VAL	CYS	VAL	LEU	ALA	HIS
Horse	ASN	PHE	ARG	LEU	LEU	GLY	ASN	VAL	LEU	ALA	LEU	VAL	VAL	ALA	ARG
Kangaroo	ASN	PHE	LYS	LEU	LEU	GLY	ASN	ILE	ILE	VAL	ILE	CYS	LEU	ALA	GLU

Figure 1

2. In Data Table 1, notice that the abbreviated names of the amino acids in human hemoglobin are printed.
3. In the appropriate spaces in Data Table 1, write the abbreviated name of each amino acid in chimpanzee hemoglobin that is *different from* that in human hemoglobin. If there are no differences, leave the spaces blank.
4. For the remaining organisms, write the abbreviated names of the amino acids that do *not* correspond to those in human hemoglobin. **Note:** *Always be sure that you compare the amino acid sequence of each organism with that of the human and not the organism on the line above.*
5. Use Figure 1 to complete Data Table 2.

Part B. Inferring Evolutionary Relationships from Differences in Amino Acid Sequences

1. Another commonly studied protein is cytochrome *c*. This protein, consisting of 104 amino acids, is located in the mitochondria of cells. There it functions as a respiratory enzyme. Examine Figure 2. Using human cytochrome *c* as a standard, the amino acid differences between humans and a number of other organisms are shown.

Species Pairings	Number of Differences
Human–chimpanzee	0
Human–fruit fly	29
Human–horse	12
Human–pigeon	12
Human–rattlesnake	14
Human–red bread mold	48
Human–rhesus monkey	1
Human–screwworm fly	27
Human–snapping turtle	15
Human–tuna	21
Human–wheat	43

Figure 2

Name ______________________ Class __________ Date __________

2. Using Figure 2, construct a bar graph on Graph 1 to show the amino acid differences between humans and other organisms.

3. Now examine Figure 3. In this figure the cytochrome *c* of a fruit fly is used as a standard in comparing amino acid differences among several organisms. Construct a bar graph on Graph 2 to show these differences.

Species Pairings	Number of Differences
Fruit fly–dogfish shark	26
Fruit fly–pigeon	25
Fruit fly–screwworm fly	2
Fruit fly–silkworm moth	15
Fruit fly–tobacco hornworm moth	14
Fruit fly–wheat	47

Figure 3

Observations

Data Table 1

	87	88	89	90	91	92	93	94	95	96	97	98	99	100	101	102	103	104	105	106
Human	THR	LEU	SER	GLU	LEU	HIS	CYS	ASP	LYS	LEU	HIS	VAL	ASP	PRO	GLU	ASN	PHE	ARG	LEU	LEU
Chimpanzee																				
Gorilla																				
Rhesus monkey																				
Horse																				
Kangaroo																				

Data Table 1 (Continued)

	107	108	109	110	111	112	113	114	115	116
Human	GLY	ASN	VAL	LEU	VAL	CYS	VAL	LEU	ALA	HIS
Chimpanzee	____	____	____	____	____	____	____	____	____	____
Gorilla	____	____	____	____	____	____	____	____	____	____
Rhesus monkey	____	____	____	____	____	____	____	____	____	____
Horse	____	____	____	____	____	____	____	____	____	____
Kangaroo	____	____	____	____	____	____	____	____	____	____

Data Table 2

Organisms	Number of Amino Acid Differences	Positions In Which They Vary
Human and chimpanzee		
Human and gorilla		
Human and rhesus monkey		
Human and horse		
Human and kangaroo		

Name ______________________ Class ____________ Date ____________

Graph 1

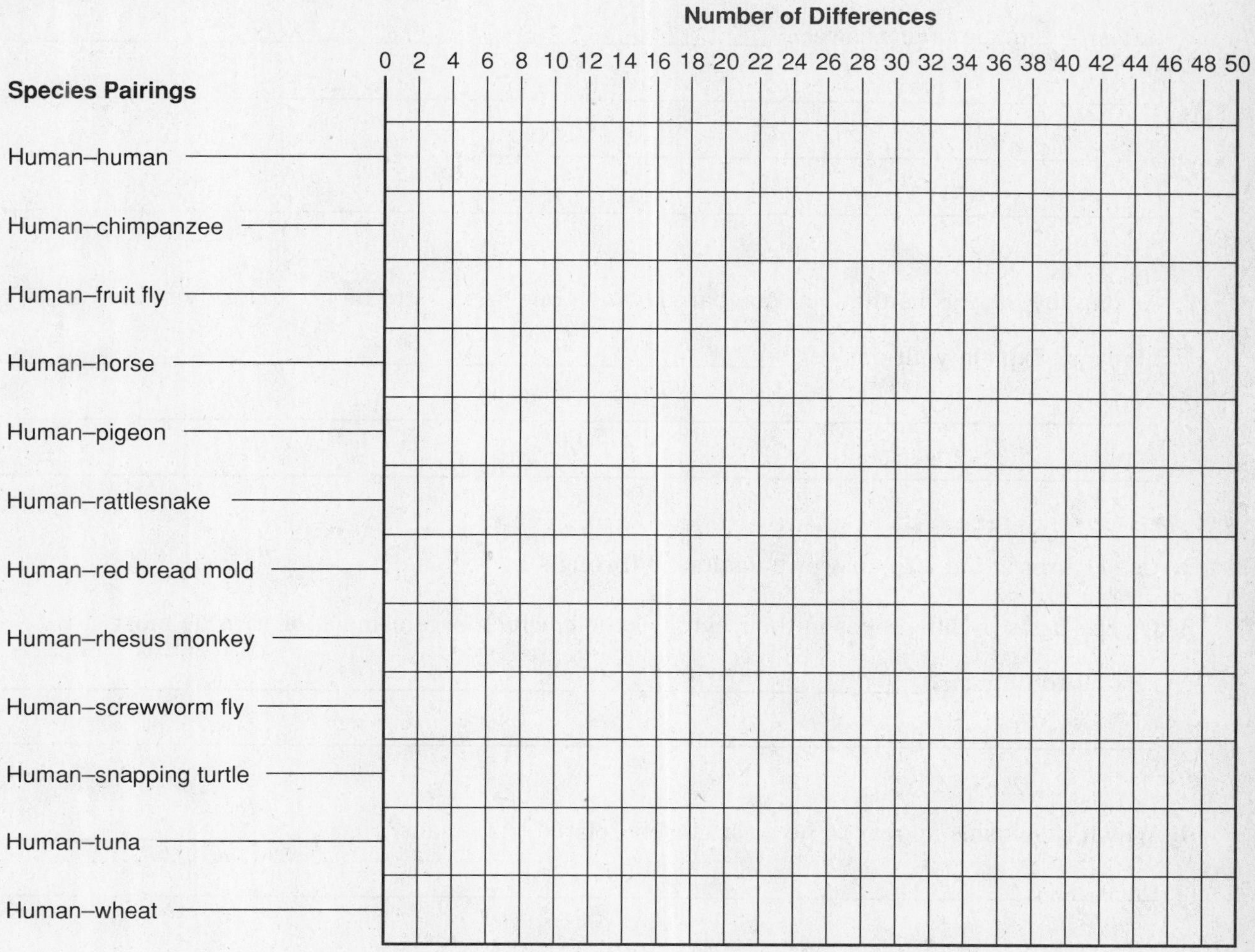

Number of Differences
0 2 4 6 8 10 12 14 16 18 20 22 24 26 28 30 32 34 36 38 40 42 44 46 48 50
Species Pairings
Human–human
Human–chimpanzee
Human–fruit fly
Human–horse
Human–pigeon
Human–rattlesnake
Human–red bread mold
Human–rhesus monkey
Human–screwworm fly
Human–snapping turtle
Human–tuna
Human–wheat

Graph 2

Number of Differences
0 2 4 6 8 10 12 14 16 18 20 22 24 26 28 30 32 34 36 38 40 42 44 46 48 50
Species Pairings
Fruit fly–fruit fly
Fruit fly–dogfish shark
Fruit fly–pigeon
Fruit fly–screwworm fly
Fruit fly–silkworm moth
Fruit fly–tobacco hornworm moth
Fruit fly–wheat

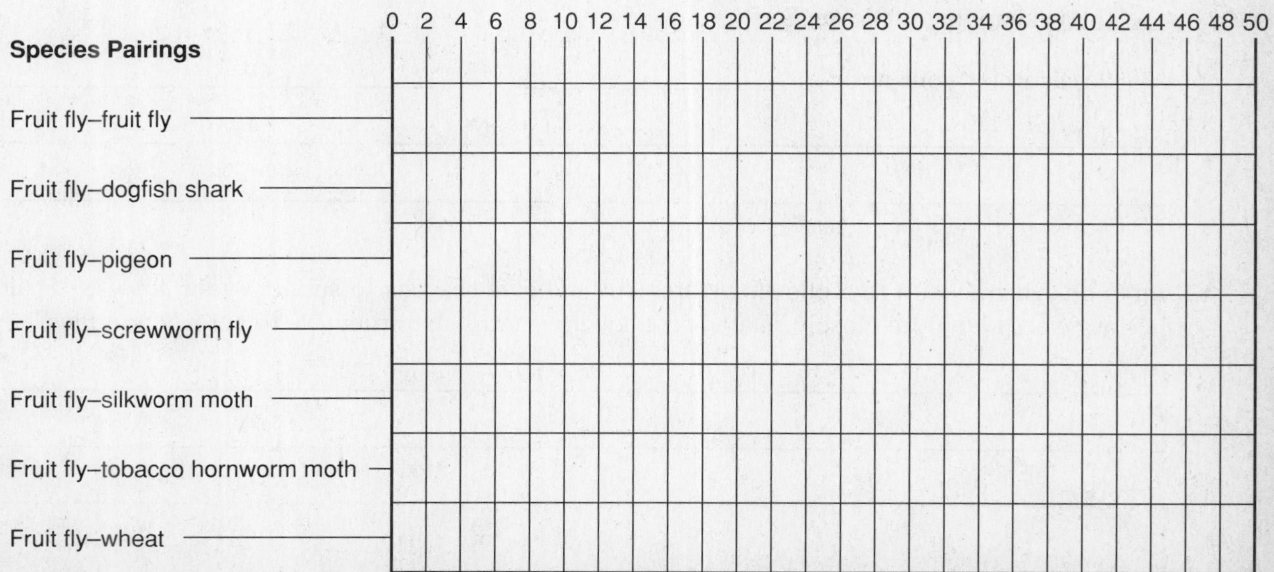

Analysis and Conclusions

Part A. Use Figure 1 to answer questions 1 and 2.

1. On the basis of the hemoglobin similarity, what organisms appear to be most closely related to humans? Explain your answer. ______________________________

____

____

____

2. Among the organisms that you compared, which one appears to be least closely related to humans? Explain your answer. ______________________________

____

____

Part B. Use Figures 2 and 3 to answer questions 3 through 9.

3. On the basis of differences in their cytochrome *c*, which organisms appear to be most closely related to humans? ______________________________

____

4. Which organisms appear to be least closely related to humans? ______________________________

____

5. Check the pair of organisms that appears to be most closely related to each other:

   _______ snapping turtle–tuna
   _______ snapping turtle–rattlesnake
   _______ snapping turtle–pigeon

   Give a reason for your answer. ______________________________

   ____

   ____

6. Agree or disagree with the following statement and give reasons to support your answer. "Fruit flies appear to be more closely related to silkworm moths than they are to screwworm flies."

   ____

   ____

   ____

Name ________________ Class ________ Date ________

7. Name the pair of organisms that appears to be equally related to humans on the basis of cytochrome *c* similarity. ________________

8. Is it possible that the organisms in question 7 could be equally related to humans but not equally related to each other? Explain your answer. ________________

9. Agree or disagree with the following statement. "Fruit flies and humans have about the same evolutionary relationship to wheat." Explain your answer. ________________

Critical Thinking and Application

1. There is a difference of only one amino acid in one chain of the hemoglobin of humans and gorillas. What might have caused this difference? ________________

2. If the amino acid sequences in the proteins of two organisms are similar, why will their DNA also be similar? ________________

3. Many biologists believe that the number of differences between the proteins of different species indicates how long ago the species diverged from common ancestors. Why do these biologists believe that humans, chimpanzees, and gorillas diverged from a common ancestor only a few million years ago? ____

____

____

____

4. Other proteins can be used to establish degrees of evolutionary relatedness between organisms. Would you expect to find roughly the same number of differences in the amino acid chains when comparing organisms? Explain your answer. ______________________

____

____

____

Going Further

Conduct an investigation on the means by which amino acid sequences are used to construct *phylogenetic,* or *evolutionary, trees.* Then, using an appropriate reference, make a large bulletin-board drawing showing an example of a phylogenetic tree.

Name ______________________ Class ____________ Date ____________

26

A Human Adaptation

Pre-Lab Discussion

Suppose that for some reason the only food source available to human beings was starch. Obviously, people who could digest starch would have a better chance of surviving than those who could not digest starch.

In human saliva, there is an enzyme called ptyalin that begins the digestion of starch in the mouth. If some people had more ptyalin in their saliva, they would be better suited to the new environment than those people who had very little ptyalin.

In this investigation, you will use a test to show the change of starch into the simple sugar glucose, determine how long it takes for your ptyalin to change starch into glucose, and find out whether other students in your class have the same amount of ptyalin. You will also determine if there is a genetic variation in your class population and whether the variation has a possible adaptive value.

Problem

What test is used to show the change of starch into the simple sugar glucose? Do people have similar amounts of ptyalin? If there is a genetic variation within a population, will this variation have an adaptive value?

Materials *(per group)*

Glass-marking pencil
10 test tubes
Test tube rack
50-mL graduated cylinder
Iodine solution
Water
2 150-mL beakers
Drinking straw
Starch solution
Medicine dropper
Glass stirring rod

Safety

Put on a laboratory apron if one is available. Put on safety goggles. Handle all glassware carefully. Always use special caution when working with laboratory chemicals, as they may irritate the skin or cause staining of the skin or clothing. Never touch or taste any chemical unless instructed to do so. Use care when collecting and working with saliva. Note all safety alert symbols next to the steps in the Procedure and review the meanings of each symbol by referring to the symbol guide on page 10.

Procedure

1. With a glass-marking pencil, number the 10 test tubes from 1 to 10. Place them in a test tube rack.
2. Using the graduated cylinder, add 2 mL of iodine solution to each of the 10 test tubes.
3. To prepare a 6% saliva solution, add 17 mL of water to a beaker. Collect 1 mL of saliva by salivating through a drinking straw into the graduated cylinder. Pour the saliva into the water in the beaker and mix the solution.
4. Place 7 mL of starch solution in another beaker. To this solution, add 1 mL of the 6% saliva solution. Mix the solution by swirling the beaker. Note the exact time that you mixed the starch-and-saliva solution.
5. After 3 minutes, use a medicine dropper to remove a small amount of the starch-and-saliva solution from the beaker.
6. Add 2 drops of the starch-and-saliva solution to the first test tube containing the iodine solution. Be sure to return the rest of the contents of the medicine dropper to the starch-and-saliva solution.
7. Record the color of the iodine solution after adding the starch-and-saliva solution. **Note**: *In the presence of starch, the iodine solution will be blue-black in color. If no starch is present, because it has been broken down by ptyalin, the iodine solution's color will not change.*
8. At 3-minute intervals, add 2 drops of the starch-and-saliva solution to each of the 9 remaining test tubes of iodine solution.

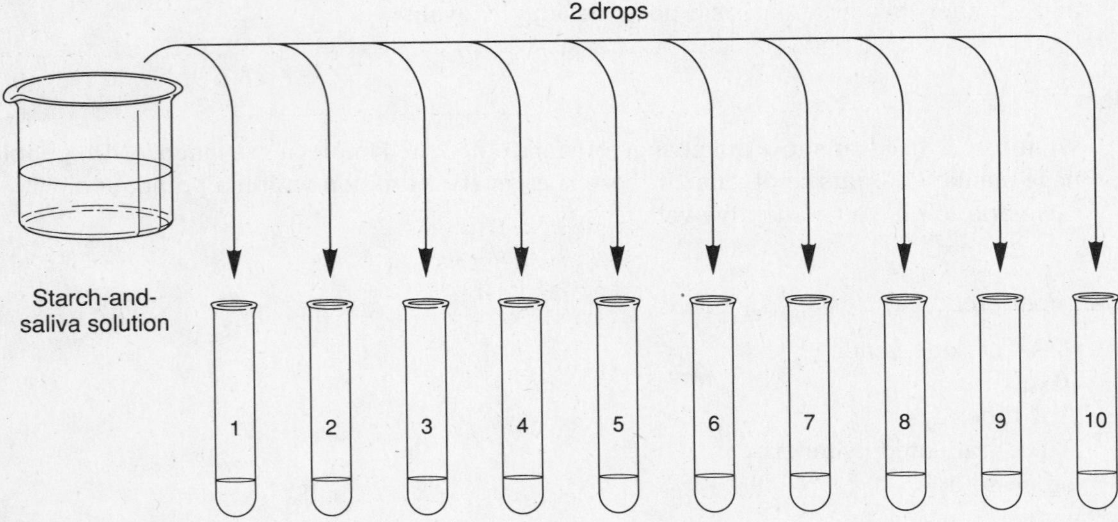

Figure 1

9. Record the color of each test tube in the proper place in Data Table 1. By observing the color of the solution, determine if the test tube contains starch or glucose and record this information as well.
10. In Data Table 2, record the total number of student groups that found a change in the starch solution within each given time interval.
11. Prepare a line graph of the results recorded in Data Table 2 using the graph provided in Observations.

Name ______________________ Class ____________ Date ____________

Observations

Data Table 1 Results of Group

3-minute Intervals	Color of Iodine Solution	Starch or Glucose Present?
0	Blue-black	Starch
3		
6		
9		
12		
15		
18		
21		
24		
27		
30		

Data Table 2 Results of Class

3-minute Intervals	Number of Student Groups Observing Change in Iodine
0	
3	
6	
9	
12	
15	
18	
21	
24	
27	
30	

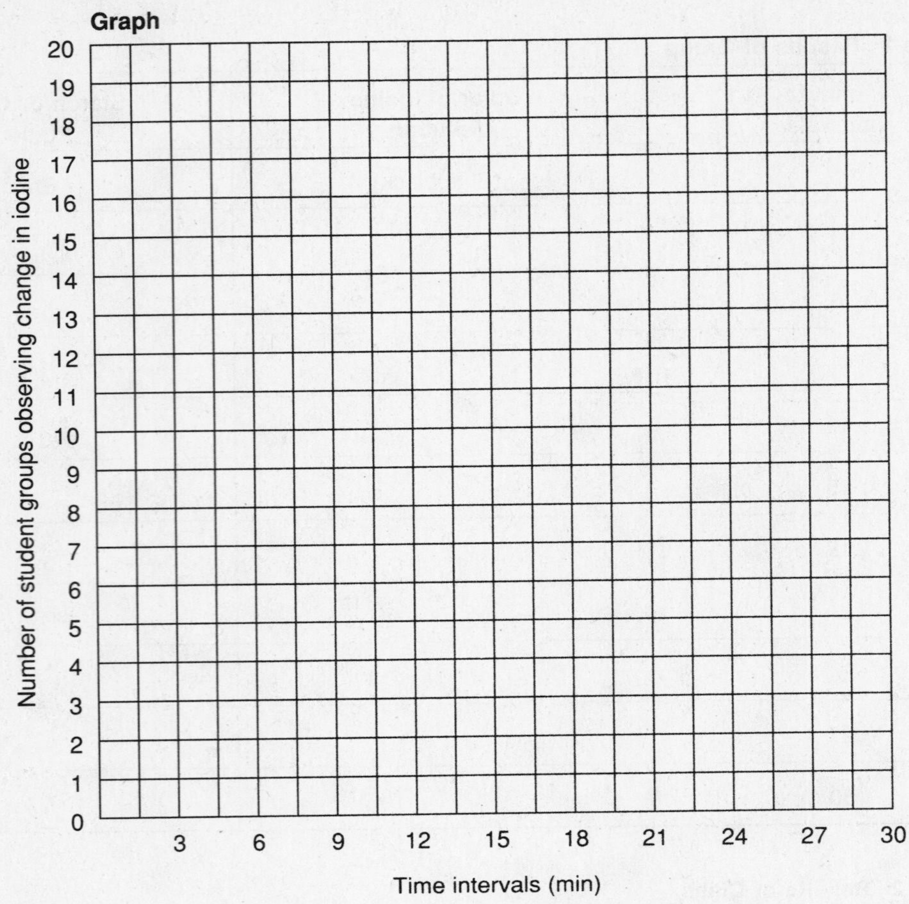

1. In your group, how long did it take the saliva to digest the starch?

____

2. How long did it take the majority of the student groups to digest the starch?

____

3. What is the shape of the graph you have constructed? ______________________

____

Analysis and Conclusions

1. If, indeed, starch were the only food source available to humans, which students would be more likely to survive? Explain your answer. ______________________

____

____

Name ____________________ Class __________ Date __________

2. Do the results of your class's groups show that there are various amounts of ptyalin in saliva? Explain your answer. ____________________

3. a. Is the source of this chemical variation genetic? ____________________

b. Did this chemical variation occur before or during the experiment?

c. Do most variations occur before they are adaptive? Explain your answer.

4. Look at the shape of the graph you constructed. Suggest a reason for the shape.

Critical Thinking and Application

1. In addition to the amount of ptyalin found in saliva, list two other chemical variations occurring within the human population that are helpful to survival.

2. Describe three adaptations that would be most useful to a person who plays the piano.

3. In prehistoric times, people had to struggle every day to survive. Why might a very smart person have an advantage? ____________________

____

____

____

4. The ball-and-socket shoulder joint is an adaptation that is unique to primates. Why is this adaptation important to the survival of the human population? ____________________

____

____

____

Going Further

Gather data on several variations in humans, such as the width of the foot, the length of the little finger, or the ratio of standing height to sitting height. Graph each of the variations. Are the shapes of the graphs consistent with that of the ptyalin investigation?

Name ______________________ Class __________ Date __________

Variation Within a Population

Pre-Lab Discussion

The members of a species are not exactly alike. Small differences called variations exist in each member of a species. Variations are found in all plants and animals. Some variations may be passed on to the offspring of an organism through reproduction. Most inherited variations are neutral; that is, they do not affect the survival of the organism. Some variations are helpful. Helpful inherited variations are called adaptations. Harmful inherited variations will cause the organism to be less well-suited to its environment.

The process by which organisms with adaptations to the environment survive is called natural selection. Natural selection tends to allow well-adapted organisms to reproduce and pass the beneficial trait to their offspring. Organisms with harmful traits will die off. This process is sometimes referred to as "survival of the fittest."

In this investigation, you will observe variations that exist in two different types of plants. You will also observe variations that exist within a classroom population.

Problem

How can the variations within plant and animal populations be measured?

Materials *(per group)*

10 white kidney beans
10 maple leaves
Metric ruler
3 colored pencils

Procedure

Part A. Variation Within a Species of Plant

1. Obtain 10 white kidney beans and 10 maple leaves.
2. Measure the length of each kidney bean and maple leaf blade in millimeters. Record the measurements in Data Table 1.
3. Locate the petiole on the leaf as shown in Figure 1. Measure the length of the petiole of each leaf. Record the measurements in Data Table 1.

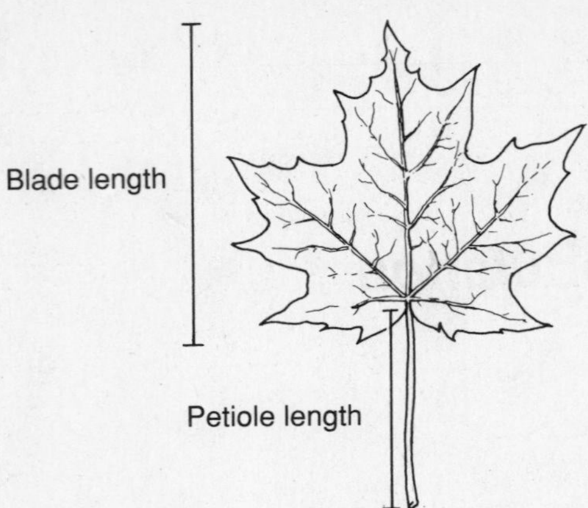

Figure 1

4. Report the measurements for each of the plants to your teacher. Your teacher will record your measurements on the chalkboard.
5. Using the data from all groups in your class, record the range in lengths for the kidney beans, leaf blades, and petioles. Record your findings in Data Tables 2, 3, and 4, respectively. Fill in the first row of each table with the graduated lengths, beginning with the shortest length and ending with the longest.
6. Record the total class count for the lengths of the kidney beans, leaf blades, and petioles in the second row of the table in Data Tables 2, 3, and 4.
7. Using the data in Data Table 2, construct a line graph for the kidney bean lengths on Graph 1 provided in Observations.
8. Using the data in Data Tables 3 and 4, construct line graphs for the leaf blade lengths and the petiole lengths on Graph 2 provided in Observations. Use a different colored pencil to graph each set of data and be sure to include a key for each graph.

Part B. Variation Within a Classroom Population

1. Measure your hand span. The measurement should be made from the top of the thumb to the tip of the little finger as shown in Figure 2. Round off the measurement to the nearest centimeter. Record your hand span in the class chart on the chalkboard.

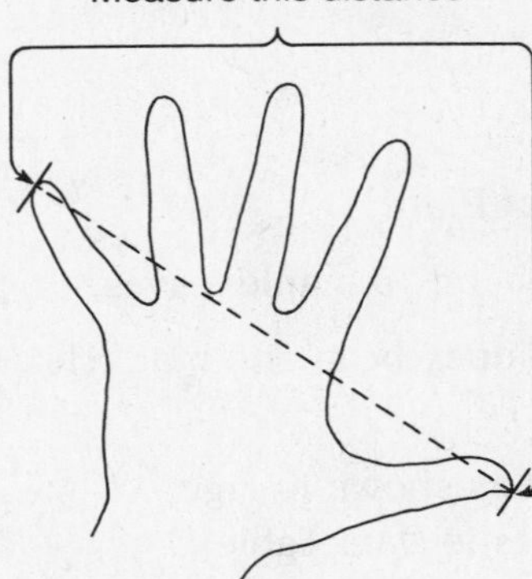

Figure 2

2. After all your classmates have recorded their hand-span measurements in the class chart, transfer the class results to Data Table 5. Your results will show the total number of hands having the same hand span.
3. Construct a line graph of your results using Graph 3 provided in Observations.

Name ______________________ Class ____________ Date ____________

Observations

Data Table 1 Species Length Measurements

	Length of Samples (mm)									
	1	2	3	4	5	6	7	8	9	10
Kidney beans										
Leaf blades										
Petioles										

Data Table 2 Comparison of Kidney Bean Lengths (mm)

Length of kidney beans																
Total class count																

Data Table 3 Comparison of Leaf Blade Lengths (mm)

Length of leaf blades																
Total class count																

Data Table 4 Comparison of Petiole Lengths (mm)

Length of petioles																
Total class count																

Data Table 5 Comparison of Hand-span Lengths (mm)

	Measurement of Hand Spans (cm)															
	15	16	17	18	19	20	21	22	23	24	25	26	27	28	29	30
Total number of hand spans with the same measurement																

Graph 1

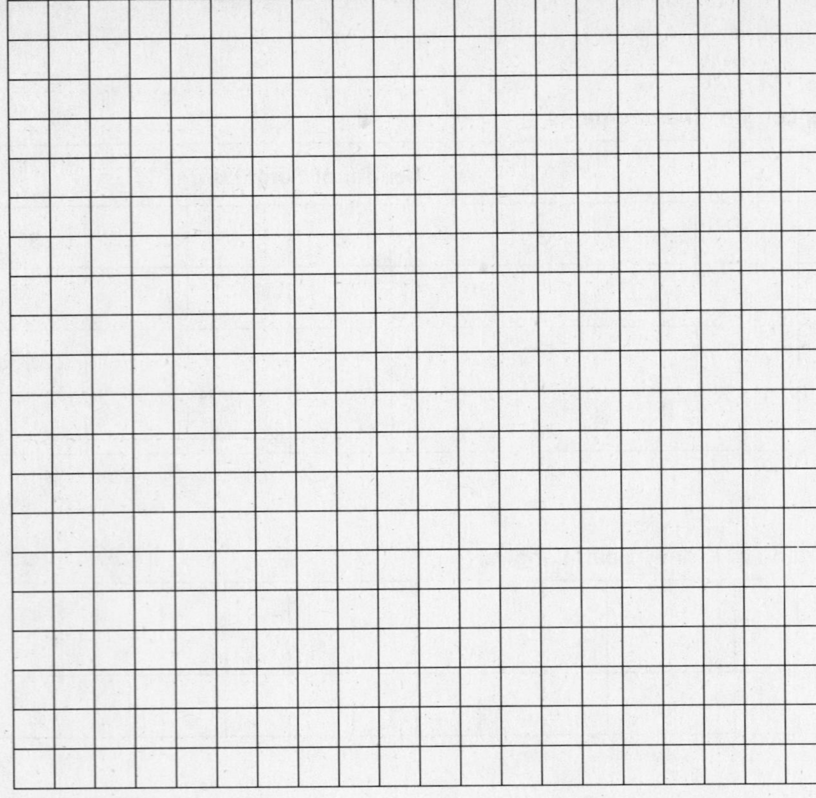

Graph 2

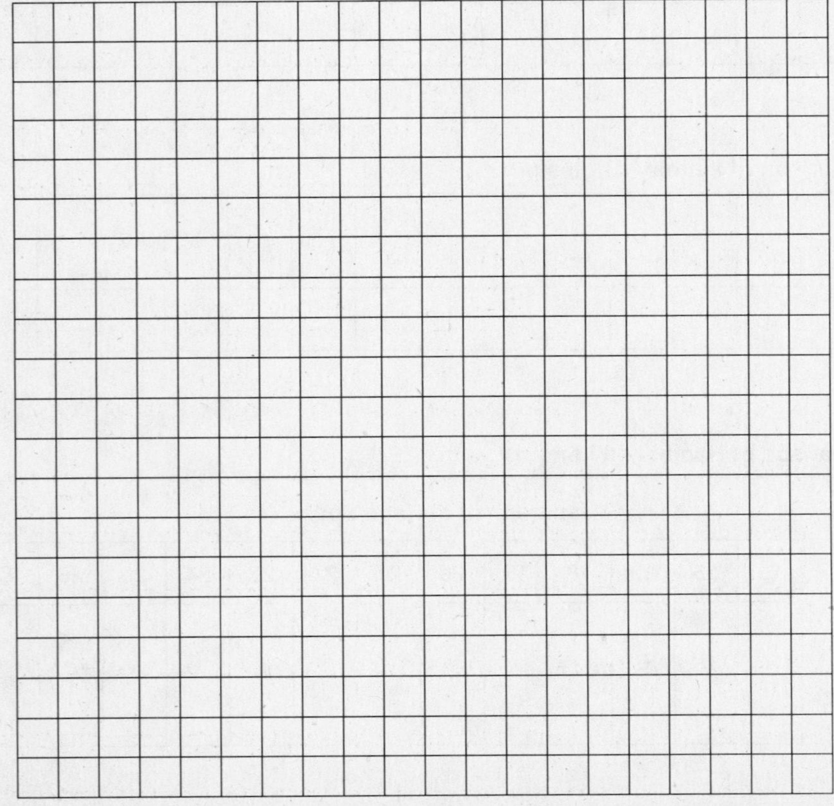

Name ______________________ Class ____________ Date ____________

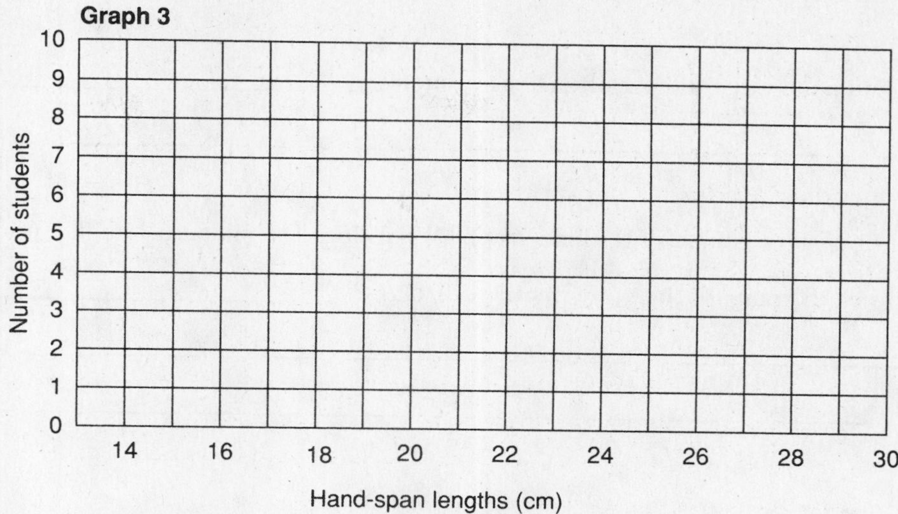

Analysis and Conclusions

1. What is meant by the term variation? ______________________

2. In what length range are most of the kidney beans? Most of the leaf blades? Most of the petioles? ______________________

3. In what length range are the fewest beans? The fewest leaf blades? The fewest petioles?

4. What is the general shape of each line graph? ______________________

5. What do the shapes of the graphs indicate about the lengths of the kidney beans, the lengths of the leaf blades, and the lengths of the petioles? ______________________

6. List two other variations that can be observed in the beans, leaf blades, and petioles.

7. Are the hand spans of all students in your class the same? ____________________

____

8. Which hand span occurs most often? Least often? ____________________

____

9. What is the general shape of the line graph? What does the shape of the line graph indicate about the hand spans of the students in your class? ____________________

____

____

Critical Thinking and Application

1. Why would having many seeds in a pod be a more useful adaptation to a bean plant than having only a few seeds? ____________________

____

____

2. Why would having a very short root system be a harmful adaptation to a desert cactus plant?

____

____

____

3. How is the white coat color of certain arctic animals a helpful adaptation in the winter?

____

____

4. List two ways in which a large hand span might be a useful human adaptation.

____

____

Going Further

Investigate variations that occur in other types of plants. Make your measurements and graph the results as you did in this laboratory investigation. Do you think that all organisms of the same species exhibit variations for all of their traits? Explain your answer.

Name ______________________ Class __________ Date ____________

Classifying Organisms

Pre-Lab Discussion

Even early biologists developed classification systems in order to understand an organism and explain its relationships to other forms of life. The basis of the classification system in current use was developed in the 1700s by the Swedish botanist Carolus Linnaeus (1707–1778). Linnaeus used structural similarities as the basis of his system. Organisms were first placed in one of two large groups—the plant kingdom or the animal kingdom. Although all animals share certain similarities and all plants share certain similarities, individual members of each kingdom are quite different. Therefore, Linnaeus divided the two large groups (kingdoms) into smaller groups, and these groups, in turn, into still smaller groups.

In the Linnaean system, the largest groups are *kingdoms,* which are divided into smaller groups called *phyla* (singular, *phylum*). Phyla are then divided into *classes,* and classes are divided into many *orders.* Orders, in turn, are divided into *families,* families into *genera* (singular, *genus*), and genera into one or more *species.* Though most biologists generally classify organisms within these categories, they do not always agree on the criteria for placing an organism in a particular category. Because an organism can be observed from many viewpoints, the same organism may be classified differently by different biologists.

In this investigation, you will group and regroup a number of pictured objects according to similar characteristics. You will also classify a variety of monkeys and apes (primates) on the basis of their similarities and differences.

Problem

How can organisms be classified?

Materials *(per student)*

Scissors

Safety

Be careful when handling sharp instruments. Note all safety alert symbols next to the steps in the Procedure and review the meanings of each symbol by referring to the symbol guide on page 10.

Procedure

Part A. Grouping Objects

1. Use scissors to cut along the dotted lines around each pictured object in Figure 4 on page 201. **CAUTION:** *Be especially careful if you are working with sharp pointed scissors.*

2. Place the group of objects you cut out in front of you. Then choose some characteristic to separate the objects into two groups of approximately equal numbers.

3. In Observations, make a branching chart similar to the one in Figure 1 to show how you divided the objects. Write the criterion you used to divide the group on the lines in your chart corresponding to line A1 and A2. Also list the numbers of the objects that belong on lines A1 and A2.

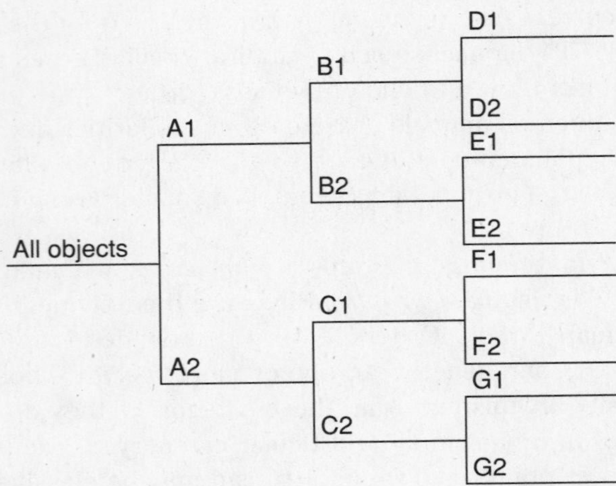

Figure 1

4. Next choose some characteristics to divide the objects in group A1 into two subgroups. Write their characteristics and numbers on your chart on the lines that correspond to lines B1 and B2 in Figure 1. Separate group A2 into two subgroups, and write their characteristics and numbers on the lines that correspond to lines C1 and C2.

5. Continue dividing each group until each pictured object is separated from all of the others. On your chart, continue to list the criterion you used for dividing each group. **Note:** *Your chart may not contain all of the branches shown in Figure 1. Also, you may need to extend your chart by adding branches not shown in Figure 1.*

Name ______________________ Class __________ Date __________

Part B. Grouping Living Organisms

1. When classifying organisms, biologists must be very observant. Often only slight differences separate the members in one classification group from those in another. Study Figure 2, which shows some traits used for identifying primates.

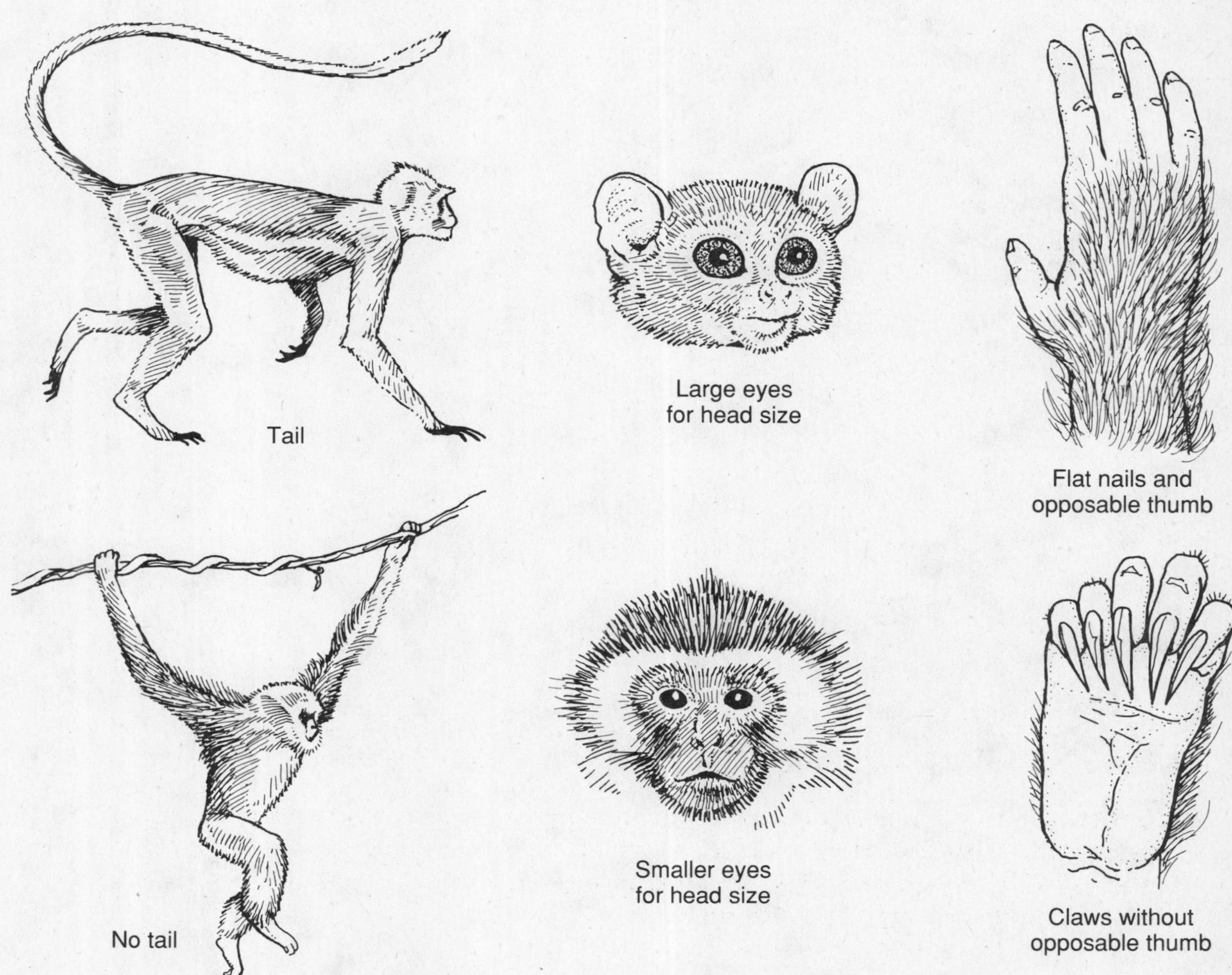

Figure 2

2. Use the traits shown in Figure 2 and other traits to separate the primates in Figure 3 into two groups of approximately equal numbers. Use the space in Observations to make a branching chart similar to the one in Figure 1 to show how you separated the primates. Write the characteristics that you used to make your first division on the lines that correspond to lines A1 and A2 in Figure 1.

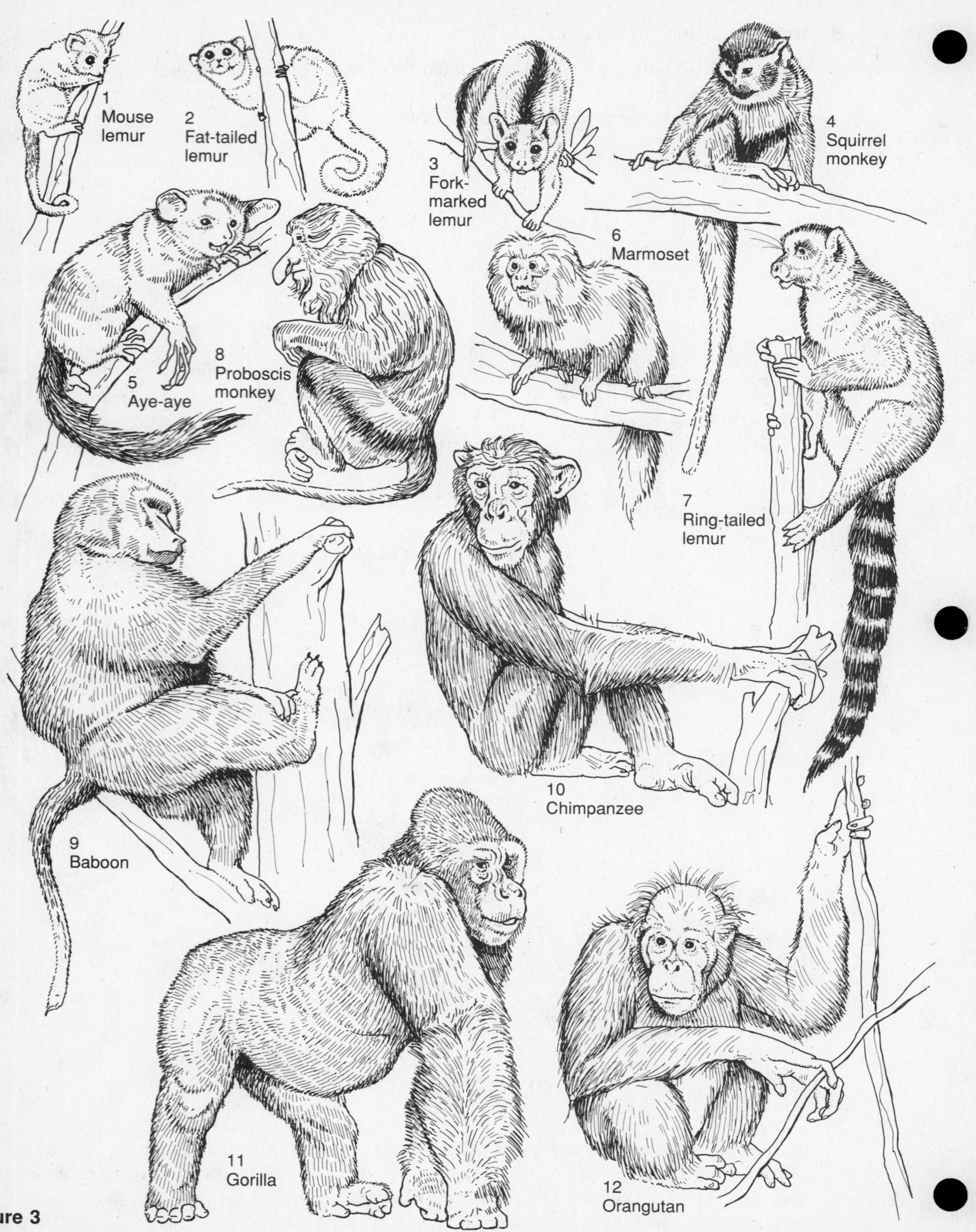
1
Mouse
lemur
2
Fat-tailed
lemur
3
Fork-
marked
lemur
4
Squirrel
monkey
5
Aye-aye
6
Marmoset
7
Ring-tailed
lemur
8
Proboscis
monkey
9
Baboon
10
Chimpanzee
11
Gorilla
12
Orangutan

Figure 3

Name ________________________ Class ____________ Date ____________

3. Continue to divide the primates into smaller groups until each is separated from all of the others. As you divide the primates into subgroups, complete your chart by listing the characteristics you used to separate them.

Observations

Branching Chart for Classifying Objects in Figure 4

Primate Branching Chart

Analysis and Conclusions

1. How is this investigation similar to the way in which biologists classify organisms?

____

____

2. Were either of your classification systems different from those developed by classmates? Why might this be possible? ____

____

____

Name ______________________ Class ____________ Date ____________

3. What characteristics did you find most useful for classifying the primates?

4. What characteristics of the primates would not be especially useful for classifying them?

5. Why should terms like *tall, short, large,* or *small* be avoided when describing traits of organisms?

6. How does classification help you to better understand organisms?

Critical Thinking and Application

1. Trees are usually identified by the characteristics of their leaves. Suggest two ways in which trees could be identified during winter when they have no leaves.

2. Suppose you wanted to identify and classify all birds that came to a particular bird feeder during a spring day. What are some common characteristics you would use in classifying the birds?

3. Almost immediately after the invention of the microscope, taxonomists began to use it in their work. Why might a microscope be useful to a taxonomist? ______________________

____

____

4. Describe three ways in which pieces of clothing are classified in a department store.

____

____

____

Going Further

1. Devise a system to classify the different kinds of shoes your classmates wear. Make a chart similar to Figure 1 to show the different categories and characteristics used to group the shoes.
2. Make a list of ten or more household appliances. Exchange your list with a classmate. Classify the appliances into groups based on similar characteristics. Make a chart similar to Figure 1 to show the different categories and characteristics used to group the appliances.

Name ______________________ Class __________ Date __________

Use Figure 4 to complete step 1 of Part A.

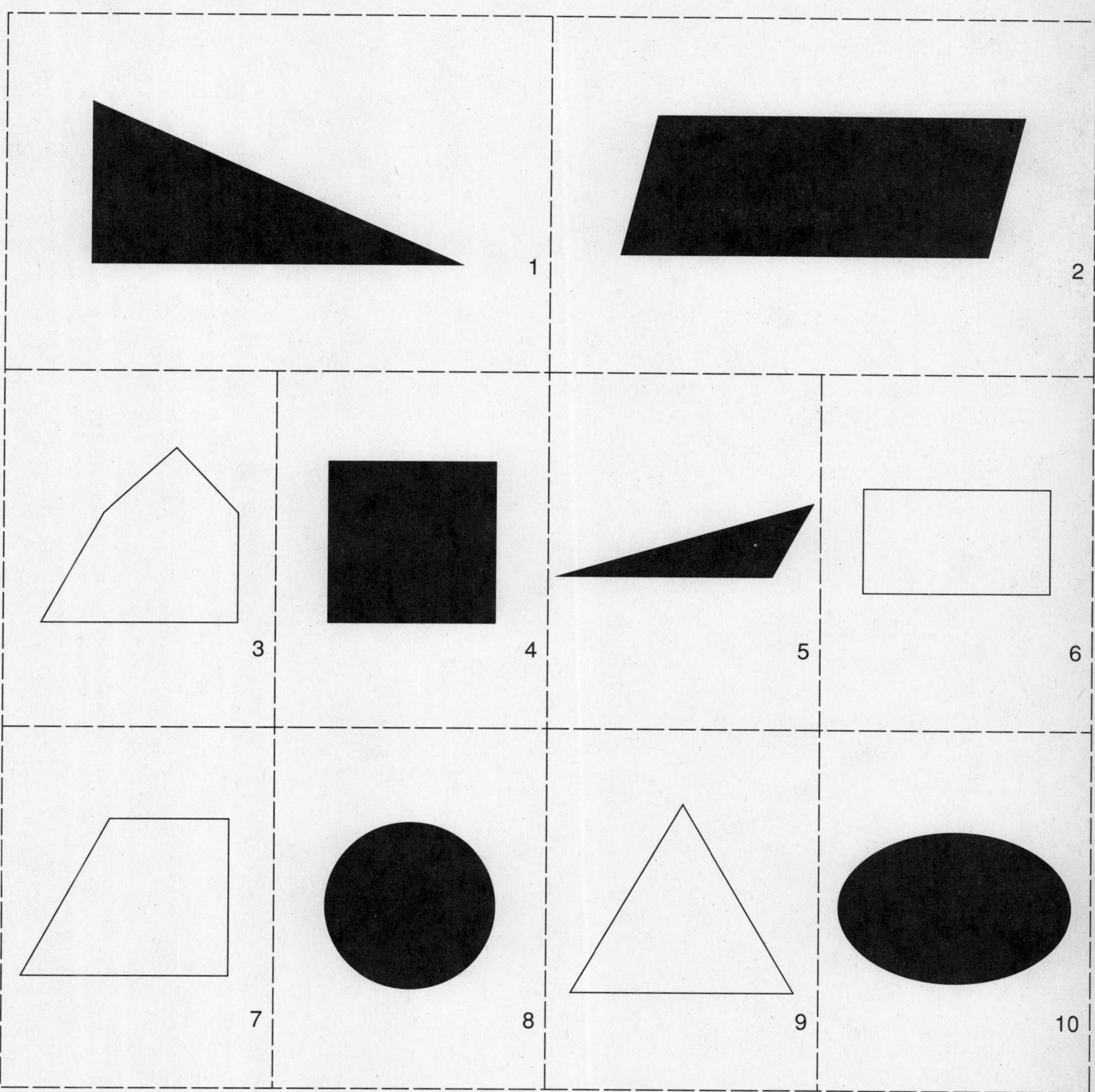

Figure 4

Name ______________________ Class __________ Date __________

Using and Constructing a Classification Key

Pre-Lab Discussion

Suppose you find a large colorful wildflower while walking through the woods. Chances are the flower has already been named and classified, but how can you learn its identity? As an aid to help others identify unknown organisms, biologists have developed classification keys.

Many classification keys have been developed to help identify wildflowers and many other kinds of plants and animals. Although these keys may vary in purpose and complexity, they have certain features in common. These classification keys are often called *dichotomous keys*. The word dichotomous comes from the word *dichotomy*, meaning "two opposite parts or categories." A dichotomous classification key presents the user with two opposite statements about some trait of an organism. By choosing the statement that best describes the unknown organism, the user is led to further pairs of statements. By going from one set of statements to another, the name of the organism or its classification group is finally determined.

In this investigation, you will use a classification key to identify several organisms. You will then write a classification key for another group of organisms.

Problem

How can a classification key be used to identify organisms?

Materials *(per student)*

No special materials are needed

Procedure
Part A. Using a Classification Key

1. Examine the drawing of salamander 1 in Figure 1.

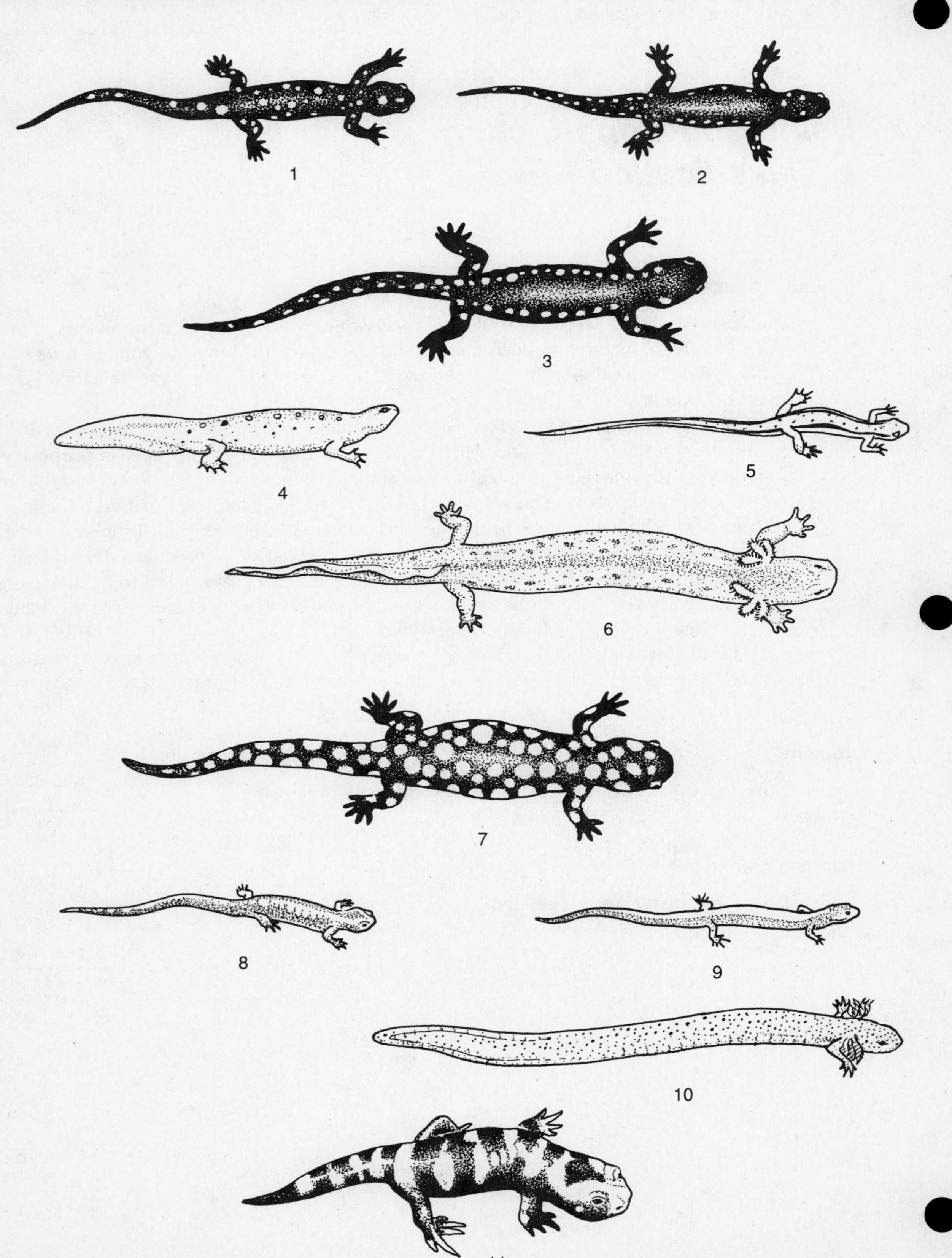

Figure 1

Name ______________________ Class ____________ Date ____________

2. Read statements 1a and 1b in the classification key in Figure 2. One of these statements describes salamander 1; the other statement does not. Follow the directions in the statement that describes salamander 1 and continue following the correct statement directions until salamander 1 has been identified.

1	**a**	Hind limbs absent	*Siren intermedia*, siren
	b	Hind limbs present	Go to 2
2	**a**	External gills present in adults	*Necturus maculosus*, mud puppy
	b	External gills absent in adults	Go to 3
3	**a**	Large size (over 7 cm long in Figure 1)	Go to 4
	b	Small size (under 7 cm long in Figure 1)	Go to 5
4	**a**	Body background black, large white spots irregular in size and shape completely covering body and tail	*Ambystoma tigrinum*, tiger salamander
	b	Body background black, small round white spots in a row along each side from eye to tip of tail	*Ambystoma maculatum*, spotted salamander
5	**a**	Body background black with white spots	Go to 6
	b	Body background light color with dark spots and/or lines on body	Go to 7
6	**a**	Small white spots on a black background in a row along each side from head to tip of tail	*Ambystoma jeffersonianum*, Jefferson salamander
	b	Small white spots scattered throughout a black background from head to tip of tail	*Plethodon glutinosus*, slimy salamander
7	**a**	Large irregular black spots on a light background extending from head to tip of tail	*Ambystoma opacum*, marbled salamander
	b	No large irregular black spots on a light background	Go to 8
8	**a**	Round spots scattered along back and sides of body, tail flattened like a tadpole	*Triturus viridescens*, newt
	b	Without round spots and tail not flattened like a tadpole	Go to 9
9	**a**	Two dark lines bordering a broad light middorsal stripe with a narrow median dark line extending from the head onto the tail	*Eurycea bislineata*, two-lined salamander
	b	Without two dark lines running the length of the body	Go to 10
10	**a**	A light stripe running the length of the body and bordered by dark pigment extending downward on the sides	*Plethodon cinereus*, red-backed salamander
	b	A light stripe extending the length of the body, a marked constriction at the base of the tail	*Hemidactylium scutatum*, four-toed salamander

Figure 2

3. Write the scientific name and the common name of salamander 1 on the appropriate line provided in Observations.

4. Repeat steps 1 through 3 with each salamander until all the animals have been identified.

Part B. Constructing a Classification Key

1. Study Figure 3, which shows some common North American wildflowers. As you study the drawings of various flowers, note different characteristics in flower shape, number of petals, and leaf number and shape.

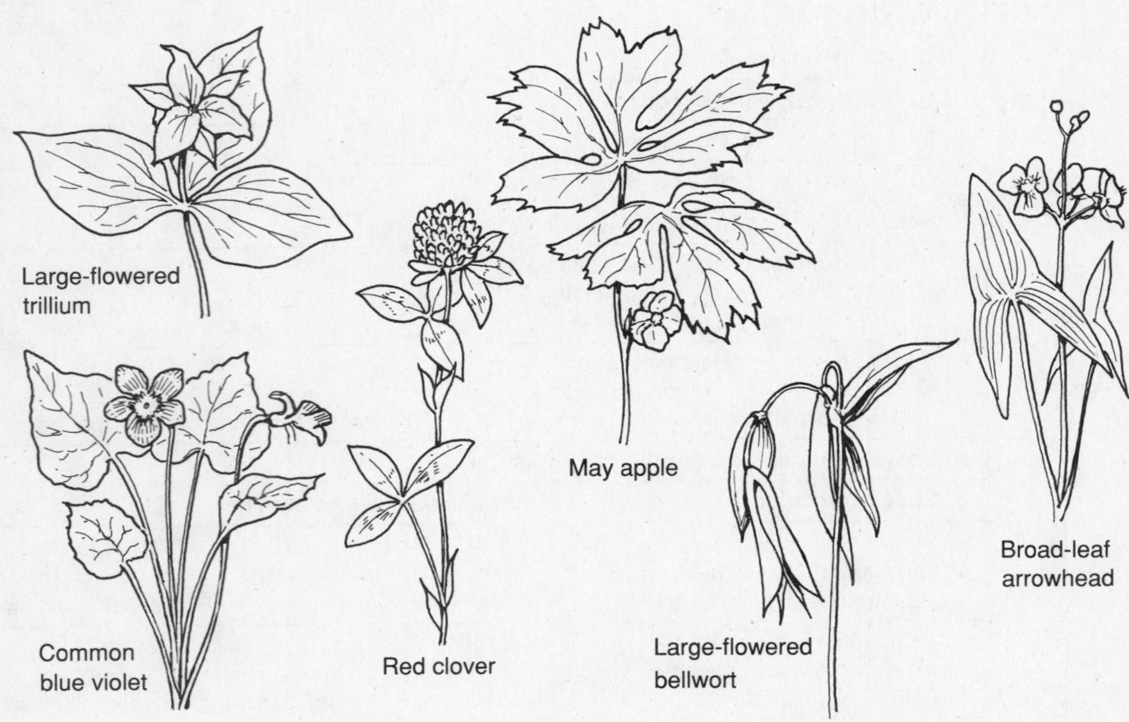

Figure 3

2. In the space provided in Observations, develop a classification key to identify each wildflower. Use the key to salamanders as a model for developing your wildflower key.

3. Check the usefulness of your wildflower key by letting another student see if he or she can use it to identify each pictured flower.

Observations

Part A. Using a Classification Key

Write the scientific and common names of each salamander in Figure 1 on the line that corresponds to its number.

1. ______________________________

2. ______________________________

3. ______________________________

4. ______________________________

5. ______________________________

6. ______________________________

Name ______________________ Class __________ Date __________

7. ______________________________

8. ______________________________

9. ______________________________

10. ______________________________

11. ______________________________

Part B. Constructing a Classification Key

Use the space below to construct a wildflower classification key.

Wildflower Classification Key

Analysis and Conclusions

1. As you used the classification key to identify the salamanders, did you go from general to specific characteristics or from specific to general characteristics? ______________________________

2. What two groupings do the scientific names of the salamanders represent?

3. Was the classification key you constructed exactly like those of other students? Explain why or why not. ______________________________

4. If you were using actual wildflowers, what other characteristics could you use to identify them? ____________________

____

Critical Thinking and Application

1. Do you think that there may be some closely related species of organisms that cannot be identified with a classification key? Explain your answer. ____________________

____

____

____

2. Why do you think biological classification keys always present two, rather than some other number, of choices at each step? ____________________

____

____

____

3. What types of problems would scientists have today if Carolus Linnaeus had not developed his classification and naming system for organisms? ____________________

____

____

____

4. Explain what is meant by the statement "Classification systems are the inventions of humans; diversity is the product of evolution." ____________________

____

____

Going Further

Use the key presented in this investigation as a model to construct a classification key to identify one or more of the following:

1. Instruments of a band or orchestra
2. Balls used in different kinds of sports (tennis ball, basketball, football, bowling ball, golf ball, and so on)
3. The students in your biology class

Name ______________________ Class ____________ Date ____________

Making Coacervates

Pre-Lab Discussion

How did life on Earth begin? According to a hypothesis by Russian scientist Alexander Oparin, all life developed gradually from materials found in the oceans on primitive Earth. According to Oparin, prehistoric oceans probably consisted of a rich mixture of organic chemicals, including proteins and carbohydrates. If certain kinds of proteins and carbohydrates are mixed together with water, *coacervates,* or droplets showing lifelike characteristics, may form. Coacervates are not alive. However, in a manner similar to cells, coacervates appear to ingest materials, grow, and reproduce. Because of this, scientists have hypothesized that coacervates may have been the precursors of cells.

In this investigation, you will produce coacervates and observe their behavior.

Problem

How does a mixture of protein and carbohydrate molecules act like a living organism?

Materials *(per group)*

1 medium-sized test tube
1 rubber stopper that fits the test tube
1% gelatin solution
1% gum arabic solution
1% hydrochloric acid solution
pH test paper
2 medicine droppers
Microscope
Glass slide
Coverslip
10-mL graduated cylinder
Glass stirring rod

Safety

Put on a laboratory apron if one is available. Put on safety goggles. Handle all glassware carefully. Always use special caution when working with laboratory chemicals as they may irritate the skin or cause staining of the skin or clothing. Never touch or taste any chemical unless you are instructed to do so. Always handle the microscope with extreme care. You are responsible for its proper care and use. Use caution when handling glass slides as they can break easily and cut you. Note all safety alert symbols next to the steps in the Procedure and review the meanings of each symbol by referring to the symbol guide on page 10.

Procedure

1. Obtain a test tube and a rubber stopper that fits the test tube. Use the graduated cylinder to pour 6 mL of gelatin solution into the test tube. Next add 4 mL of gum arabic solution to the test tube. Mix the two solutions by *gently* inverting the test tube several times. **Note:** *Mix the solutions gently. Do not shake the test tube as this will hinder the formation of coacervates.*

2. Coacervates will form only under specific environmental conditions. One important condition is the acidity of the environment. Unstopper the test tube and dip a glass stirring rod into the mixture. Touch a drop of the mixture onto a piece of pH paper. Compare the color of your pH paper to information given on the package of test papers or information supplied by your teacher. Record the pH under Trial 1 in the Data Table.

3. The cloudiness or clearness of the mixture may indicate the presence or absence of coacervates. Hold the test tube up to a light source. Record if the mixture is cloudy or clear under Trial 1 in the Data Table.

4. Use a medicine dropper to put 2 drops of the mixture on a clean glass slide. Place a coverslip over the drops and examine them under low power. Coacervates, as shown in Figure 1, will look like droplets of material with tiny bubbles inside. **Note:** *You may not see coacervates until the pH (acidity) is adjusted, so do not get discouraged.* Observe the number, size, and movement of any coacervates present and record this information under Trial 1 in the Data Table. Clean the slide and coverslip.

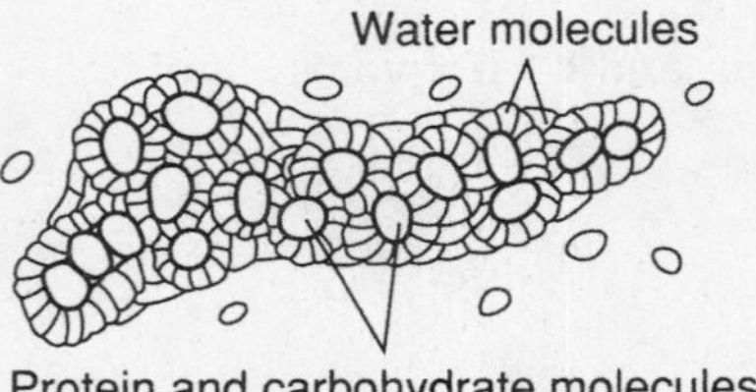

Figure 1

5. Use a second medicine dropper to change the pH of the mixture by adding 3 drops of weak hydrochloric acid solution. **CAUTION:** *Hydrochloric acid çan cause burns. If the acid touches your skin or clothes, immediately wash the area with water and notify your teacher.* Stopper the test tube and gently invert it once or twice to mix the solution.

6. Repeat steps 2 through 4, recording all information under Trial 2 in the Data Table.

7. Repeat step 5 four more times, adding 3 drops of weak hydrochloric acid each time. Test the pH mixture, observe and record the degree of cloudiness, and make observations of coacervates under the microscope. Record this information under Trials 3 through 6 in the Data Table.

Name ______________________ Class __________ Date __________

Observations

Data Table

	Observation of Coacervates					
	Trial					
	1	2	3	4	5	6
pH						
Cloudy/clear						
Microscopic observations						

Analysis and Conclusions

1. At what pH do coacervates appear to be the largest and most abundant?

____

2. At what pH or pHs do coacervates appear to be the smallest and least abundant?

____

3. How does the cloudiness of the solution relate to the presence or absence of coacervates?

____

____

4. List two ways in which coacervates seem to be living organisms. ______________

____

____

5. List two ways in which coacervates seem to be nonliving substances.

____

____

Critical Thinking and Application

1. The word "coacervate" comes from the Latin word meaning "heap." How is this name suitable for describing the droplets you observed?

2. Some scientists feel strongly that life may have evolved on other planets as well as on Earth. Why might they think so?

3. Do you think your coacervate investigation might have worked if you had substituted other proteins and carbohydrates? How could you test your hypothesis?

Going Further

Some scientists believe that microspheres may be another type of substance that may have been a precursor of the living cell. Conduct library research on the production and characteristics of microspheres. Refer to the work of American biochemist Sidney Fox.

Identifying Bacteria

Pre-Lab Discussion

Bacteria are divided into different groups based on a number of characteristics. One way to distinguish bacteria is by shape. Bacteria that resemble spheres are called *cocci. Bacilli* are bacteria that look like rods. Bacteria that resemble sprials are called *spirilla.*

Another way to distinguish bacteria is by their ability to accept a specific type of stain called a Gram stain. The Gram-staining method uses two stains—a dark purple stain and a light pink stain. Bacteria that are Gram-positive retain the dark purple stain. Bacteria that loose the dark purple stain and show only the light pink stain are called Gram-negative. The difference in staining characteristics is due to differences in the bacterial cell wall.

In this investigation, you will identify bacteria by their shape, mobility, and ability to accept a Gram stain.

Problem

How are different types of bacteria identified?

Materials *(per group)*

Microscope
Bunsen burner
Inoculating loop
Test tube rack
Glass-marking pencil
Flint striker or matches
2 glass slides
Slide holder
Staining tray
Wash bottle
Bibulous paper

Prepared slides of different-shaped bacteria
Cultures of:
- *Escherichia coli*
- *Bacillus subtilis*

Gram stain set of 4 dropper bottles:
- Crystal violet
- Gram's iodine
- 95% ethyl alcohol
- Safranin

Safety

Put on a laboratory apron if one is available. Put on safety goggles. Handle all glassware carefully. Use extreme care when working with heated equipment or materials to avoid burns. Always use special caution when working with laboratory chemicals, as they may irritate the skin or cause staining of the skin or clothing. Never touch or taste any chemical unless instructed to do so. Use special caution when working with bacterial cultures. Always handle the microscope with extreme care. You are responsible for its proper care and use. Use caution when handling glass slides as they can break easily and cut you. Note all safety symbols next to the steps in the Procedure and review the meanings of each symbol by referring to the symbol guide on page 10.

Procedure

Part A. Identifying Bacteria by Shape

1. Obtain prepared slides of different-shaped bacteria from your teacher. Bacteria have three characteristic shapes: round, or *coccus;* rod-shaped, or *bacillus;* and spiral-shaped, or *spirillum*. These three basic shapes can be arranged singly, in pairs (*diplo-*), or in chains (*strepto-*), and in clusters (*staphylo-*).
2. Observe a prepared bacteria slide under high power. Make a drawing of what you observe in the appropriate place in Observations. Write the name of the bacteria. Observe how the bacteria on this slide are arranged and record this information in Data Table 1.
3. Repeat step 2 with the other prepared slides obtained from your teacher.

Part B. Gram-Staining Bacteria

1. Obtain tubes of bacterial cultures A and B. **CAUTION:** *Use extreme care when working with bacterial cultures. Avoid spills. If a spill does occur, immediately call your teacher for assistance. Do not try to clean it up by yourself.* Place the tubes in an upright position in a test tube rack. With a glass-marking pencil, label one glass slide A and the other glass slide B.
2. Put on safety goggles. Light the Bunsen burner. **CAUTION:** *Use extreme care when working with or near an open flame. Tie back loose hair and clothing.*
3. Pick up test tube A and make a sterile transfer from the test tube to a clean glass slide in the following manner. Remove the plug and pass the mouth of the test tube through the flame of the Bunsen burner. Hold the inoculating loop in the flame until it is red hot. Allow the loop to cool for a few seconds. Pick up a loopful of bacterial culture from test tube A. Reflame the mouth of the test tube and replace the plug. Return the test tube to the test tube rack. Touch the loopful of bacterial culture to the center of slide A. Spread the bacteria into a circle about the size of a dime, as shown in Figure 1A. The area of bacteria on the slide is called a smear. Set slide A aside to air-dry. Flame the inoculating loop until it is red hot. **CAUTION:** *Do not touch the bacterial smear to determine if it is dry. Always avoid direct contact with bacterial cultures.*
4. Repeat step 3 making a transfer from test tube B to slide B.
5. Fasten a slide holder onto the end of slide A. Pass the slide, *smear side up,* three times through the flame of the Bunsen burner, as shown in Figure 1B. **CAUTION:** *Do not touch the hot glass slide. Handle it only with a slide holder.* Set the slide aside to cool. Repeat this procedure with slide B. Turn off the Bunsen burner. Passing the slide through the flame kills the bacteria and attaches, or fixes, the bacteria to the slide.
6. Place both slides in the center of a staining tray. Find the dropper bottle marked "crystal violet." Flood the smears on both slides with crystal voilet, as shown in Figure 1C. **CAUTION:** *Handle the crystal violet with care. It is a stain and is difficult to remove from hands and clothing.* Leave the crystal violet on the smears for 60 seconds.

Name ______________________ Class ____________ Date ____________

7. As shown in Figure 1D, use a wash bottle filled with water to rinse all excess crystal violet from the slides. **CAUTION:** *Do not let the water flow directly onto the smear.*

8. Place the slides between two sheets of bibulous paper, as shown in Figure 1E. Close the sheets and carefully press on the slides to blot them dry. **CAUTION:** *To avoid breaking the slides, use only gentle pressure.*

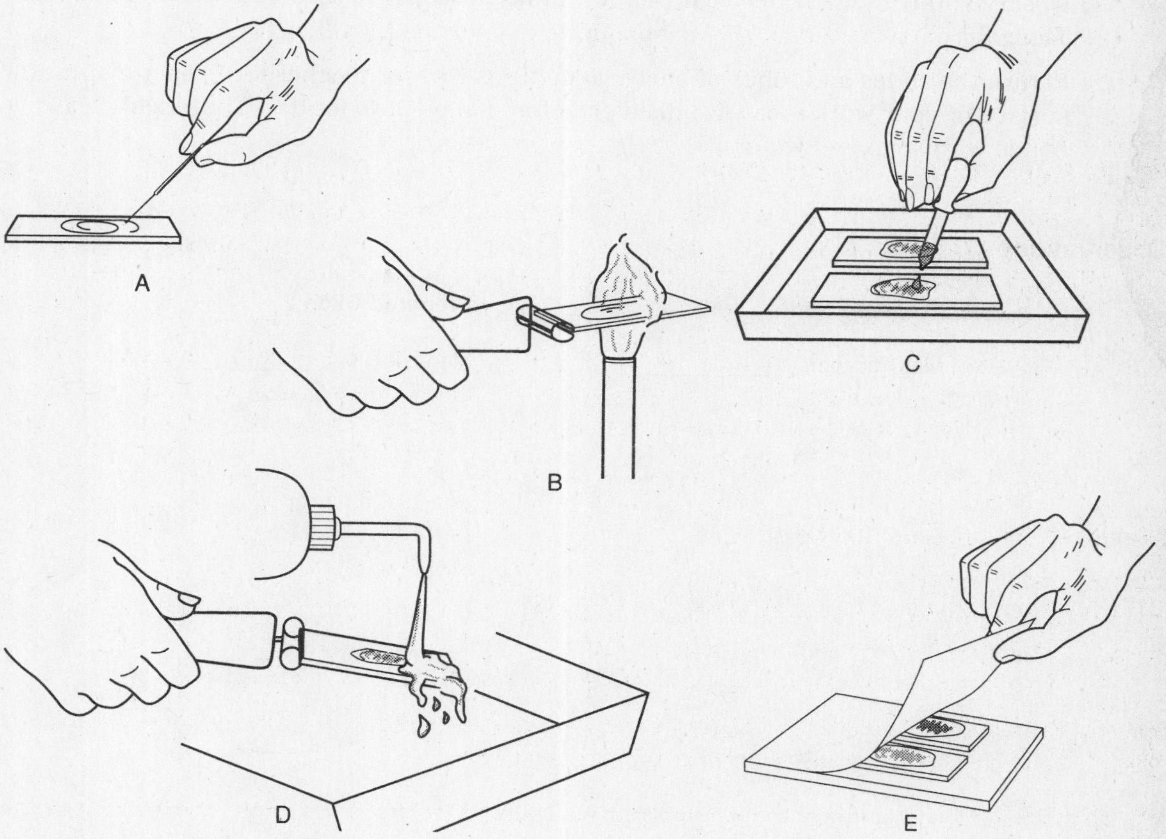

Figure 1

9. Steps 2 through 8 are the first steps in making a Gram stain. The result up to this step is called a *simple stain.*

10. Read steps 11 through 15 before you proceed with the Gram stain. The Gram stain is a differential stain; that is, it has one of two possible results. Gram-positive (Gram+) bacteria are stained violet or blue. Gram-negative (Gram-) bacteria are stained pink or red. You may want to stain only one slide at a time.

11. Replace the slide, smear side up, in the staining tray. Find the dropper bottle marked "Gram's iodine." Flood the smear with Gram's iodine. **CAUTION:** *Handle the Gram's iodine with care. It is a stain and is difficult to remove from hands and clothing.* Leave the Gram's iodine on the smear for 30 seconds.

12. With a wash bottle, wash away all excess Gram's iodine from the slide. **CAUTION:** *Do not let the water flow directly onto the smear.*

13. Find the dropper bottle marked "95% ethyl alcohol." Flood the smear with alcohol. Leave the alcohol on the smear for 10 seconds. Use a wash bottle to rinse away any excess alcohol. **CAUTION:** *Do not let the water flow directly onto the smear.* The alcohol removes the crystal violet from the Gram-negative bacteria. The Gram-negative bacteria are now colorless.

14. Find the dropper bottle marked "safranin." Flood the smear with safranin stain. **CAUTION:** *Handle the safranin with care. It is a stain and is difficult to remove from hands and clothing.* Leave the safranin on for 45 seconds. Wash away any excess safranin with a wash bottle. **CAUTION:** *Do not let the water flow directly onto the smear.* The safranin colors the Gram-negative bacteria pink or red.

15. Dry the slide between two sheets of bibulous paper.

16. Repeat steps 11 through 15 with the remaining slide.

17. Observe both slides under high power of a microscope. Record your observations in Data Table 2.

18. Return the slides and tubes of bacterial cultures to your teacher for proper disposal. Wipe the surface of your work area with disinfectant and allow it to air-dry. Thoroughly wash your hands with soap and water.

Observations

Prepared Slide 1

Magnification ____________

____

Prepared Slide 2

Magnification ____________

____

Prepared Slide 3

Magnification ____________

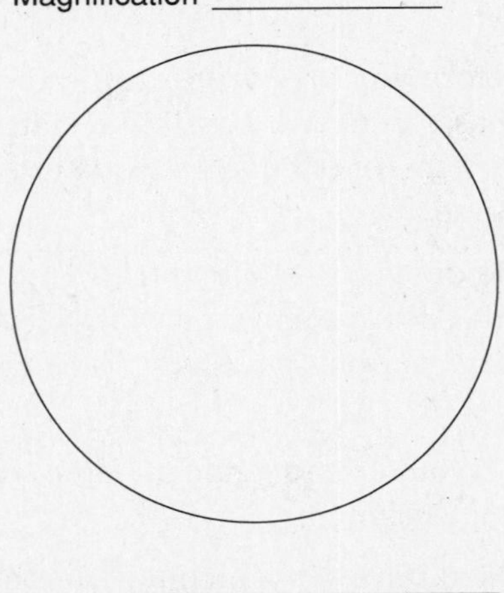

____

Name ______________________ Class ____________ Date ____________

Data Table 1

Bacterium Name	Basic Shape	Arrangement (✓) Single	Pair (diplo-)	Chain (strepto-)	Cluster (staphylo-)	Other (draw)

Data Table 2

Bacterium	Basic Shape	Color	Gram-Positive or Gram-Negative?
A *(Escherichia coli)*			
B *(Bacillus subtilis)*			

Analysis and Conclusions

1. Why do you think you flame the slide with the smear facing away from the flame?

____

____

2. Why do you think you do not run water directly onto the smear?

____

____

3. What are the two reasons for staining bacteria? ______________________

____

____

4. Does a bacterium's Gram reaction seem to be directly related to its shape? How could you prove your hypothesis? ______________________

____

____

Critical Thinking and Application

1. Penicillin works by interfering with a bacterium's ability to build a cell wall. Using this information, explain why Gram-positive bacteria are more susceptible to destruction by penicillin than Gram-negative bacteria are. (*Hint:* Review the structural differences between Gram-positive and Gram-negative bacteria in your text.) ______________________________

2. You are given an unknown bacterial culture by your teacher. After completing the Gram-staining technique used in this investigation, you observe both pink and purple bacteria in the same smear. What could be a possible explanation for this? ______________________________

3. Occasionally after a flood or the breaking of a large water pipe, people living in the area are advised to boil their drinking water. Why is this done? ______________________________

4. It has been only in recent years that scientists have discovered fossilized bacteria. Why might this discovery have taken so long to occur? ______________________________

Going Further

1. Using the techniques for Gram staining that you used in this investigation, prepare a smear of bacteria that have been isolated from bacterial cultures prepared from various foods. Devise a system of research and chart your results.
2. Some scientists now think they have isolated a fourth bacterial shape—the bent rod. Conduct library research on this new finding and share your information with the class.

Name ______________________ Class __________ Date __________

32

Controlling Bacterial Growth

Pre-Lab Discussion

Chemical substances that either kill bacteria or inhibit bacterial growth are called *antimicrobial agents*. Antimicrobial agents are of three basic types: *antiseptics*, or chemicals used to inhibit the growth of or kill bacteria on living tissues; *disinfectants*, or chemicals used to inhibit the growth of or kill bacteria on nonliving things; and *antibiotics*, or chemical substances produced by living organisms, which inhibit the growth of bacteria.

The effectiveness of each type of antimicrobial agent is influenced by many factors. Some of these factors include the environmental conditions in which the agent is applied, the chemical properties of the agent, how long the agent has been stored, and the rate of deterioration of the agent.

In this investigation, you will test the effectiveness of disinfectants and antibiotics in inhibiting the growth of bacteria.

Problem

How can the growth of bacteria be controlled?

Materials *(per group)*

Glass-marking pencil
Bunsen burner
Flint striker or matches
2 sterile cotton swabs
Beaker of water
Test tube rack
Sterile filter-paper disks
Sterile forceps
Distilled water
Metric ruler
Transparent tape
Culture of *Escherichia coli*
2 sterile nutrient agar plates
3 disinfectants chosen from the following: chlorine bleach, household cleaner, household disinfectant, lye, phenol
3 antibiotic disks chosen from the following: aureomycin, chloromycetin, penicillin, streptomycin, tetracycline, terramycin

Safety

Put on a laboratory apron if one is available. Put on safety goggles. Handle all glassware carefully. Tie back loose hair and clothing when using the Bunsen burner. Always use special caution when working with laboratory chemicals and bacterial cultures. Note all safety symbols next to the steps in the Procedure and review the meanings of each symbol by referring to the symbol guide on page 10.

Procedure

Part A. Inoculating a Sterile Nutrient Agar Plate

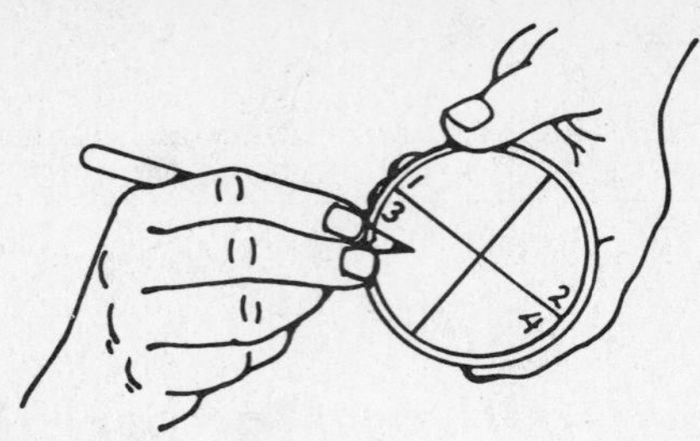

Figure 1

1. Obtain two sterile nutrient agar plates. Carefully turn each plate over and lay it on your worktable. **CAUTION:** *Be very careful not to open the the petri dishes of sterile agar while handling them.*

2. With a glass-marking pencil, mark the bottom of each petri dish shown in Figure 1. Draw two lines at right angles to each other so that the dish is divided into four equal areas, or quadrants. Number the quadrants on each dish 1 through 4. **Note:** *Place the numbers near the edges of the dishes.* Near the top center of each dish write your initials. Carefully turn the petri dishes right side up.

3. Obtain a test tube containing a bacterial culture of *Escherichia coli.* Place the test tube in a test tube rack. **CAUTION:** *Use extreme care when working with bacterial cultures. Avoid spills. If a spill does occur, immediately call your teacher for assistance.* Obtain two sterile cotton swabs. Carefully read steps 4 through 8 and study Figure 2 before you proceed.

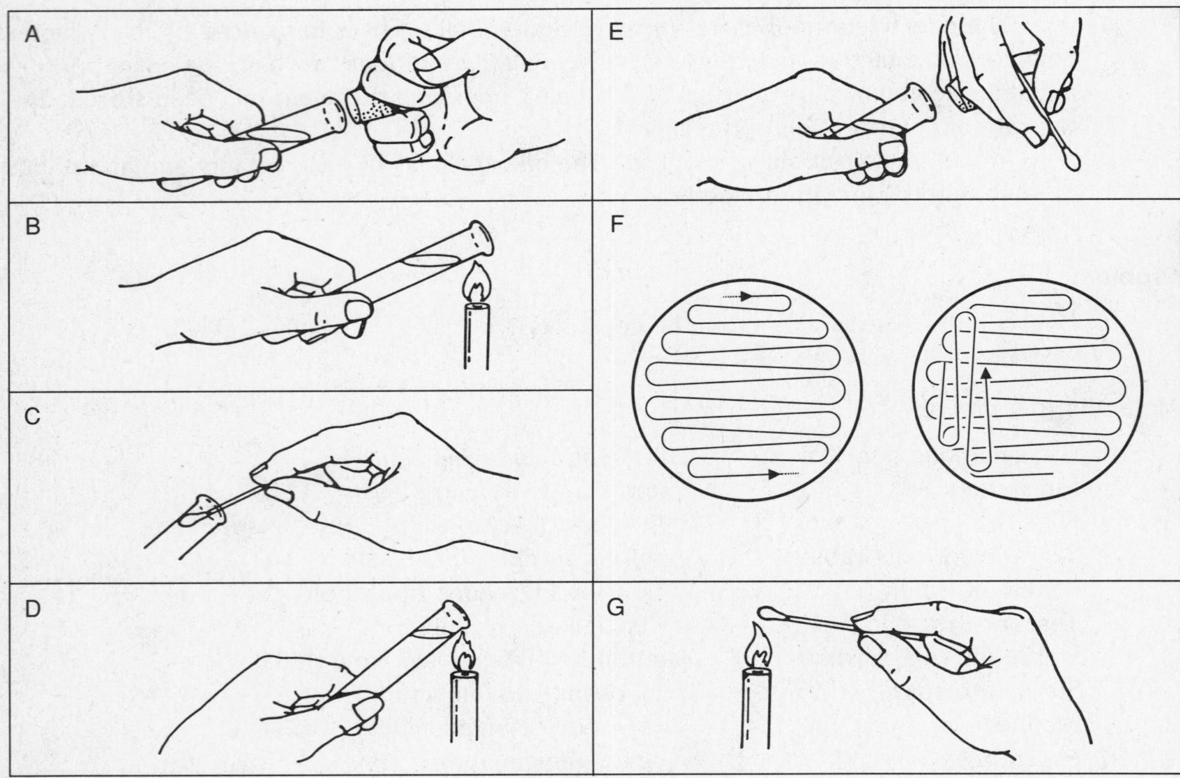

Figure 2

4. Put on your safety goggles and light the Bunsen burner. **CAUTION:** *Use extreme care when working with or near an open flame. Tie back loose hair and clothing.*

5. Pick up the test tube of *E. coli.* Remove the cotton plug. **Note:** *Do not let the cotton plug come in contact with any other object.* Pass the mouth of the tube back and forth through the burner flame.

6. Insert a sterile cotton swab into the bacterial culture. **Note:** *Shake off any excess liquid into the test tube.* Remove the cotton swab. **Note:** *Do not let the cotton swab come in contact with any other object.* Pass the mouth of the tube back and forth through the burner flame. Replace the cotton plug and return the test tube to the rack.

Name ______________________ Class __________ Date __________

7. Slightly open a sterile nutrient agar plate. Place the tip of the cotton swab near the top center of the agar. Streak the agar as shown in Figure 3A. Lift the swab off the plate and turn the petri dish 90° to the right. Streak the agar again, as shown in Figure 3B. Close the petri dish.

8. Hold the top of the cotton swab in the flame of the Bunsen burner until it catches fire. Remove the swab from the flame and plunge it into a beaker of water.

9. Repeat steps 5 through 8 for the other sterile nutrient agar plate. **Note:** *Be sure to use a new sterile cotton swab for this transfer.* Turn off the Bunsen burner after you have completed the second plate.

10. The nutrient agar plates that you have just streaked, or inoculated, with bacteria will be used in part B of this investigation.

11. Return the test tube of *E. coli* to your teacher. Thoroughly wash your hands with soap and water.

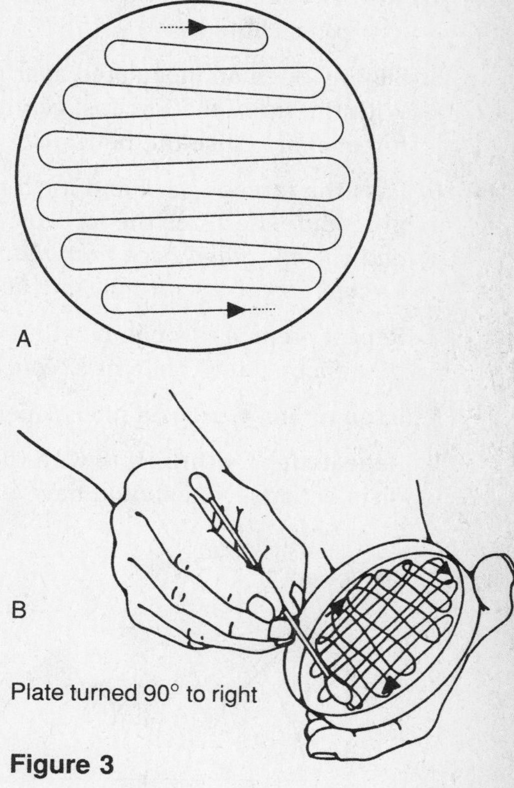

Figure 3

Part B. Controlling the Spread of Bacteria with Disinfectants and Antibiotics

1. Take one inoculated agar plate of *E. coli* that you prepared in Part A.

2. Select three disinfectants and three antibiotics, and record these selections in the Data Table. Carefully read steps 3 through 10 and study Figure 4 before you proceed.

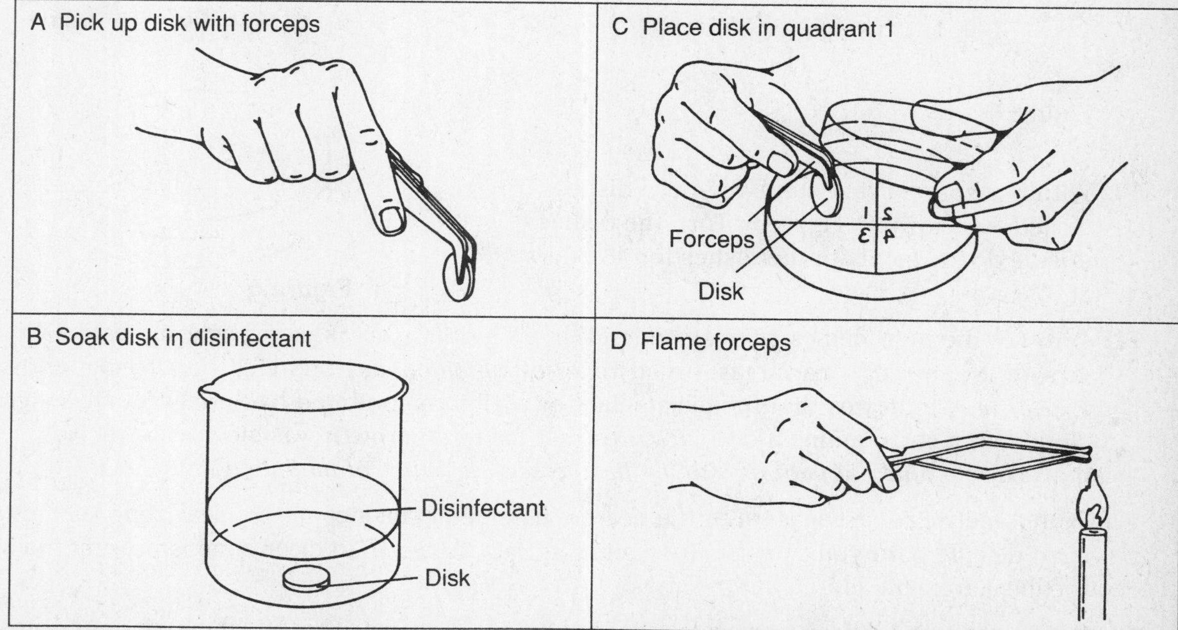

Figure 4

3. Light the Bunsen burner. **CAUTION:** *Use extreme care when working with or near an open flame. Tie back loose hair and clothing.*

4. With sterile forceps, pick up a disk of filter paper. Insert the disk into disinfectant 1. Shake off any excess liquid.

5. Slightly open an inoculated agar plate. Position the filter-paper disk in the center of quadrant 1. With the tip of the forceps, gently press the disk against the agar until it sticks. Remove the forceps and close the petri dish.

6. Pass the forceps back and forth through the flame of the Bunsen burner several times. This procedure sterilizes the forceps. **CAUTION:** *If there is alcohol on the forceps, it will burn brightly and quickly. Stand back from the Bunsen burner when burning alcohol off the forceps.* Allow the forceps to cool before picking up the next filter-paper disk.

7. Repeat steps 4 through 6 with the remaining disinfectant-soaked disks in quadrants 2 and 3 of the inoculated plate of *E. coli*. **Note:** *Remember to sterilize the forceps after each use.*

8. In quadrant 4, place a filter-paper disk soaked in distilled water.

9. Repeat steps 4 through 8 with the other inoculated agar plate using antibiotic disks instead of disinfectants. You should have two inoculated agar plates as shown in Figure 5.

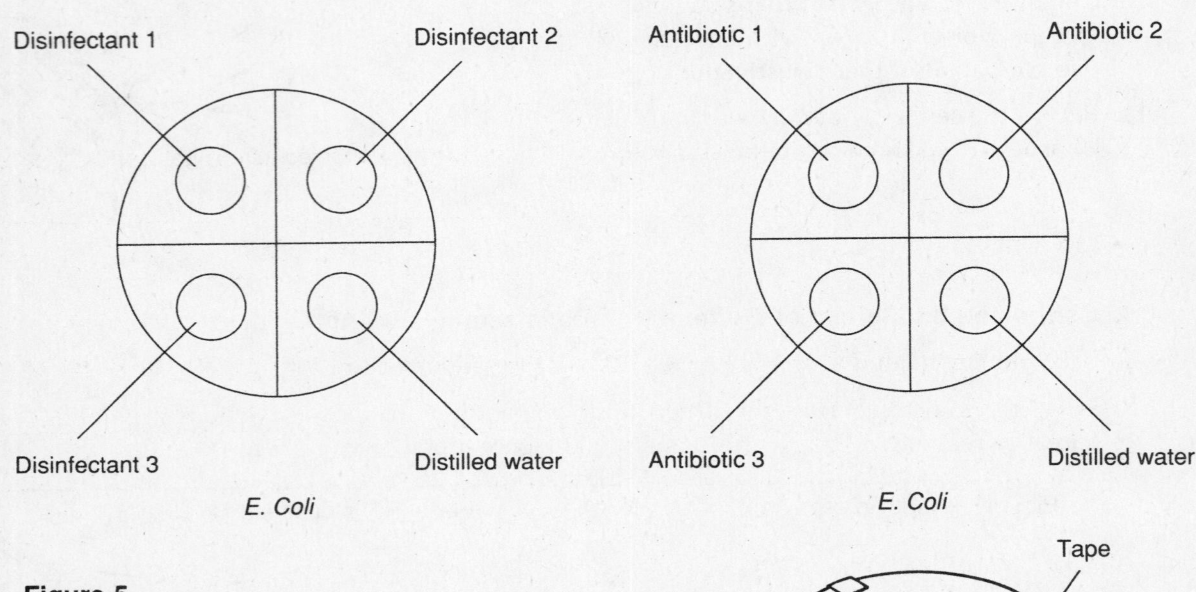

Figure 5

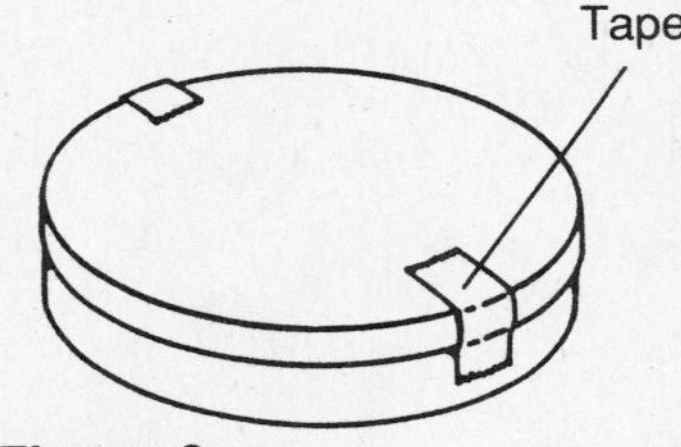

Figure 6

10. With transparent tape, tape the petri dishes closed, as shown in Figure 6. Turn the dishes upside down. Incubate the dishes for 48 hours at 37°C.

11. Observe the petri dishes after 48 hours. White or cloudy areas of the agar indicate bacterial growth. Notice any clear areas, called *zones of inhibition,* surrounding the filter-paper disks. A clear area indicates that the disinfectant or antibiotic inhibited bacterial growth. A lightly cloudy area surrounding a disk indicates that bacterial growth was slowed down. **Note:** *You may want to hold the petri dish to the light to see the zones of inhibition more clearly.*

12. With a metric ruler, measure to the nearest millimeter the size of the clear zone surrounding each disk. Record your measurements in the Data Table. If no clear zone is present, record the measurement as 0.

13. Return the petri dishes to your teacher for proper disposal. Thoroughly wash your hands with soap and water.

Name ______________________ Class ____________ Date ____________

Observations

Data Table

Effects of Disinfectants and Antibiotics on Growth of *E. Coli*	
Disinfectant	**Zone of Inhibition** (mm)
1	
2	
3	
4 distilled water	
Antibiotic	**Zone of Inhibition** (mm)
1	
2	
3	
4 distilled water	

Analysis and Conclusions

1. Why is it important not to open the sterile nutrient agar plates? ______________________

____

____

2. Why do you think it is so important to write the numbers and initials near the edges of the petri dish? ______________________

____

3. Why is it important to use sterile techniques while inoculating the agar plate?

____

____

4. What is the purpose of the disk soaked in distilled water in each inoculated petri dish?

____

____

5. What is the purpose of taping closed the lids of the petri dishes? ______________________

____

____

6. Which disinfectant was most effective in preventing the growth of *E. coli?*

____

7. Which antibiotic was most effective in preventing the growth of *E. coli?*

____

8. How do you know that any inhibition you have observed is due to the disinfectants and antibiotics on the disks? ________________________________

Critical Thinking and Application

1. Pretend that a serious staphylococcus infection has developed in the locker room of your school's gym. Assume that you are responsible for getting rid of the staph contamination. How would you do so? ________________________________

____

____

2. Scientists have observed that an antibiotic seems to lose its effectiveness against a particular population of bacteria after a prolonged period of time. What do you think is responsible for this phenomenon? ________________________________

____

3. Why are the different disinfectants not equally effective against all species of bacteria?

____

____

4. Suppose that your doctor diagnoses your condition as a bacterial infection and prescribes an antibiotic. Your doctor cautions you to take the antibiotic for 10 days even though you may feel fine after a few days. Explain why you should follow your doctor's orders.

____

____

____

Going Further

1. You may want to conduct the following experiment dealing with the origin of antibiotics. Stir a handful of soil into about 100 mL of water. Allow the soil to settle. Dip filter-paper disks into the soil water. Set the disks aside to dry. Use the procedures of this investigation to test for the presence of natural antibiotics.
2. Using the procedures presented in this investigation, test other species of bacteria—such as *B. subtilis, P. vulgaris,* and *S. lutea*—for their resistance or sensitivity to various disinfectants and antibiotics.

Name ______________________ Class __________ Date __________

Adaptations of the Paramecium

Pre-Lab Discussion

Paramecia are animallike single-celled organisms that belong to the phylum *Ciliata.* In nature, paramecia can be found swimming freely in fresh water. The paramecium, like any other living organism, has certain metabolic requirements for survival. The activities that fulfill these requirements are accomplished by certain cell organelles. These cell structures, like the tissues of more complex multicellular organisms, perform specialized tasks. Paramecia are relatively large protozoans with easily identifiable organelles, making them especially useful for study in the biology laboratory.

In this investigation, you will examine the structure and function of the various organelles of the paramecium. You will also observe adaptations of a paramecium that help it survive.

Problem

How is a paramecium structured for survival?

Materials *(per group)*

Culture of *Paramecium caudatum*
Methyl cellulose solution
Carmine solution
Yeast suspension
Congo red solution
Glass slide
Coverslip
Cotton ball
Microscope
Toothpick or glass rod
Medicine dropper

Safety

Put on a laboratory apron if one is available. Put on safety goggles. Handle all glassware carefully. Always use special caution when working with laboratory chemicals, as they may irritate the skin or cause staining of the skin or clothing. Never touch or taste any chemical unless you are instructed to do so. Always handle the microscope with extreme care. You are responsible for its proper care and use. Use caution when handling glass slides as they can break easily and cut you. Note all safety alert symbols next to the steps in the Procedure and review the meanings of each symbol by referring to the symbol guide on page 10.

Procedure

Part A. Paramecium Structure and Movement

1. With a toothpick or glass rod, place a ring of methyl cellulose on a clean glass slide, as shown in Figure 1.

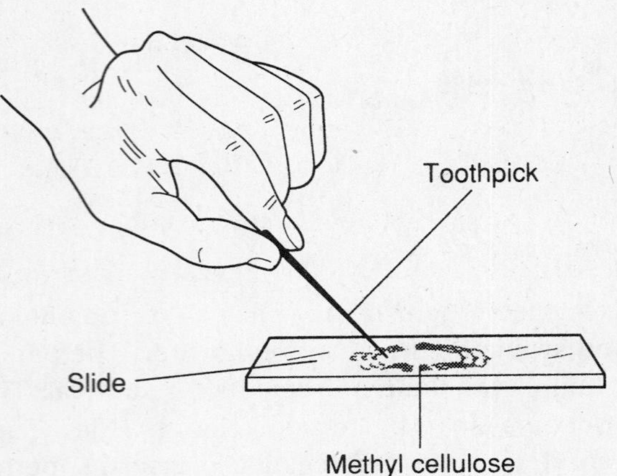

Figure 1

2. Use a medicine dropper to place a drop of paramecium culture within the ring of methyl cellulose. The methyl cellulose is thick like syrup and will slow the paramecia down to make observations easier. **CAUTION:** *Be careful not to spill the methyl cellulose. It can irritate your skin.*
3. Pull a few fibers from the cotton ball and place them over the paramecia and methyl cellulose.
4. Cover the paramecium culture with a coverslip.
5. Observe the paramecium culture under the low-power objective of the microscope. Look for slipper-shaped paramecia that are relatively motionless or trapped near a cotton fiber.
6. Carefully switch to the high-power objective.
7. Adjust the diaphragm to permit a low level of light to enter the lens. This will allow a better contrast for observing organelles.
8. Observe one paramecium for a few minutes.
9. In the space provided in Observations, draw a single paramecium as you observe it in the microscope. Include the organelles that you see. Indicate which end is the posterior (rear) end and which is the anterior (front) end.
10. Refer to Figure 2 to label the organelles you see in your specimen.

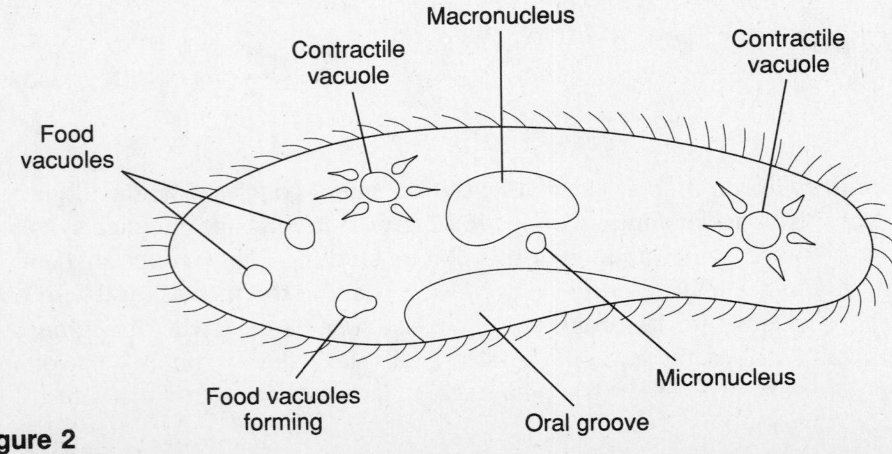

Figure 2

Name ______________________________ Class ______________ Date ______________

11. Observe the motion of the paramecium and answer question 1 in Observations.
12. Observe how the paramecium reacts when it bumps into a cotton fiber. Answer question 2 in Observations.
13. Look for hairlike structures on the outer surface of the organism. These are called cilia. Answer questions 3 and 4 in Observations.
14. Carefully lift the coverslip and add a drop of carmine solution to the paramecium culture. **CAUTION:** *Carmine solution is a dye that can stain your clothing and skin. Be careful.*
15. Replace the coverslip and refocus a single paramecium under high power.
16. Observe the way in which the particles of the carmine solution are moved around the paramecium. Answer question 5 in Observations.
17. Carefully wash and dry the slide and coverslip.

Part B. Ingestion and Digestion in the Paramecium

1. Prepare a second slide of paramecia and methyl cellulose.
2. Add a drop of yeast suspension that has been stained with Congo red to the paramecium culture on the slide. Congo red is a pH indicator. It turns blue in acidic conditions, pink in neutral conditions, and red in basic conditions. **CAUTION:** *Congo red is a dye that can stain your clothing and skin. Be careful.*
3. Cover the slide with a coverslip.
4. Locate a single paramecium under low power and then focus under high power.
5. Look for red yeast cells being drawn into the paramecium.
6. Look for the formation of a food vacuole at the end of the paramecium's gullet. See Figure 2. Continue your observations of the paramecium feeding for several minutes. Answer questions 6 and 7 in Observations.
7. Observe what happens to the food vacuole when it becomes full. Answer question 8 in Observations.
8. Continue observing "older vacuoles" full of yeast. Note any color change in the yeast. Answer questions 9 and 10 in Observations.
9. While observing the food vacuoles, also look for star-shaped organelles toward the anterior and posterior ends of the paramecium. These are called contractile vacuoles. Notice that the yeast cells are not in contact with the contractile vacuoles.
10. Continue observing the contractile vacuoles. Note that the contractile vacuoles occasionally shrink and almost disappear.
11. Time the intervals between contractions of this vacuole. Answer question 11 in Observations.
12. Carefully wash and dry the slide and coverslip.

Observations

Paramecium

1. Describe the motion of the paramecium as it moves throughout the microscope field.

2. How does a paramecium react when it bumps into a cotton fiber?

3. Describe the movement of the cilia on a paramecium.

4. Are the cilia located only on the perimeter of the paramecium?

5. How are the particles of the carmine solution affected by the paramecium?

6. How did the yeast cells end up in the gullet of the paramecium?

7. Describe how a food vacuole forms as yeast cells are ingested by a paramecium.

Name ______________________ Class __________ Date __________

8. What happens to the food vacuole when it becomes full? ______________________

9. What is the initial color of the yeast cells in the food vacuole? ______________________

10. What color do the yeast cells eventually change to after a period of time in a vacuole?

11. What was the time interval between contractions of the contractile vacuole?

Analysis and Conclusions

1. Why is methyl cellulose necessary for observing paramecia? ______________________

2. What is the function of the cilia on a paramecium? ______________________

3. What does a color change in the food vacuoles containing the yeast cells indicate?

4. What is the function of the food vacuole? ______________________

5. Paramecia are constantly taking in water. What natural processes cause the water to move in and out of a paramecium? ______________________

6. What is the function of the contractile vacuole? ______________________

Critical Thinking and Application

1. The trachea in your throat is lined with ciliated cells. Based on your observations of the paramecium's cilia, explain how the function of the cilia in your throat is similar to the function of the cilia of the paramecium. ______________________________

____

____

2. If a paramecium were to live in a saltwater environment, would it need to have a contractile vacuole? Explain your answer. ______________________________

____

____

____

____

3. Although protozoans are unicellular organisms, parts of their cells exhibit a high degree of specialization. List three cell parts of a paramecium that show specialization within the cell. Relate each cell part in the organism to an organ in the human body.

____

____

____

Going Further

1. Changes in the environment may produce changes in paramecium behavior. Set up an experiment to observe a paramecium's response to changes in: the salt concentration of water; the pH of water; the temperature of water; and light.
2. Cilia are embedded in the pellicle, or outer membrane, of the paramecium. Embedded in the pellicle are trichocysts, which normally are not visible. To find this organelle, add a drop of weak acetic acid to a slide of paramecium culture. The trichocysts will protrude from the pellicle and appear as clusters of long fibers or hairs. Report your findings to your teacher and class.

Name ______________________ Class __________ Date __________

Comparing Protists

Pre-Lab Discussion

Kingdom *Protista* includes algal protists, animallike protists, and fungal protists. Protists are often thought of as simple organisms, but a closer examination reveals that they possess a variety of specialized structures. Protists are eukaryotic, which means they have a nucleus enclosed in a membrane. Protists also possess other organelles that perform different functions. Although most protists are one-celled organisms, a few are multicellular. Protists are usually found in damp or watery environments.

In this investigation, you will study several algal and animallike protists.

Problem

What are some characteristics of protists?

Materials *(per group)*

Culture of mixed diatoms
Culture of *Euglena gracilis*
Culture of *Amoeba proteus*
Culture of *Stentor coeruleus*
Microscope
2 glass slides
2 coverslips
Medicine dropper
Methyl cellulose
Index card
Scissors
Toothpick

Safety

Put on a laboratory apron if one is available. Put on safety goggles. Handle all glassware carefully. Always use special caution when working with laboratory chemicals, as they may irritate the skin or cause staining of the skin or clothing. Never touch or taste any chemical unless instructed to do so. Always handle the microscope with extreme care. You are responsible for its proper care and use. Use caution when handling glass slides as they can break easily and cut you. Note all safety symbols next to the steps in the Procedure and review the meanings of each symbol by referring to the symbol guide on page 10.

Procedure

Part A. Observing Algal Protists

1. Prepare a wet-mount slide of a mixed diatom culture.

2. Observe the diatoms under the low-power objective of a microscope. Note the variety of colors, shapes, and sizes. Locate an area containing two or three different diatoms.

3. Switch to the high-power objective and focus on one individual diatom at a time. All diatoms have a two-part cell wall, or shell, made of silica. Observe the ridges and patterns in the diatoms' shells.

4. Notice the grooves in the shell. These grooves connect with pores that open to the outside environment. Look for other structures within the diatoms. In the appropriate place in Observations, draw a sketch of two diatoms that you observed. Record the magnification of the microscope.

5. Look for movement of the diatoms. Answer question 1 in Observations.

6. Prepare a wet-mount slide of *Euglena gracilis* culture. Add a drop of methyl cellulose to the culture before placing the coverslip on the slide. Methyl cellulose will slow down the euglena's movement. **CAUTION:** *Be careful not to spill the methyl cellulose. It can irritate your skin.*

7. Mount the slide onto the microscope and focus under low power. Locate several euglenas.

8. Switch to high power and focus on one euglena. Observe the shape of the euglena and distinguish its anterior (front) end from its posterior (rear) end.

9. Note the chloroplasts, which are greenish structures that enable a euglena to make its own food.

10. Locate the euglena's outer covering, or pellicle. The pellicle is flexible and allows the euglena to change its shape. Notice the two flagella, or taillike structures. The longest flagellum helps the organism move. Find the flask-shaped groove called the reservoir, where the flagella are attached.

11. Find the contractile vacuole, which collects extra water and discharges it from the cell. Locate the nucleus.

12. Locate the red eyespot, which is sensitive to light. From an index card, cut a strip the size of a glass slide. Then cut a slit about 1.5 mm long in the center of the card. Place the card with the slit under the slide on the microscope stage. After several minutes, carefully remove the card and examine the euglenas again. Answer question 2 in Observations.

13. In the appropriate place in Observations, draw and label one euglena. Label the following parts: pellicle, flagellum, reservoir, chloroplasts, contractile vacuole, eyespot, and nucleus. Record the magnification of the microscope.

14. Carefully wash and dry your slides and coverslips.

Name ______________________ Class ____________ Date ____________

Part B. Observing Animallike Protists

1. Prepare a wet-mount slide of *Amoeba proteus* culture.

2. Mount the prepared slide under the microscope and focus under low power. Use the diaphragm to cut down the amount of light entering the lens. An ameba is easier to observe in low light. Move the slide back and forth to search for the organism. Ameba are somewhat transparent and very irregular in shape.

3. Locate one ameba and switch to high power. Observe the organism closely for several minutes. Locate the two kinds of cytoplasm: endoplasm, a clear outer layer near the cell membrane; and ectoplasm, the granular interior cytoplasm that contains organelles.

4. Notice the movement of the ameba's cytoplasm. Classified as a sarcodine, the ameba moves with psuedopodia (false feet), or cell extensions.

5. Observe the organelles such as the nucleus and the contractile vacuole.

6. Carefully remove the coverslip from the slide and add a drop of *Stentor* culture. Mix the two cultures with a toothpick and carefully replace the coverslip. Observe what happens when an ameba comes in contact with a *Stentor.*

7. In the appropriate place in Observations, draw and label one ameba. Label the following structures: ectoplasm, endoplasm, pseudopod, nucleus, contractile vacuole, and food vacuole. Answer questions 3 and 4 in Observations.

8. Prepare a wet-mount slide of *Stentor coeruleus* culture.

9. Mount the slide under the microscope and focus under low power. Examine the size and shape of the *Stentor.*

10. Switch to the high-power objective. Notice the narrow posterior end of the *Stentor.* It may be attached to debris or algae particles. Normally the organism is free-swimming, but when feeding it attaches its posterior end to an object.

11. Locate the hairlike cilia. The *Stentor* is a ciliate protist that moves by means of cilia. Observe the organism for a few minutes. Notice if it changes shape.

12. Note the shape and location of the mouth opening and the cilia that surround the mouth. The *Stentor* sucks in food by setting up water currents with these cilia.

13. Find the gullet, where food is taken into the organism. Also locate the long thin nucleus.

14. In the appropriate place in Observations, draw and label one *Stentor.* Label the following structures: cilia, gullet, and nucleus.

15. Carefully wash and dry your slides and coverslips.

Observations

Magnification ______________

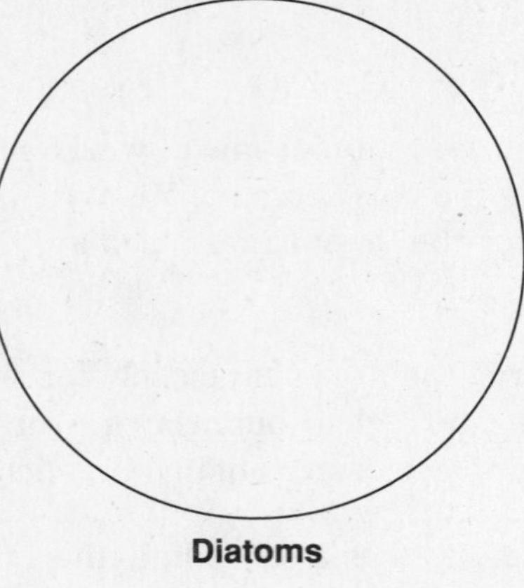

Diatoms

Magnification ______________

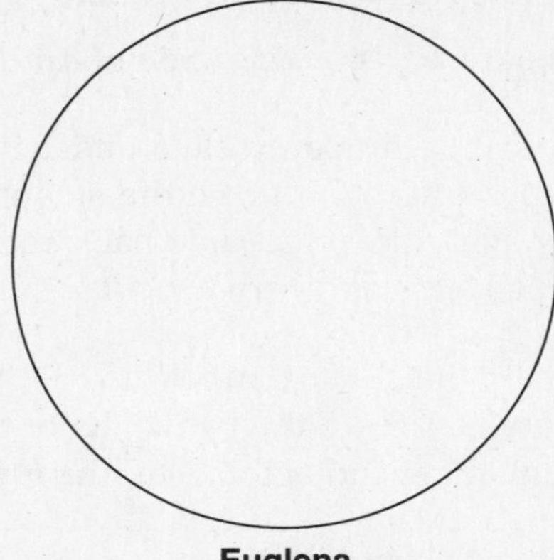

Euglena

Magnification ______________

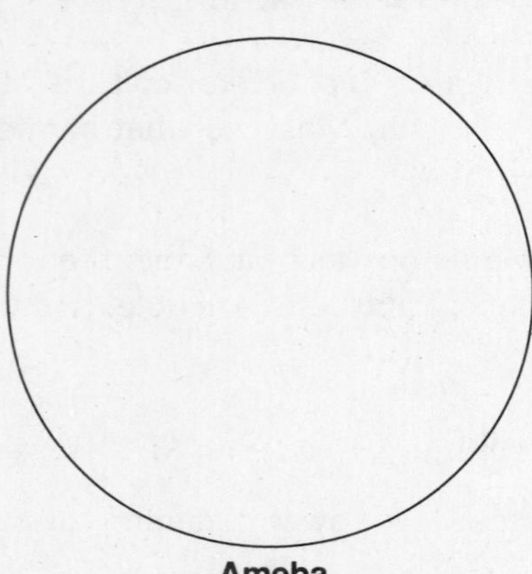

Ameba

Magnification ______________

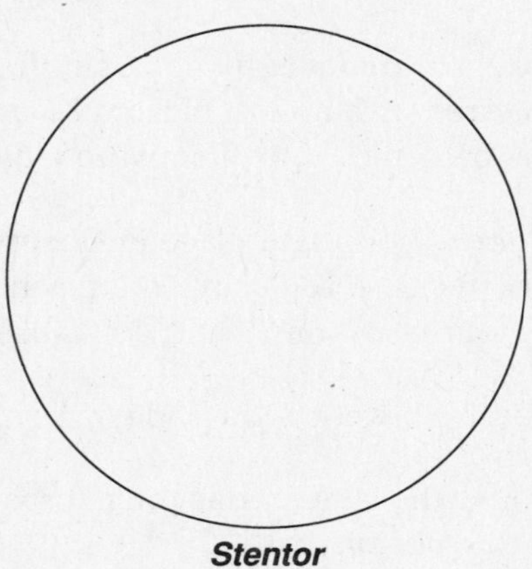

Stentor

1. Describe any kind of diatom movement that you discovered. ______________________________

____

2. Where did you find the euglenas after removing the index card? ______________________________

____

3. Does an ameba retain a constant shape? Explain your answer. ______________________________

____

4. Describe what happens when an ameba meets some food. ______________________________

____

5. Which protist moved the slowest? ______________________________

Name ____________________ Class __________ Date __________

Analysis and Conclusions

1. Why do you think diatoms show colors of green, yellow, and brown?

2. How do the eyespot and the chloroplasts work together to help the euglena survive?

3. What characteristics of an animal does the euglena possess?

4. What are the functions of cilia in *Stentor*?

Critical Thinking and Application

1. Some of the white blood cells in your body show ameboid movement. What does this mean?

2. Why is it important for a photosynthetic organism such as a diatom to live near the surface of a body of water?

3. Euglenas are generally autotrophic organisms. Why does this characteristic make the euglena a good choice for use in the biology laboratory?

4. Why are ground-up diatoms sometimes added to powdered cleansers?

Going Further

Obtain a sample of pond water, or prepare a hay infusion for culturing protists. To make a hay infusion, boil a few pieces of timothy hay stalks in about 200 mL of spring water. **CAUTION:** *Wear safety goggles.* Transfer the mixture to a jar, add a few grains of boiled rice, and allow it to stand for two days. If you add pond water instead, let the culture stand for several days. The culture will become cloudy and then will become clear. When it is clear, make several slides from drops of water taken from different levels of the pond water or the hay infusion. Examine the culture or pond water for protists and identify them. Attempt to outline a food web based on the organisms found.

Name ______________________ Class __________ Date __________

The Structures of Fungi

Pre-Lab Discussion

Fungi include a variety of organisms that are widely distributed in many different habitats. Fungi are heterotrophs that secrete enzymes to break down materials outside their body. Then they absorb these products of digestion and use them as food. Most fungi are saprophytes that live on dead organisms; some are parasites that feed on living organisms; and a few are symbionts that live in association with other organisms.

Most fungi have a "body" made up of tangled filaments called a *mycelium*. A mycelium is, in turn, made up of filaments called *hyphae*. The mycelium attaches to a food source, and the hyphae, although microscopic, present a large surface area through which food can be absorbed.

In this investigation, you will observe and identify the structures and characteristics of several different types of fungi.

Problem

What structures and characteristics of fungi can be identified in a slime mold, mushroom, bracket fungus, and lichen?

Materials *(per group)*

Physarum polycephalum culture and prepared slides
Hand lens
Microscope
Agaricus campestris
Bracket fungus
2 glass slides
2 coverslips
Forceps
Medicine dropper
Lichen
Probe

Safety

Put on a laboratory apron if one is available. Put on safety goggles. Handle all glassware carefully. Be careful when handling sharp instruments. Always handle the microscope with extreme care. You are responsible for its proper care and use. Use caution when handling glass slides as they can break easily and cut you. Note all safety alert symbols next to the steps in the Procedure and review the meanings of each symbol by referring to the symbol guide on page 10.

Procedure

Part A. Observing a Slime Mold

1. Use a hand lens to observe a culture of *Physarum polycephalum.* Record your observations.
2. Obtain a prepared slide of *Physarum polycephalum* and observe it under the low-power objective of a microscope. Record your observations.

Part B. Examining Basidiomycetes—Mushroom and Bracket Fungus

1. Obtain an *Agaricus campestris* mushroom.
2. Examine the structure of the mushroom. Locate the stalk, the umbrella-shaped cap, and the rows of gills on the underside of the cap. Record your observations.
3. Using forceps, carefully remove one gill from the mushroom cap.
4. Place the gill in a drop of water on a glass slide and cover with a coverslip.
5. Use the microscope to observe the gill under both low and high power. Each gill has thousands of extensions called basidia. Under the microscope most of the cells of the gill will appear as rectangles. On the edge of the gill, locate the club-shaped basidia and the attached basidiospores. In *Agaricus,* each basidium contains four spores. (You may not see all four spores on each basidium.) Record your observations.
6. Obtain a specimen of bracket fungus, a basidiomycete that grows on trees.
7. Use a hand lens to examine the underside of the fungus. Record your observations.

Part C. Identifying Parts of a Lichen

1. Obtain a lichen specimen.
2. Use a hand lens to examine your lichen. Powdery masses on the surface of the lichen may be spore cases. Record your observations.
3. Place a small section of the lichen on a glass slide. Add a drop of water.
4. Gently tease the lichen apart with a probe. Add a coverslip.
5. Observe the lichen under low power.
6. Switch to high power and again examine the lichen. Note the greenish cells enmeshed in the whitish filaments. The greenish cells may be a green alga or a cyanobacterium. The white filaments are the hyphae of the fungus. Record your observations.

Observations

Part A. Observing a Slime Mold

1. Describe the appearance of the slime mold. ______________________________

____

2. What characteristic of an animal did you observe in the slime mold?

____

Name ______________________ Class __________ Date __________

3. What characteristic of a fungus did you observe in the slime mold?

____

____

Part B. Examining Basidiomycetes—Mushroom and Bracket Fungus

1. Draw and label the parts of a mushroom that you observed.

***Agaricus* Structures**

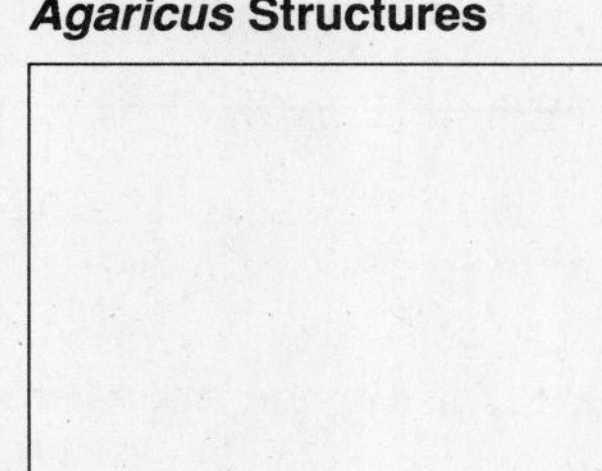

2. Draw and label several basidia with basidiospores from the gill you observed under the microscope.

Magnification __________

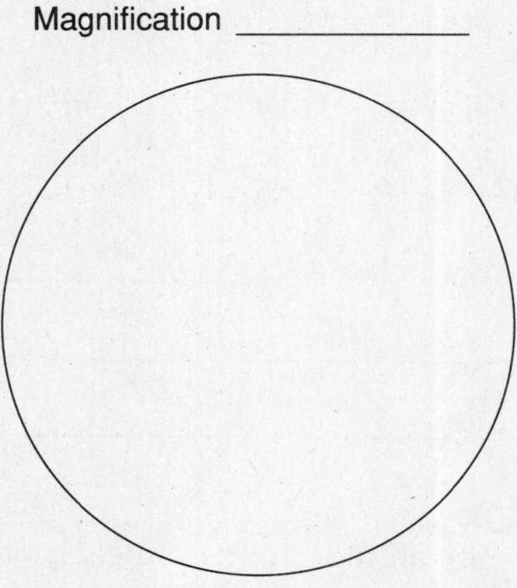

***Agaricus* Basidia**

3. Did you observe the mycelium of the mushroom? Explain your answer.

____

____

4. Did you observe the mycelium of the bracket fungus? Explain your answer.

____

____

5. Describe the appearance of the bracket fungus. ______________________________

____

Part C. Identifying Parts of a Lichen

1. Describe the appearance of your lichen. ______________________________

____

2. Draw and label the parts of the lichen you observed using the microscope.

Magnification ____________

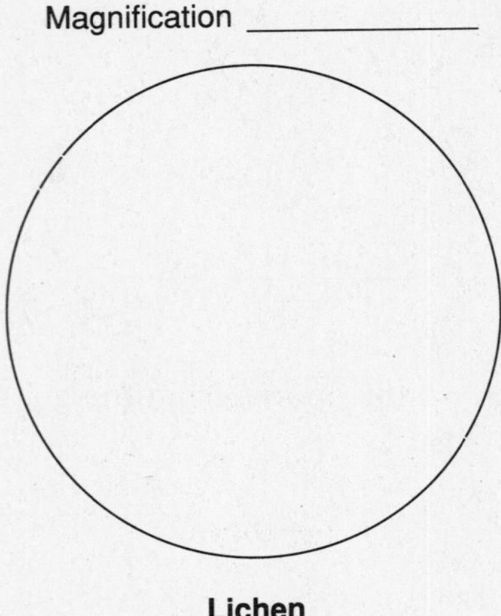

Lichen

Analysis and Conclusions

1. Why are slime molds difficult organisms to classify? ______________________________

____

____

2. Why would you not always be able to see all four spores on each basidium of a mushroom gill? ____

____

3. Which parts of a mushroom are normally eaten? ______________________________

____

Name ______________________ Class __________ Date __________

4. Which fungus in this investigation is a parasite? Explain your answer.

5. Which fungus is actually a symbiotic relationship between two different organisms? How does each organism benefit?

Critical Thinking and Application

1. Why is it an advantage for a mushroom to develop under ground and produce many spores?

2. Lichens are very sensitive to sulfur dioxide, an air pollutant. How might the study of lichens benefit human health?

3. *Agaricus campestris* is an edible mushroom. Is it safe to eat all types of mushrooms? Explain your answer.

Going Further

1. Purchase a mushroom-growing kit from a biological supply company and grow your own mushrooms. Once growth starts, mushrooms should continue to appear for several weeks. Examine and describe the growth stages of the emerging mushrooms. Make a drawing of the life cycle of a mushroom.
2. Go on a fungi scavenger hunt. Locate, but do not disturb, examples of slime molds, mushrooms, bracket fungi, and lichens in your neighborhood.

Name ______________________ Class __________ Date __________

36

Mold Growth

Pre-Lab Discussion

Fungi provide us with food and return important nutrients to the environment for future use. But when fungi appear on our food as mold, they are definitely not welcome!

One of the most commonly purchased food items in a grocery store is bread. Bread has a relatively short shelf-life (the time it takes for food to spoil). And bread is often spoiled by the presence of mold. Food producers have developed chemical additives that help to extend the shelf-life of many foods. For example, to extend the shelf-life of bread, producers often add calcium propionate to bread to prevent the growth of mold.

In this investigation, you will test the resistance of different types and brands of bread to the growth of mold.

Problem

Which type of bread shows the greatest resistance to mold growth?

Materials *(per group)*

Petri dishes
Forceps
Medicine dropper
Glass-marking pencil
Bread—a variety of different types and brands

Safety

Put on a laboratory apron if one is available. Handle all glassware carefully. Note all safety alert symbols next to the steps in the Procedure and review the meanings of each symbol by referring to the symbol guide on page 10.

Procedure

Use the type of bread provided by your teacher to perform the following procedure.

1. Use forceps to place a 1/4 slice of bread in a petri dish.
2. Use the medicine dropper to moisten the bread with 20 drops of water.
3. Close the petri dish and place it in a warm location along with the petri dishes from other lab groups. Label the petri dish with a glass-marking pencil or place it on a sheet of paper with your group's name and class.

4. Observe the bread daily for a week. Record your observations in Data Table 1.
5. At the end of the week report your observations to the class and complete Data Table 2.

Observations

Data Table 1

Day	Appearance of Bread
1	
2	
3	
4	
5	

Data Table 2

Brand	Day on Which Mold Was First Observed

Name ______________________ Class ____________ Date ____________

1. Calcium propionate is an additive put in bread to prevent mold growth. Using the list of ingredients on the bread packages, which breads contained calcium propionate?

2. Which bread was the first to show the growth of mold?

3. How did the appearance of the mold change during the week?

4. Why were all the petri dishes placed in the same location?

Analysis and Conclusions

1. Why did mold grow on the bread in the closed petri dishes?

2. What conditions favor the growth of mold?

3. Would you expect homemade breads or commercially produced breads to have a longer shelf-life? Explain your answer.

Critical Thinking and Application

1. Many dehydrated foods do not contain chemical additives. Would you expect mold to grow on dehydrated foods? Explain your answer.

2. Some molds that grow on bread are green. Can bread mold produce its own food through photosynthesis? Explain your answer. ______________________________

____

Going Further

Chemical additives are used in many foods. Using reference materials, identify the purpose of adding each of the following chemicals to foods: ammonium sulfate, calcium sulfate, monocalcium phosphate, potassium bromate, ethoxylated mono- and diglycerides.

Name ______________________ Class ____________ Date ____________

 37

Characteristics of Green Algae

Pre-Lab Discussion

Green algae are members of phylum Chlorophyta and contain chlorophylls *a* and *b* within their chloroplasts. It is these chlorophylls that give the algae their green color. Green algae may live as single cells or they may be multicellular. Green algae are found mainly in moist areas on land and in fresh water. Some species, however, live in the oceans.

In this investigation, you will observe some characteristics of several different types of green algae.

Problem

What are some characteristics of green algae?

Materials *(per group)*

Microscope	Cultures of:
4 glass slides	*Chlamydomonas*
4 coverslips	*Volvox*
4 medicine droppers	*Ulothrix*
	Ulva

Safety

Put on a laboratory apron if one is available. Handle all glassware carefully. Use caution when handling laboratory cultures. Always handle the microscope with extreme care. You are responsible for its proper care and use. Use caution when handling glass slides as they can break easily and cut you. Note all safety alert symbols next to the steps in the Procedure and review the meanings of each symbol by referring to the symbol guide on page 10.

Procedure

Part A. Unicellular Green Algae

1. Prepare a wet-mount slide of a *Chlamydomonas* culture.
2. Place the slide on the stage of the microscope and focus under the low-power objective, locating several *Chlamydomonas.*
3. Switch to the high-power objective and focus on one *Chlamydomonas.* Observe the shape of the *Chlamydomonas.* Notice the two flagella, or taillike structures. Locate the large cup-shaped chloroplast that surrounds the centrally located nucleus.

4. Find the small organelle called the pyrenoid at the base of the chloroplast. The pyrenoid synthesizes and stores starch. Note the eyespot, which enables the *Chlamydomonas* to sense bright light or darkness.

5. In the appropriate place in Observations, sketch one *Chlamydomonas*. Include the following labels on your sketch: cell wall, flagellum, chloroplast, nucleus, pyrenoid, and eyespot if visible. Record the magnification.

Part B. Colonial Green Algae

1. Prepare a wet-mount slide of a *Volvox* culture.

2. Place the slide on the stage of the microscope and observe the *Volvox* colony under the low-power objective.

3. In the appropriate place in Observations, sketch several cells in the *Volvox* colony. Include the following labels on your sketch: cell wall, chloroplast, cytoplasm, and flagellum. Record the magnification.

4. Switch to the high-power objective. Note any differences in the cells of the colony.

Part C. Filamentous Green Algae

1. Make a wet-mount slide of a *Ulothrix* culture. *Ulothrix* is a filamentous green alga. With the exception of the cell at the base, the cells of each filament are similar. The cell at the base, called the holdfast cell, is modified to hold the plant to stones and other objects.

2. Place the slide on the stage of the microscope and observe a filament of *Ulothrix* under the low-power objective.

3. Switch to the high-power objective. Note the green, horseshoe-shaped chloroplast.

4. In the appropriate place in Observations, sketch a portion of the *Ulothrix* filament. Include the following labels on your sketch: cell wall, chloroplast, and nucleus. Record the magnification.

Part D. Multicellular Green Algae

1. Prepare a wet-mount slide of an *Ulva* culture.

2. Place the slide on the stage of the microscope and focus under the low-power objective.

3. Switch to the high-power objective. To determine the number of cell layers present in *Ulva,* slowly focus up and down while observing the slide under the high-power objective.

4. In the appropriate place in Observations, sketch two or three cells of *Ulva*. Include the following labels on your sketch: cell wall, cytoplasm, chloroplast, and nucleus. Record the magnification.

5. Carefully clean and dry the slides and coverslips you have used.

Observations

1. How does a *Volvox* colony move? ______________________________

____

2. Were any of the cells in the *Volvox* colony different from the others? If so, how were they different? ______________________________

____

3. How many cell layers does *Ulva* have? ______________________________

Name ______________________ Class __________ Date __________

High-power objective

Magnification ____________

Chlamydomonas

High-power objective

Magnification ____________

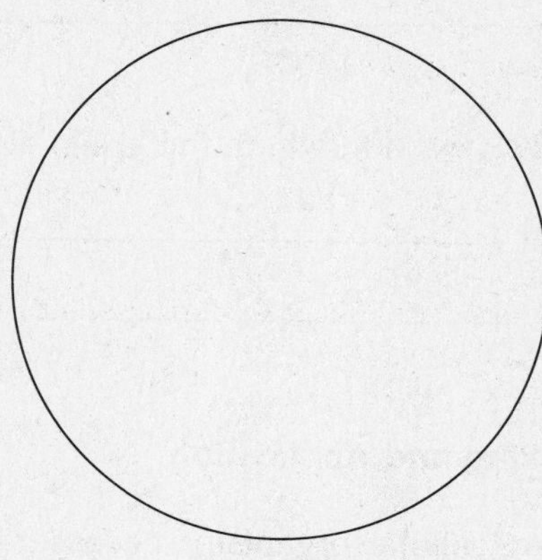

Volvox

High-power objective

Magnification ____________

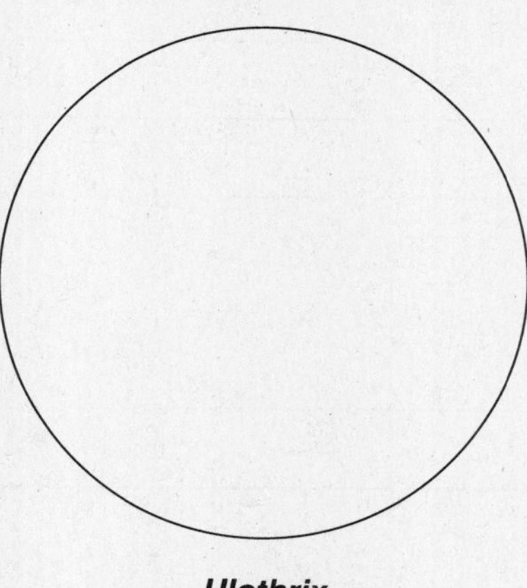

Ulothrix

High-power objective

Magnification ____________

Ulva

Analysis and Conclusions

1. How is *Chlamydomonas* animallike? How is it plantlike? ______________________

____

____

2. How is *Volvox* animallike? How is it plantlike? ______________________

____

3. List three ways in which the green algae you observed in this investigation are different.

4. List two ways in which the green algae you observed in this investigation are similar.

Critical Thinking and Application

1. Are unicellular organisms necessarily simple organisms? Explain your answer.

2. Why are algae not able to grow on land to any great extent?

3. What is the advantage of a holdfast to an alga, such as *Ulva,* that lives in coastal waters?

4. Suppose a new species of unicellular green algae containing chlorophyll and an eyespot is discovered. Would you expect this type of algae to have either flagella or cilia? Explain.

Going Further

1. Obtain samples of red and brown algae for observation. How are red and brown algae similar to green algae? How are they different?
2. Survey some of the foods in your local supermarket. Which contain products made from algae? Identify the products. To which group of algae do they belong? What role do the algal products play? Prepare a report or poster on your findings to be presented to your class.

Name ______________________ Class __________ Date __________

Examining an Algal Bloom

Pre-Lab Discussion

When pollutants, such as phosphate-containing detergents and fertilizers, are dumped into ponds and lakes, the algae population greatly increases. This sudden massive growth of algae is called an *algal bloom* and appears as a thick, green covering on the surface of the water.

Although the algae on the water's surface flourish, those lying below the surface do not. These algae die because they are shaded from the sun by the masses of algae above them. The dead algae sink to the bottom of the pond or lake, taking their nutrients with them. Thus a large amount of organic matter accumulates on the bottom.

Because the dead organic matter serves as food for bacteria, there is a tremendous increase in the growth of these organisms. As the bacteria grow, they use up the oxygen in the water. Eventually, fishes and other aquatic organisms suffocate and die. This oxygen-depletion process is called *eutrophication.*

In this investigation, you will observe the ability of some detergents and fertilizers to cause algal blooms.

Problem

How is the growth of algae affected by chemical pollutants?

Materials *(per group)*

Green algae culture
Pond water
1% pollutant solution
Microscope
Glass slide
Coverslip
Medicine dropper
Glass-marking pencil
3 250-mL Erlenmeyer flasks
100-mL graduated cylinder
10-mL graduated cylinder
3 cotton balls
Sheet of translucent tissue paper

Safety

Put on a laboratory apron if one is available. Put on safety goggles. Handle all glassware carefully. Always use special caution when working with laboratory chemicals, as they may irritate the skin or cause staining of the skin or clothing. Never touch or taste any chemical unless instructed to do so. Always handle the microscope with extreme care. You are responsible for its proper care and use. Use caution when handling glass slides as they can break easily and cut you. Note all safety alert symbols next to the steps in the Procedure and review the meanings of each symbol by referring to the symbol guide on page 10.

Procedure

Part A. Preparing Algal Cultures

1. With the glass-marking pencil, mark one flask "Control," the second flask "0.5% pollution," and the third flask "1% pollution."
2. Measure out 100 mL of pond water and place it in the flask marked Control. Use the 10-mL graduated cylinder to add 5 mL of the green algae culture to the flask. **Note:** *To make sure that there is an equal amount of algae in each 5-mL sample, gently swirl the container containing the green algae culture before pouring each sample.*
3. Obtain the 1% pollutant solution from your teacher. Measure out 50 mL of the pollutant solution and place it in the flask marked 0.5% pollution. Then add 50 mL of pond water and 5 mL of green algae to the flask.
4. Measure out 100 mL of the 1% pollutant solution and place it in the flask marked 1% pollution. Then add 5 mL of the green algae culture to the flask.
5. Gently swirl the flasks to mix the algae with the other solutions. Plug the flasks with cotton balls.
6. If a fluorescent light is available, place the flasks about 60 centimeters from the light for one week. If a fluorescent light in not available, place the flasks in a window. To protect the algae from heat, cover the flasks with a sheet of translucent tissue paper.

Part B. Observing Algal Cultures

1. After one week, observe the cultures in the flasks. Look for a greenish tint or cloudiness in the flasks.
2. Gently swirl the Control flask and make a wet-mount slide from one drop of the culture.
3. Place the slide on the stage of the microscope and focus under low power. Count the number of algal cells in each of the three fields of view shown in Figure 1. Record the numbers in the appropriate places in the Data Table. Record the average number of algal cells.

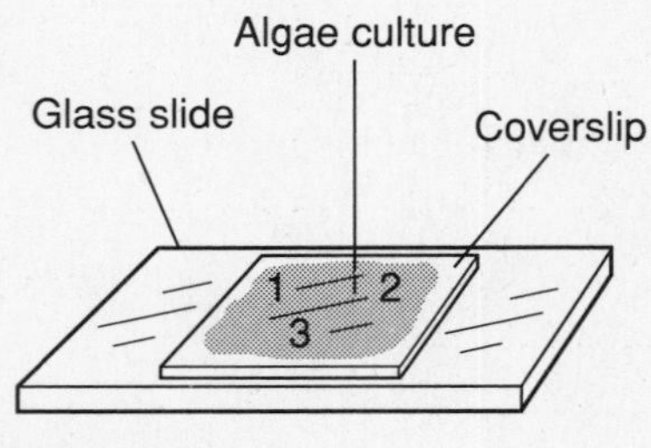

Figure 1

Name ______________________ Class ____________ Date ____________

4. Repeat steps 2 and 3 with the two other flasks. Record the information in the appropriate places in the Data Table.
5. Using the graph provided in Observations, plot the average number of algal cells found in the Control, in the 0.5% pollution solution, and in the 1% pollutant solution. Connect the points on the graph.
6. Place your data in the data table provided by your teacher on the chalkboard.

Observations

Data Table

Field of View	Number of Algal Cells		
1	Control	0.5% Pollution	1% Pollution
2			
3			
Average			

Graph

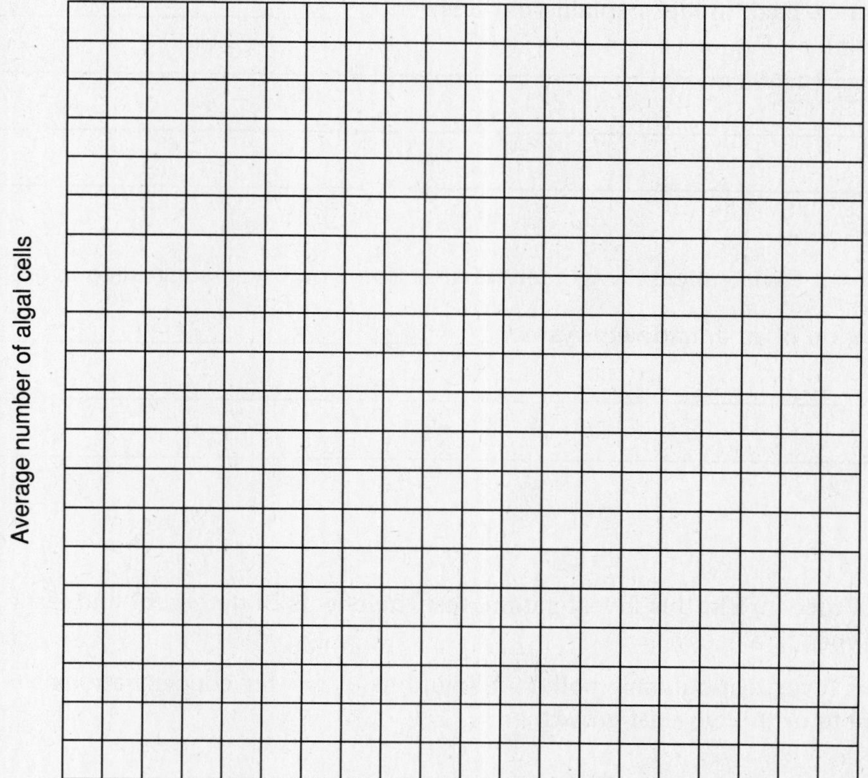

Analysis and Conclusions

1. Why is a light source necessary in this investigation? ______________________________

____

2. According to your data, which pollution solution caused the least algal growth?

____

3. According to your data, which pollution solution caused the greatest algal growth?

____

Critical Thinking and Application

1. How might algae be an indicator of the amount of water pollution in a body of water?

____

____

____

2. Do you think it would be an easy task to bring a body of water that has undergone eutrophication back to life? Explain your answer. ______________________________

____

____

____

3. If you were an environmental engineer, what action would you suggest to avoid the eutrophication of an aquatic ecosystem? ______________________________

____

____

Going Further

1. Using the Procedure in this investigation, test the effects of detergent and fertilizer pollutants on other types of algae.
2. Repeat this investigation using pollutant solutions of greater concentrations to determine if they promote or inhibit algal growth.

Name ______________________ Class ____________ Date ____________

Comparing Marine and Land Nonvascular Plants

Pre-Lab Discussion

Nonvascular plants that live in water are called *algae*. Algae can be unicellular or multicellular. Most are multicellular. Some species of multicellular algae are commonly seen along the shoreline or washed up on the beach. These algae are called seaweeds. Seaweeds come in a variety of colors, including green, brown, and red. The brown alga *Fucus* is one of the more familiar seaweeds and is often seen attached to rocks and wharf pilings at the intertidal zone. At the intertidal zone, the seaweed is exposed at low tide and under water at high tide.

Nonvascular plants that live on land include the liverworts and the mosses. These plants are generally found in moist environments because they lack the water-conducting tubes that are found in higher plants.

In this investigation, you will compare a marine nonvascular plant (alga) and a land nonvascular plant (moss).

Problem

How do marine and land nonvascular plants compare?

Materials *(per group)*

Large shallow bowl
Sea water
Microscope
Dissecting microscope
Hand lens
Dissecting tray
Scissors
Scalpel
Glass slide
Coverslip
Dissecting needle
Forceps
Brown alga, *Fucus*
Moss, *Polytrichum*
Prepared slide of moss protonema
Prepared slide of moss antheridia and archegonia

Safety

Put on a laboratory apron if one is available. Always handle the microscope with extreme care. You are responsible for its proper care and use. Use caution when handling glass slides as they can break easily and cut you. Be careful when handling sharp instruments. Note all safety alert symbols next to the steps in the Procedure and review the meanings of each symbol by referring to the symbol guide on page 10.

Procedure

Part A. Adaptive Structures and Reproduction of *Fucus*

1. Place the *Fucus* in a large shallow bowl. Cover the plant with sea water.
2. Carefully remove a piece of *Fucus* from the bowl and place in on the dissecting tray.

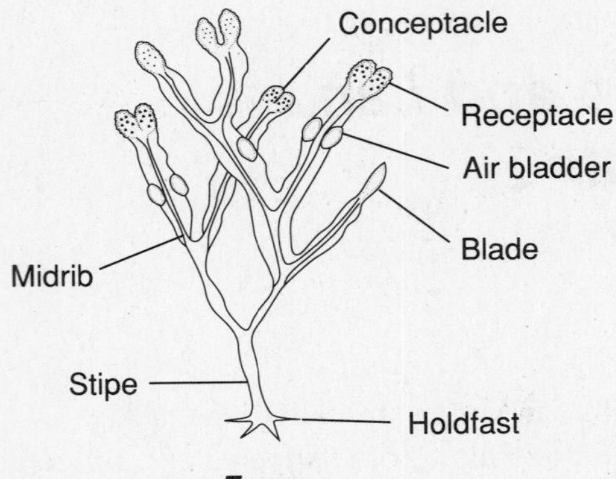

Figure 1 ***Fucus***

3. Look for the main stem. At its base, notice a tough, fibrous pad of tissue called the holdfast.
4. Feel the stems in their midregions until you come across a small lump. This is the air bladder. Air bladders sometimes come in pairs on either side of the midrib.
5. Using the scissors, cut out a small section of the stem containing an air bladder. Place it back in the bowl and observe what happens.
6. Look at the leaflike structures on the seaweed. Find the flattened forked stem tips. Special cells called apical cells located at the tips divide by mitosis and produce the forked branching pattern.
7. Examine the stem tips for swollen areas called receptacles. Receptacles contain eggs and sperm.
8. Look for the most swollen receptacles because they are the most mature and are the best ones for examination. Orange-yellow receptacles contain gametes that are ready to be dispersed.
9. With a hand lens, notice the tiny bumps on the surface of a receptacle. For a closer look, use the dissecting microscope.
10. In order to examine the gametes, you will have to open the receptacles and look inside. Using a scalpel, carefully cut out a very thin cross section of the receptacle. See Figure 2. **CAUTION:** *Use extreme care when using a sharp instrument.*

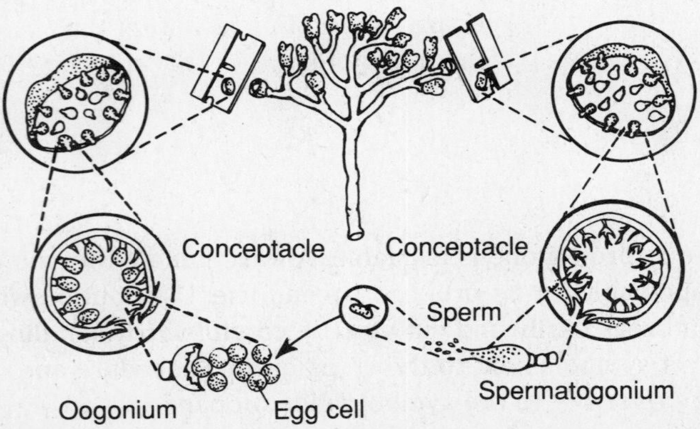

Figure 2 **Life Cycle of *Fucus***

Name ______________________ Class __________ Date __________

11. With the forceps, place the thin section of the receptacle on the glass slide and prepare a wet mount.

12. Observe the section of the receptacle under the low power of the microscope. **Note:** *It may take a little practice to cut the section thin enough so that it can be viewed under the microscope.*

13. Inside the receptacle, look for several small, round chambers that come into contact with the receptacle. These chambers are called conceptacles. Notice whether a conceptacle has an opening to the outside wall. A female conceptacle has several round oogonial sacs, each of which contains 8 egg cells. A male conceptacle has spermatogonial sacs containing many orange dots. Each dot is a sperm cell. In *Fucus,* fertilization and development are external. In the appropriate place in Observations, sketch and label the conceptacle. Record the magnification.

Part B. Adaptive Structures and Reproduction of *Polytrichum* Moss

1. Examine a small clump of *Polytrichum* moss under the dissecting microscope. Use Figure 3 to identify the structures of a moss.

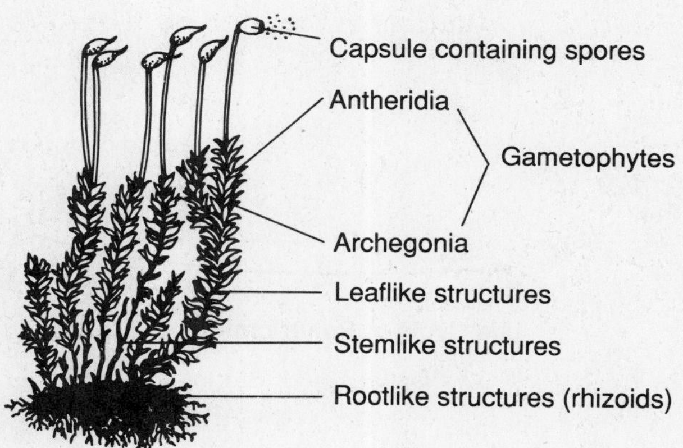

Figure 3

2. Carefully separate a sporophyte from a gametophyte. A sporophyte consists of one stalk with a capsule on it. The green "leafy" part is the female gametophyte. Notice that the capsule has a cap on it.

3. To examine the contents of the capsule, carefully remove the cap. Then place a drop of water on a glass slide and squeeze the contents of the capsule into the water. Cover with a coverslip and locate the capsule's contents under low power. Then observe under high power. **CAUTION:** *When turning to the high-power objective lens, always look at the objective from the side of the microscope so that the objective lens does not hit or damage the slide.* In the appropriate place in Observations, sketch what you see. Record the magnification.

4. Examine a prepared slide of moss protonema. A protonema is a mass of tangled green filaments that germinate and grow from a spore.

5. Examine a prepared slide of moss antheridia and archegonia. These reproductive organs are found in the upper tips of the gametophytes.

6. Study Figure 4, which shows the life cycle of a moss. Compare the similarities and differences in the life cycles of *Fucus* and *Polytrichum.*

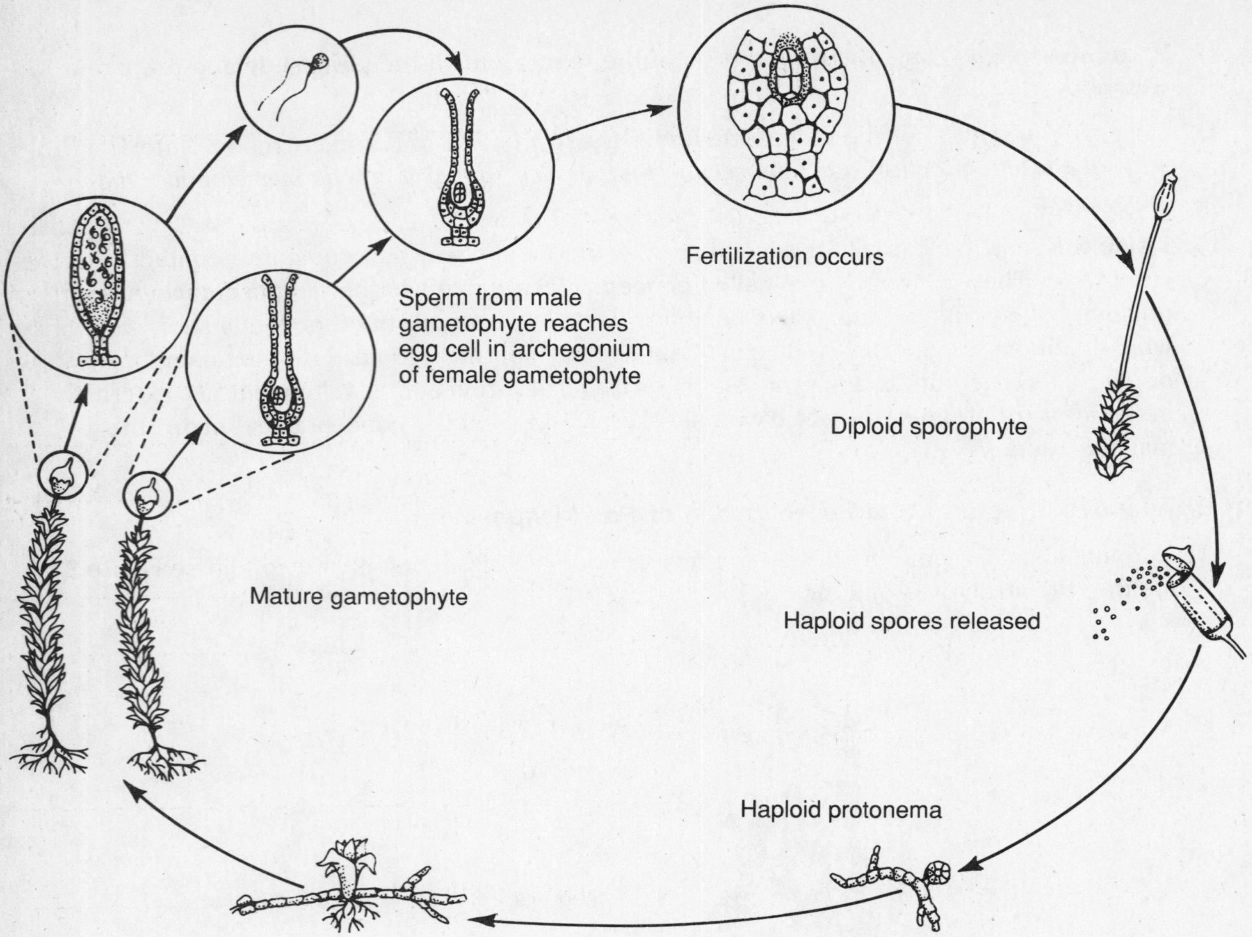

Sperm from male gametophyte reaches egg cell in archegonium of female gametophyte

Fertilization occurs

Diploid sporophyte

Mature gametophyte

Haploid spores released

Haploid protonema

Life Cycle of Polytrichum

Figure 4

Observations

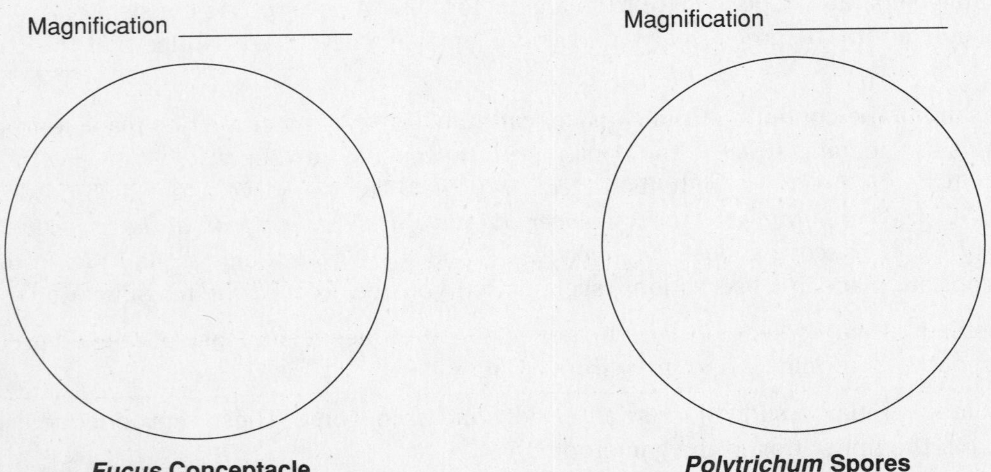

Fucus **Conceptacle** | ***Polytrichum*** **Spores**

Name ______________________ Class __________ Date __________

1. What structure enables *Fucus* to attach itself to rocks and wharf pilings?

2. What happens when you put the *Fucus'* air bladder in water?

3. Is the conceptacle male or female? Explain your answer.

4. What are the structures inside the moss capsule called?

5. Does the protonema contain any structures? If so, what might they be?

6. What is the female gametophyte called? The male gametophyte?

Analysis and Conclusions

1. What is the adaptive value of the holdfast on an alga?

2. What is the adaptive value of the air bladder on an alga?

3. Of what value is the flat, branching system of seaweed?

4. How does a moss capsule disperse its contents? ______________________________

____

5. Compare the similarities and differences in the life cycles of *Fucus* and *Polytrichum*.

____

____

____

____

____

____

Critical Thinking and Application

1. What structural differences allow the moss to be more successful on land than the alga?

____

____

____

2. Why is moss not usually found growing in areas of little rainfall? ______________

____

____

3. Why do moss plants grow close to the ground? ______________________________

____

____

4. Gardeners often add sphagnum moss to soil. How might this be helpful?

____

____

Going Further

1. Using references in the library, find out about the kelp (seaweed) forests that grow off the California coast. What types of kelps grow there? How large are they? What other organisms live within the kelp community? What advantages do the kelps provide to other organisms? What effects would harvesting the kelps have on the environment?
2. Examine the external and reproductive structures of another land nonvascular plant, the liverwort *Marchantia*. Sketch and label what you see.

Name ______________________ Class __________ Date __________

Special Characteristics of Ferns

Pre-Lab Discussion

Ferns belong to a group of plants that are "true" land plants. These plants have specialized tissues called *vascular tissues* that transport water and the products of photosynthesis throughout the plant. For this reason, these plants are called vascular plants. Vascular plants can grow much larger than plants without vascular tissues. When comparing cellular organization, ferns lie between mosses and more complex vascular plants, such as trees.

In this investigation, you will observe some of the characteristics of ferns. You will also examine how ferns reproduce.

Problem

What are some special characteristics of ferns?

Materials *(per group)*

Whole bracken fern sporophyte plant
Bracken fern frond with sori
Dissecting microscope or hand lens
2 glass slides
2 coverslips
Medicine dropper
Dissecting tray
Scalpel or razor blade
Forceps
Ethyl alcohol
Microscope

Safety

Put on a laboratory apron if one is available. Handle all glassware carefully. Always handle the microscope with extreme care. You are responsible for its proper care and use. Use caution when handling glass slides as they can break easily and cut you. Always use special caution when working with laboratory chemicals, as they may irritate the skin or cause staining of the skin or clothing. Never touch or taste any chemical unless instructed to do so. Be careful when handling sharp instruments. Note all safety alert symbols next to the steps in the Procedure and review the meanings of each symbol by referring to the symbol guide on page 10.

Procedure

Part A. Characteristics of the Sporophyte Fern

1. Examine a whole sporophyte fern plant. Use Figure 1 to identify the structures of a fern.

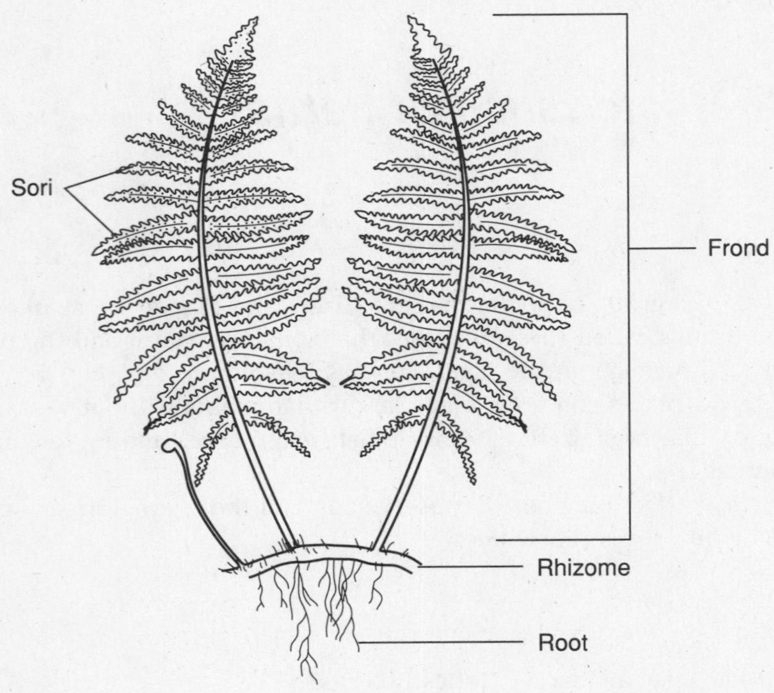

Figure 1

2. Locate the horizontal stem structure called the rhizome.
3. Notice how the fern leaves, or fronds, arise from the rhizome. Note the shape of the frond and the presence of veins in the stem part.
4. Examine the underside of the frond for brownish-yellow spots. These spots are called sori (singular, sorus).
5. With the scalpel, cut off a 0.5-cm piece of rhizome from the fern. **CAUTION:** *Use extreme care when using a sharp instrument.* Place the rhizome in the dissecting tray. While holding the rhizome with the forceps, cut a very thin cross section of the rhizome with the scalpel.
6. Prepare a wet-mount slide of the rhizome cross section. If it is too thick, place it on a glass slide without the coverslip. Place the slide onto the stage of the microscope and observe it under low power. Note the epidermis and the vascular tissue: xylem (water and mineral conduction) surrounded by phloem (organic material transport). In the space provided in Observations, sketch and label the cross section of the rhizome. Label the following structures on the sketch: epidermis, xylem, and phloem. Record the magnification of the microscope.

Part B. Characteristics of the Gametophyte Fern

1. Obtain the fern frond that contains sori.
2. Use the dissecting microscope or hand lens to examine a single sorus.

Name ________________________ Class ____________ Date ____________

3. Place a drop of water on a clean glass slide. Then, as shown in Figure 2, use the scalpel to gently scrape a sorus into the drop of water on the slide. **CAUTION:** *Use extreme care when using a sharp instrument.*

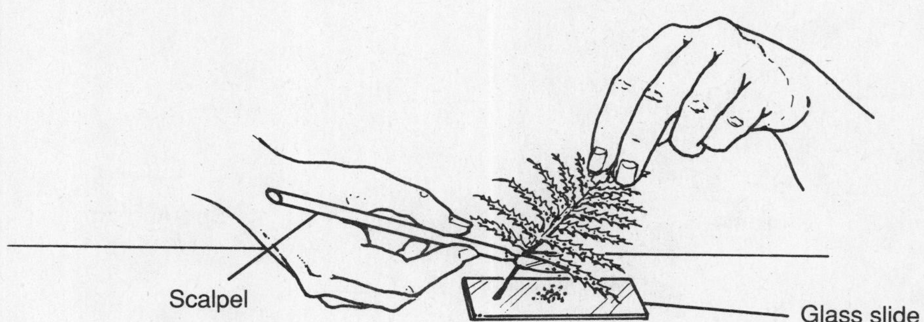

Figure 2

4. Examine the sorus under the low power of the dissecting microscope. Locate the club-shaped structures that were scraped from the sorus. These structures are called sporangia and contain spores. In the appropriate place in Observations, sketch a single sporangium. Record the magnification of the microscope.

5. Prepare a second slide of only spores by removing a few sporangia from a sorus onto a clean, dry slide. Add one drop of ethyl alcohol to the sporangia on the slide. **CAUTION:** *Be very careful when working with laboratory chemicals.*

6. Examine the sporangia under the low power of the dissecting microscope. Observe what happens to the spore cases of the sporangia. In the appropriate place in Observations, sketch a few of the spores that were ejected from the sporangia.

7. Examine the life cycle of a fern in Figure 3.

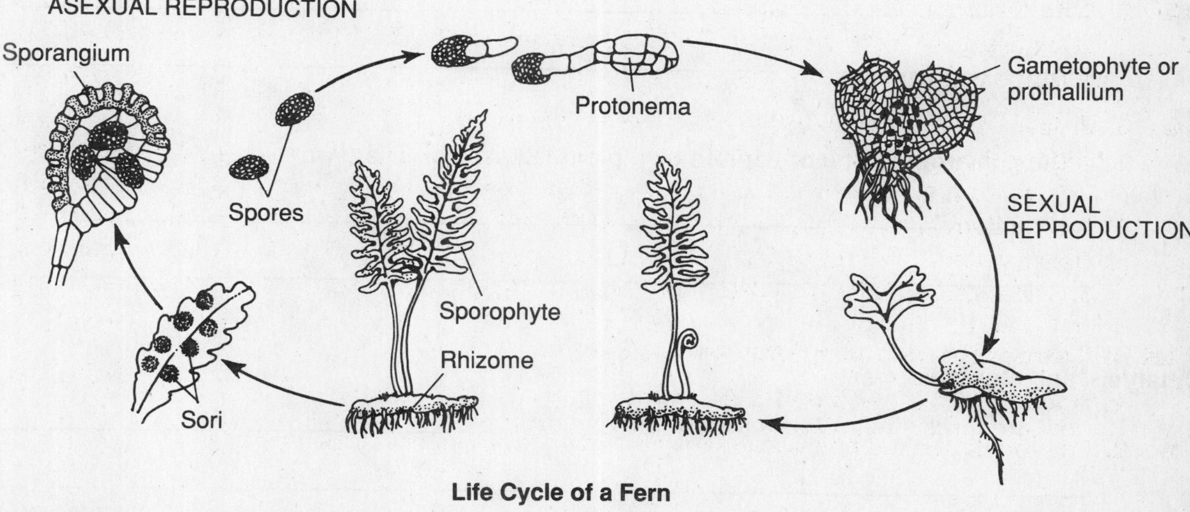

Life Cycle of a Fern

Figure 3

Observations

Magnification ______________

Rhizome

Magnification ______________

Sporangium

Magnification ______________

Spores

1. How are the vascular tissues arranged in the fern rhizome? ______________________________

2. Is the gametophyte plant haploid or diploid? Explain your answer.

Analysis and Conclusions

1. How are ferns similar to mosses and liverworts? ______________________________

Name ______________________ Class __________ Date __________

2. How are ferns similar to woody trees? ______________________

3. Fern fronds are difficult to snap or tear from the main part of the plant. Explain why this is so.

4. Why are ferns larger than mosses and liverworts? ______________________

5. Describe the life cycle of a fern. ______________________

Critical Thinking and Application

1. Why are ferns found in a greater range of habitats than mosses or liverworts?

2. Considering the large number of spores that are produced by a single frond of a fern plant, why do you think there are not more fern plants growing on Earth?

3. Why is it advantageous for the spores to be located on the bottom surface of the fern fronds?

____

____

____

4. Many flowering plants have leaves that are very similar in appearance to fern fronds. Although these plants are not true ferns, some of them are called ferns. The asparagus fern is an example. How would you determine whether such a plant was a fern?

____

____

____

Going Further

1. Grow your own fern prothallia. Tightly pack peat moss into a small, clay-type plant container. Place the container of peat moss in boiling water for 5 to 10 minutes. With tongs, remove the container and set the bottom of it in a petri dish of water. The peat moss will soak up the water and provide a source of moisture. Place a few sections of fronds containing ripe sori on the peat moss for a few days. Cover the entire surface of peat moss and fronds with a glass plate to retain moisture. Set it in a shaded area. Prothallia should appear within 4 to 6 weeks.
2. After collecting several types of fern fronds, set up an exhibit to show how ferns are identified. How are fronds classified? What are the names of the leaflets of fronds? Illustrate the different locations of sori. If possible, show the locations of fertile and nonfertile leaflets.

Name ______________________ Class __________ Date __________

41

Seed-Bearing Plant Tissues

Pre-Lab Discussion

The seed-bearing plants of the subphylum Spermopsida are more complex structurally than the plants of any other grouping. Plants in this subphylum reproduce sexually by means of seeds. Based on the appearance of their seeds, seed-bearing plants are divided into two classes. The seeds of the *angiosperms,* or the flowering plants, are enclosed in protective structures called fruits. The seeds of the *gymnosperms,* or coniferous plants, are found unprotected on structures called cones.

The seed-bearing plants developed specialized plant tissues. For the most part, these tissues are similar in both angiosperms and gymnosperms. Each type of specialized tissue has a specific function. Epidermal tissue is the outer covering of plants and protects the tissues beneath it. Meristematic tissue is found at the tips of roots and stems and contains cells that undergo rapid cell division. Vascular tissues (phloem and xylem) conduct water and nutrients throughout the plant. Sclerenchyma tissue, which may contain fiber cells or stone cells, strengthens and supports parts of the plants.

In this investigation, you will identify several specialized plant tissues in seed-bearing plants.

Problem

What is the structure of some specialized plant tissue in seed-bearing plants?

Materials *(per group)*

Plant leaf
Forceps
Glass slide
Coverslip
Medicine dropper
Microscope
Prepared slides of longitudinal sections of:
- coleus stem tip
- Mature stem of *Aristolochia*
- sycamore stem

Safety

Put on a laboratory apron if one is available. Always handle the microscope with extreme care. You are responsible for its proper care and use. Use caution when handling glass slides as they can break easily and cut you. Note all safety alert symbols next to the steps in the Procedure and review the meanings of each symbol by referring to the symbol guide on page 10.

Procedure

Part A. Examining Epidermal Tissue

1. Examine the upper and lower sides of the plant leaf.
2. Carefully tear the leaf in half. Using the forceps, remove a small section of the lower epidermis, or the thin transparent covering of the leaf.
3. Prepare a wet-mount slide of the epidermis. Examine it under the low power of the microscope. Observe the shape and arrangement of the epidermal cells. In the appropriate place in Observations, sketch and label a section of the epidermis. Record the magnification of the microscope.

Part B. Examining Meristematic Tissue

1. Examine the prepared slide of the coleus stem tip under the low power of the microscope. Locate the tip of the stem and then find an area that contains many cells in various stages of mitosis.
2. Switch to the high-power objective. **CAUTION:** *When switching to the high-power objective lens, always look at the objective from the side of the microscope so that the objective lens does not hit or damage the slide.* In the appropriate place in Observations, sketch and label some of the cells you observe under high power. Record the magnification of the microscope.

Part C. Examining Vascular Tissue

1. Examine the prepared slide of *Aristolochia* (Dutchman's pipe vine) under the low power of the microscope.
2. Switch to the high-power objective. Locate the sieve-tube element, which is a long tubelike structure, and the companion cells next to the sieve-tube element. Together these structures compose the phloem. In the appropriate place in Observations, sketch and label the sieve-tube element and a few companion cells. Record the magnification of the microscope.
3. Examine the prepared slide of the sycamore stem under the low power of the microscope.
4. Switch to the high-power objective. Locate a few wide tubular cells called vessel elements, which compose the xylem. Observe the pitted appearance of the side walls of the vessel elements. In the appropriate place in Observations, sketch and label a few vessel elements.

Part D. Examining Supporting (Sclerenchyma) Tissue

1. Examine the prepared slide of *Aristolochia* under the low power of the microscope. Locate the long narrow cells with thick cell walls. These cells are the fibers.
2. Switch to the high-power objective. In the appropriate place in Observations, sketch and label a few of these fiber cells. Record the magnification of the microscope.

Name ______________________ Class __________ Date __________

Observations

Magnification __________

Leaf Epidermis

Magnification __________

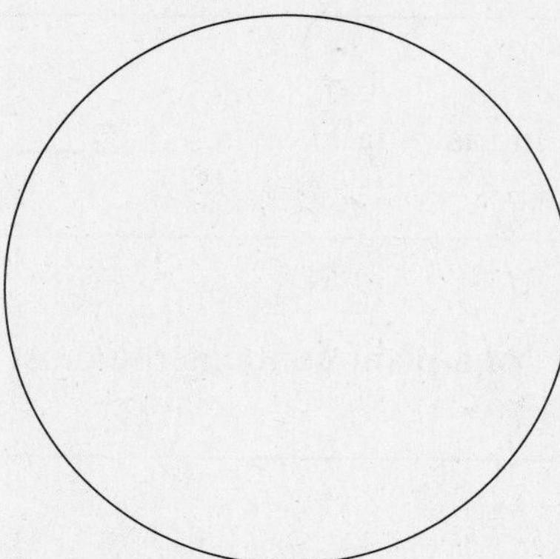

Meristematic Tissue of Coleus

Magnification __________

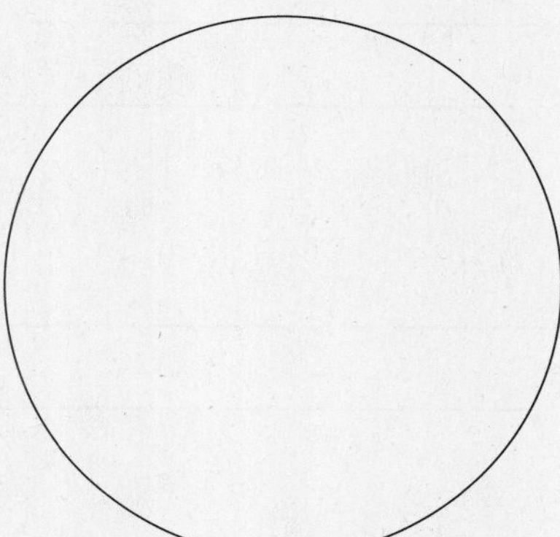

Phloem of *Aristolochia*

Magnification __________

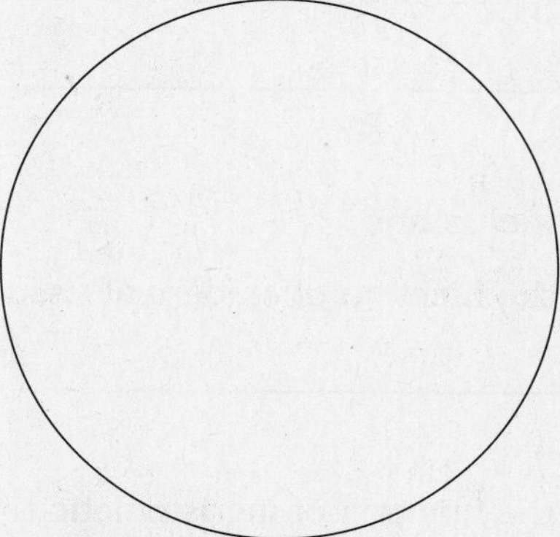

Xylem of Sycamore Stem

Magnification __________

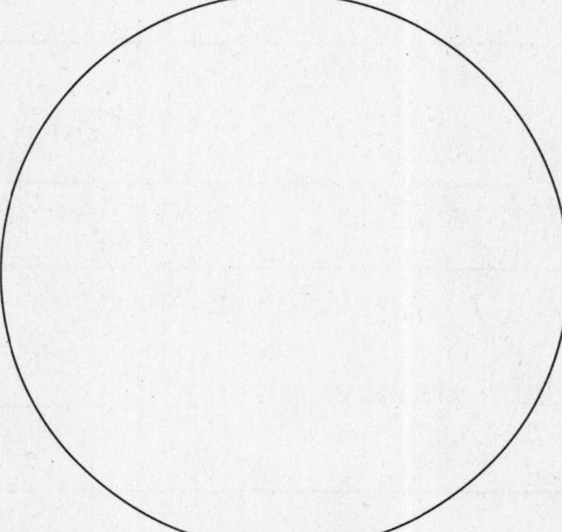

Fiber Cells of *Aristolochia*

1. How do the upper and lower sides of the leaf differ in appearance?

2. Describe epidermal tissue.

3. Describe meristematic cells.

4. What part of a plant contains the most meristematic cells?

5. What major cell structure is absent in fiber cells?

6. What part of a plant contains fiber cells?

Analysis and Conclusions

1. What is the function of epidermal tissue?

2. What is the function of meristematic tissue?

3. What is the function of the phloem?

4. What is the function of the xylem?

5. Why do the walls of the vessel elements have pits?

6. What is the function of the fiber cells?

Name ______________________ Class __________ Date __________

Critical Thinking and Application

1. Describe how each of the following characteristics of meristematic cells would be an advantage in rapid cell division.

   a. thin cell walls ______________________

   b. no vacuoles ______________________

   c. much smaller size than mature cells ______________________

2. In some forests, porcupines do much damage to young trees as they chew off the bark, which contains the phloem. Soon the trees' roots die, and then the trees themselves. Explain why this happens. ______________________

3. A cross section of a cactus leaf, or needle, shows that it has a thick epidermis. How does this adaptation help the cactus leaf to survive in hot climates? ______________________

Going Further

1. Prepare an illustrated report on the seed plants that live in one of the following environments: arctic, swamp, desert, temperate forest. Describe the special adaptations that these plants have in order to survive there. Include drawings or photographs of these plants.
2. Make wet-mount slides of the pollen grains from three different types of flowers. Examine the pollen grains under the low power and the high power of a microscope. What are some differences among these pollen grains? What are some similarities?

Name ______________________ Class __________ Date __________

Comparing Monocots and Dicots

Pre-Lab Discussion

Angiosperms, or flowering plants, are divided into two subclasses: monocots and dicots. The simplest difference between monocots and dicots is the number of *cotyledons*, or first leaves developed by the embryo plant. Monocots, such as grasses, orchids, irises, lilies, and palms, have one cotyledon. Dicots, such as cacti, roses, maples, peas, and oaks, have two cotyledons. In addition to the number of cotyledons, monocots and dicots have several other differences in structure.

In this investigation, you will observe some of the characteristics of monocots and dicots.

Problem

What are some of the structural differences between monocots and dicots?

Materials *(per group)*

2 monocot flowers with leaves
2 dicot flowers with leaves
Prepared slide of a cross section of a monocot stem
Prepared slide of a cross section of a dicot stem
Microscope

Safety

Put on a laboratory apron if one is available. Always handle the microscope with extreme care. You are responsible for its proper care and use. Use caution when handling glass slides as they can break easily and cut you. Note all safety alert symbols next to the steps in the Procedure and review the meanings of each symbol by referring to the symbol guide on page 10.

Procedure

1. Examine the leaves of the monocot and dicot specimens and observe any differences in the pattern of veins on the leaves. In the appropriate place in Observations, sketch one monocot leaf and one dicot leaf. Be sure to include the vein patterns in the sketches.

2. Count the number of petals on each of the monocot flowers and the dicot flowers.

3. Carefully examine the center of each flower. The centrally located stalklike structure is called the pistil. The pistil is surrounded by several other stalklike structures called the stamens. Each stamen contains a pollen sac. Count the number of stamens in each monocot and dicot flower.

4. Examine the prepared slide of the monocot stem under the low power of the microscope. Notice that xylem and phloem tissues are gathered into vascular bundles. Observe the arrangement of the vascular bundles. In the appropriate place in Observations, sketch the monocot stem. Record the magnification of the microscope.

5. Repeat step 4 using the prepared slide of the dicot stem. In the appropriate place in Observations, sketch the dicot stem. Record the magnification of the microscope.

Observations

Monocot Leaf

Dicot Leaf

Magnification ______________

Magnification ______________

Monocot Stem

Dicot Stem

Name ______________________ Class __________ Date __________

1. How do the veins in a monocot leaf differ from those in a dicot leaf?

____

____

____

2. Compare the number of petals on a monocot flower to the number on a dicot flower.

____

____

____

3. Compare the number of stamens on a monocot flower to the number on a dicot flower.

____

____

____

4. How does the arrangement of vascular bundles in a monocot stem differ from the arrangement in a dicot stem? ______________________________

____

____

Analysis and Conclusions

1. What are two functions of the vascular bundles found in the stems of angiosperms?

____

____

____

____

2. What are three differences between monocots and dicots? __________________

____

____

____

Critical Thinking and Application

1. A newly discovered fossil plant has 8 petals. Using this information, predict the type of venation (vein arrangement), the number of stamens, and whether the plant is a monocot or a dicot. ______________________________

2. The lima bean is the seed of the lima bean plant. Each half of the bean is a cotyledon, or seed leaf. Based on this seed structure, is this plant a monocot or a dicot? Explain your answer.

3. What characteristics of angiosperms have made them so successful in a variety of environments? ______________________________

Going Further

1. Start growing bean plants from seeds. Separate the seedlings into three groups. Allow one group of seedlings to remain undisturbed. Remove one cotyledon from each seedling in the second group. Remove both cotyledons from the seedlings in the third group. Compare the growth of each group of seedlings over a period of time. Construct a data table to present your results.
2. Make a collection of the different types of leaves found in your area. Examine each type of leaf and classify it as a monocot or a dicot. Use a reference book to determine how accurate your classifications are.

Name ______________________ Class ____________ Date ____________

43

Root and Stem Structures

Pre-Lab Discussion

The first structures to appear on a germinating seed are the roots. The initial root to grow from a seed is the primary root, which is then followed by secondary roots that branch out from the primary root. In a *taproot system,* the primary root grows longer and thicker than the secondary roots. In a *fibrous root system,* the secondary roots continue to grow, and eventually all the roots are of equal or nearly equal size.

Roots have several functions. They anchor the plant in place, absorb water containing dissolved minerals from the environment, and act as storage areas for excess food. Some roots even develop into new plants. *Adventitious roots* grow from parts of the plant other than the roots. *Aerial,* or *prop, roots* are roots that are suspended in the air.

Plant structures that grow between the roots and the leaves are called stems. Although stems usually grow above the ground in vertical positions, they can also grow under the ground in horizontal positions. All stems begin growing as soft, tubelike structures. If the stem remains soft, and usually green, for the entire life of the plant, it is a *herbaceous stem.* A *woody stem* becomes hard and often turns brown.

The primary function of stems is to conduct water and dissolved minerals from the roots to the leaves and, at the same time, to conduct food from the leaves to the rest of the plant. Stems may also function as food storage areas, supporting structures, and places for the growth of new plants.

In this investigation, you will examine the structures of roots and stems. You will also observe the structural differences between some monocot and dicot roots and stems.

Problem

What are some structures of roots and stems?

Materials *(per group)*

2-week-old radish seedlings
Hand lens or dissecting microscope
Microscope
Carrot
Methylene blue stain
Ethyl alcohol
Scalpel or single-edged razor blade
Dissecting tray
Forceps
Glass slide

150-mL beaker
Prepared slides of cross section of:

Helianthus root
Zea root
Helianthus stem
Zea stem

Safety

Put on a laboratory apron if one is available. Put on safety goggles. Handle all glassware carefully. Always handle the microscope with extreme care. You are responsible for its proper care and use. Use caution when handling glass slides as they can break easily and cut you. Always use special caution when working with laboratory chemicals, as they may irritate the skin or cause staining of the skin or clothing. Never touch or taste any chemical unless instructed to do so. Be careful when handling sharp instruments. Note all safety alert symbols next to the steps in the Procedure and review the meanings of each symbol by referring to the symbol guide on page 10.

Procedure

Part A. External and Internal Structures of a Root

1. Examine Figure 1 and identify each root as being taproot, fibrous, adventitious, or aerial.

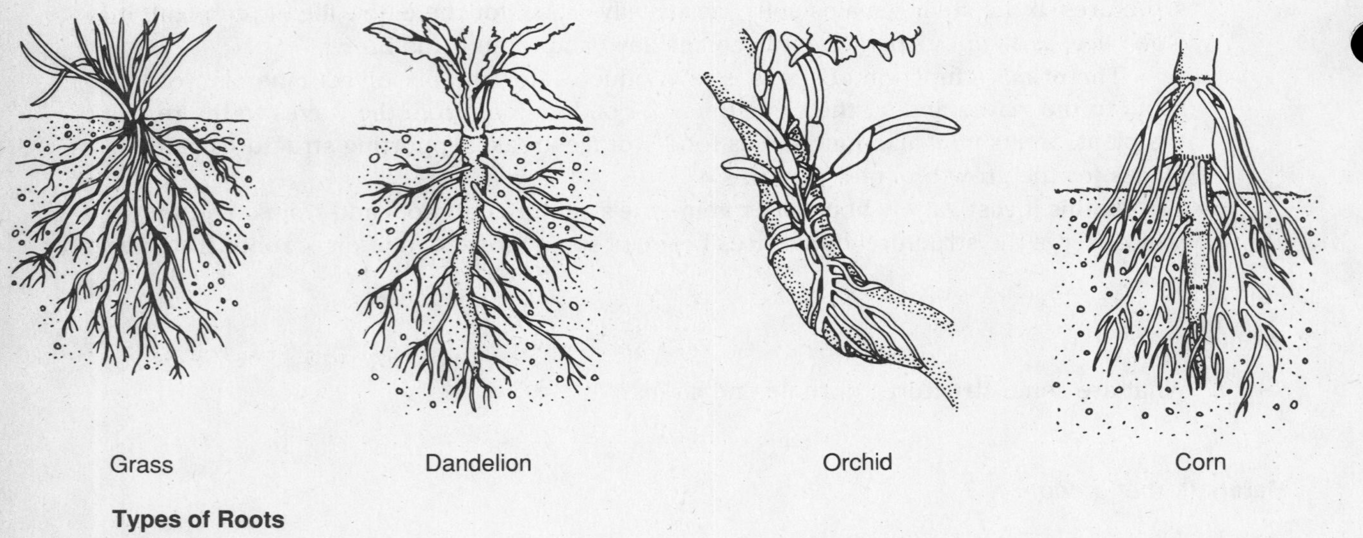

Types of Roots

Figure 1

2. Examine the two-week-old radish seedling. Note the basic structures of the seedling: leaves, stem, and roots.
3. With a hand lens or dissecting microscope, examine the delicate root hairs extending from the root.
4. Place the carrot in the dissecting tray. As shown in Figure 2, hold the carrot steady with one hand while you cut it in half lengthwise with the scalpel or single-edged razor blade. **CAUTION:** *Always cut in a direction away from yourself. Because the carrot is very hard, be careful not to let the scalpel or razor blade slip and cut you.*

Name ______________________________ Class ____________ Date ____________

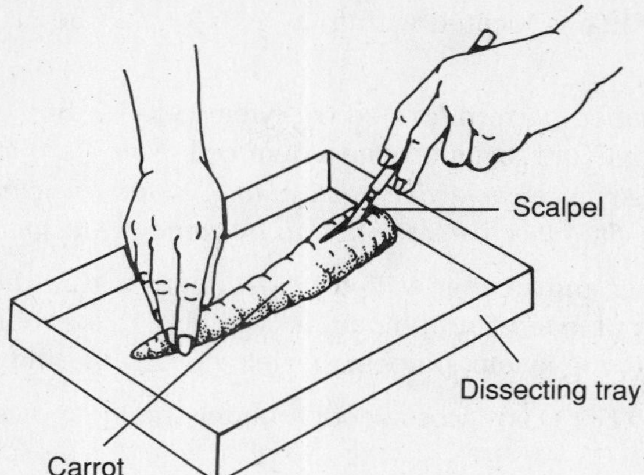

Figure 2

5. Examine the two halves of the carrot. Each half is a longitudinal section. Notice the stele in the center. The stele contains the xylem and the phloem. Surrounding the stele is the food-storing cortex. In the appropriate place in observations, sketch a longitudinal section of the carrot. Label the stele and the cortex.

6. To make a cross section of the carrot, place one of the longitudinal sections cut side down on the dissecting tray. About 4 cm from the end of the carrot, cut straight down using the razor blade. Discard the small piece. **CAUTION:** *Be careful when handling a sharp instrument.*

7. Make another cut straight down, as close as possible to the one you just made. See Figure 3. With the forceps, carefully remove the carrot cross section from the razor blade and place it on a glass slide.

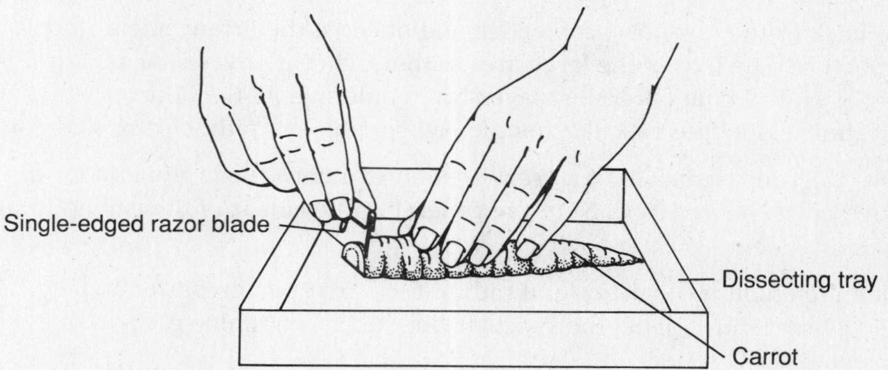

Figure 3

8. Place the slide over the mouth of the beaker. Then cover the carrot cross section completely with methylene blue stain. Allow the stain to set for 1 minute. **CAUTION:** *Methylene blue stain is a permanent stain. Be careful not to get it on your hands or clothing.*

9. After 1 minute, slowly pour alcohol over the carrot cross section until no more stain washes away.

10. Observe the cross section of the carrot with a hand lens or under a dissecting microscope. In the appropriate place in Observations, sketch the cross section of the carrot. Label the stele and the cortex.

11. Place the prepared slide of the sunflower *(Helianthus)* root cross section under the low power of the microscope. The sunflower is a dicot.

12. Locate the epidermal cells that form the outer edge of the root. Examine different areas on the glass slide. Notice some root hairs, which are extensions of single epidermal cells.

13. Find the cortex, which is located within the epidermis. The cells of the cortex are large and thin walled.

14. Locate the star-shaped pattern formed by xylem cells at the center of the root. Switch to the high-power objective and focus on one xylem cell. Note its thick cell wall. **CAUTION:** *When switching to the high-power objective, always look at the objective from the side of the microscope so that the objective does not hit or damage the slide.*

15. Observe the smaller and thinner-walled phloem cells within the arms of the star. This distinctive pattern of xylem and phloem is typical of dicot roots. In the appropriate place in Observations, label the xylem, phloem, cortex, epidermis, and root hair.

16. Examine the corn *(Zea)* root cross section under the low power of the microscope. Corn is a monocot.

17. Notice that groups of xylem cells are scattered within the central area of the root. Move the glass slide around until you find the phloem cells, which are also scattered in bunches through the central area of the root.

18. Notice that the cortex and epidermis are similar in both the sunflower root and the corn root. In the appropriate place in Observations, label the xylem, phloem, cortex, epidermis, and root hair in the corn root.

Part B. Internal Structures of Stems

1. Observe the prepared slide of a cross section of a sunflower *(Helianthus)* stem under low power of the microscope. The sunflower has a herbaceous, or nonwoody, stem. Notice that the vascular bundles are arranged in a ring within the stem. Switch to the high-power objective and focus on a single vascular bundle. Observe the thick-walled xylem cells. Notice the smaller, thinner-walled phloem cells within the bundle.

2. Switch back to the low-power objective and observe the arrangement of cells within the stem cross section. The pith is the large area within the ring of vascular bundles. Surrounding the ring is the cortex. The outermost layers are epidermis. In the appropriate place in Observations, label the vascular bundle, xylem, phloem, pith, cortex, and epidermis.

3. Observe the prepared slide of a cross section of a corn *(Zea)* stem under the low-power objective of the microscope. Note the general arrangement of the cell and the position of the xylem and phloem.

4. Examine the epidermis, cortex, and pith in the corn stem cross section. In the appropriate place in Observations, label the vascular bundle, xylem, phloem, epidermis, cortex, and pith.

Observations

Longitudinal Section of Carrot

Cross Section of Carrot

Name ______________________ Class __________ Date __________

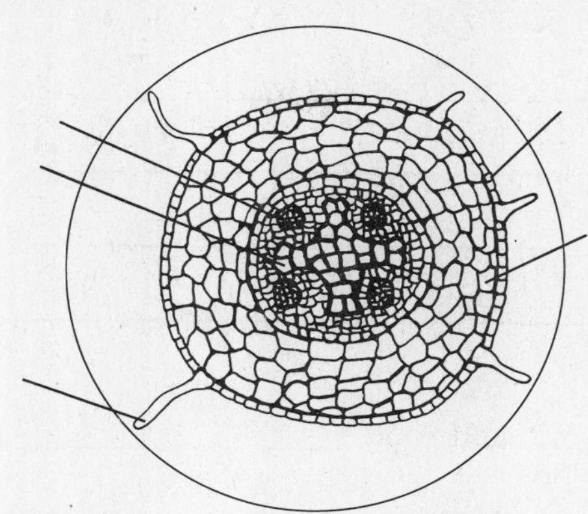
Sunflower Root

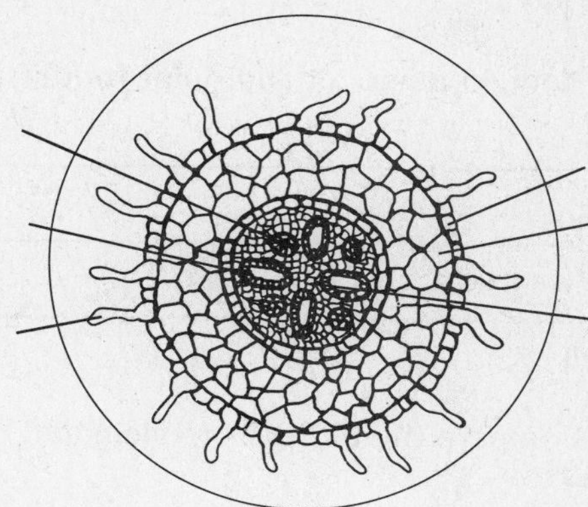
Corn Root

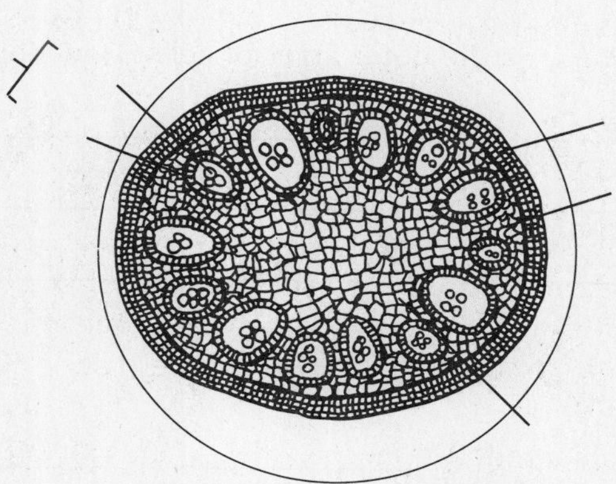
Sunflower Stem

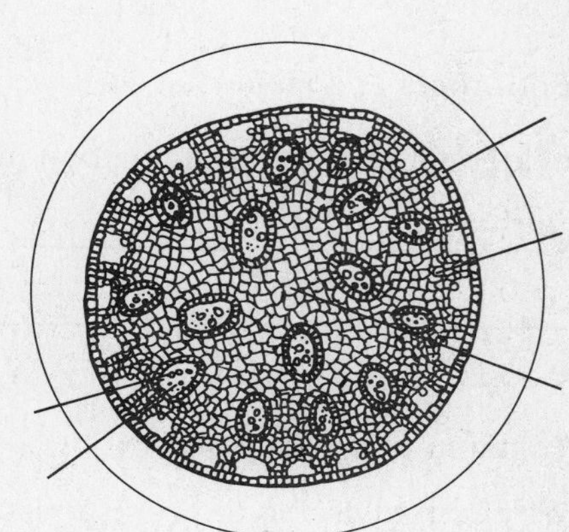
Corn Stem

Analysis and Conclusions

1. Next to the name of each plant, identify the type of root.

   a. Grass ______________________

   b. Dandelion ______________________

   c. Orchid ______________________

   d. Corn ______________________

2. Based on its root structure, is the carrot a monocot or a dicot? ______________________

   ____

3. Of the fibrous and taproot, which is best adapted for food storage? ______________________

   ____

4. Which does the shape of the root hair increase—the volume or the surface area of the root?

5. How do monocot and dicot root structures differ?

6. Compare the pattern of xylem and phloem in dicot stem and root.

7. Both roots and stems have a layer of epidermis that forms the outer layer of cells. How do the epidermal cells differ in function in roots and stems?

Critical Thinking and Application

1. Why is it advisable to prune a plant when it is transplanted?

2. Would plants with fibrous roots or taproots be more suitable for planting in a windy area? Explain your answer.

3. What types of roots, fibrous or taproot, are most commonly found on desert plants? Explain your answer.

Name ____________________ Class __________ Date __________

4. Some giant sequoia trees that have had an automobile tunnel cut through their woody stems continue to grow. How is this possible? ____________________

____

____

5. What advantages do plants with woody stems have over plants with herbaceous, or nonwoody, stems? ____________________

____

____

Going Further

1. To observe how materials are transported up through a stem, cut the stem of a white carnation in half lengthwise. Measure out 50 mL of water into each of two beakers. Add 5 drops of red food coloring to one beaker and 5 drops of blue food coloring to the other beaker. Place the beakers side by side and carefully place one half of the carnation stem in each beaker. Allow the beakers to remain undisturbed for 24 hours. After 24 hours, observe the color of the carnation flower.
2. To determine the regions of greatest growth in new roots and stems of bean seedlings, germinate some been seeds in a petri dish lined with wet filter paper. Using a waterproof pen, mark 1-mm intervals on the root when it reaches 1 cm in length. Periodically examine the roots to determine which areas show the most growth. Repeat this procedure on stems of seedlings planted in soil.

Name ______________________ Class __________ Date __________

Leaf Structures

Pre-Lab Discussion

Leaves are specialized organs found on the stems of plants. The most important function of the leaf is to carry out the process of photosynthesis. All of the necessary raw materials for photosynthesis come together in the leaf. The *xylem* conducts water to the leaf. Carbon dioxide in the air enters through the *stomata* (singular, stoma), or pores, in the leaf. The *chloroplasts* in the leaf capture the light energy needed to convert water and carbon dioxide into food.

Most leaves have large, thin, flattened sections called *blades*. The blade is attached to the stem by a thin structure called the *petiole*. Leaf blades occur in a variety of shapes and sizes. *Simple leaves* have only one blade and one petiole. *Compound leaves* have several blades, or leaflets, that are joined together and to the stem by several petioles. The leaflets of some compound leaves spread out like fingers on a hand. The leaflets of others grow in pairs along a long central petiole. The leaflets of still others are arranged on petioles, which, in turn, are arranged around a long central petiole.

In this investigation, you will identify some of the specialized structures in a leaf. You will also see how leaves vary in size and shape.

Problem

What are some structures of a leaf?

Materials *(per group)*

Dicot leaf
Monocot leaf
Conifer (pine) leaf
10 leaves (include simple and compound leaves, monocots and dicots)
Prepare slide of a cross section of privet leaf or a lilac leaf
Hand lens
Reference books
Microscope

Safety

Handle all glassware carefully. Always handle the microscope with extreme care. You are responsible for its proper care and use. Use caution when handling glass slides as they can break easily and cut you. Note all safety alert symbols next to the steps in the Procedure and review the meanings of each symbol by referring to the symbol guide on page 10.

Procedure

Part A. Observing External Leaf Structure

1. Note the color of the upper and lower surfaces of the dicot leaf.
2. Identify the blade and the petiole. Use a hand lens to observe the midrib, or large central vein that runs the length of the leaf. Look for smaller veins that branch out from the midrib to all parts of the blade. In the appropriate place in Observations, sketch the dicot leaf and label the blade, petiole, midrib, and smaller veins.
3. Observe the shape of the monocot leaf. Use a hand lens to examine the leaf veins. Note that the lower edge of the leaf blade surrounds the stem, forming a sheath. In the appropriate place in Observations, sketch the monocot leaf and label the blade, sheath, stem, and veins.
4. Examine the shape of the pine leaf. Pine leaves are found in clusters that are enclosed by a scaly sheath. Pine leaves grow in groups of two, three, and five needles. In the appropriate place in Observations, sketch the pine leaf and label the blades and the sheath.

Part B. Classifying Leaves

1. Using reference books, identify each of the 10 different leaves. Record each leaf name in the appropriate place in the Data Table.
2. Classify each leaf as simple or compound. Compound leaves may be further classified into palmately compound leaves, which have several leaflets that arise from the tip of the petiole, and pinnately compound leaves, which have leaflets that run along opposite sides of an extension of the petiole. Use Figure 1 to identify each leaf type. Record the leaf type for each leaf in the appropriate place in the Data Table.

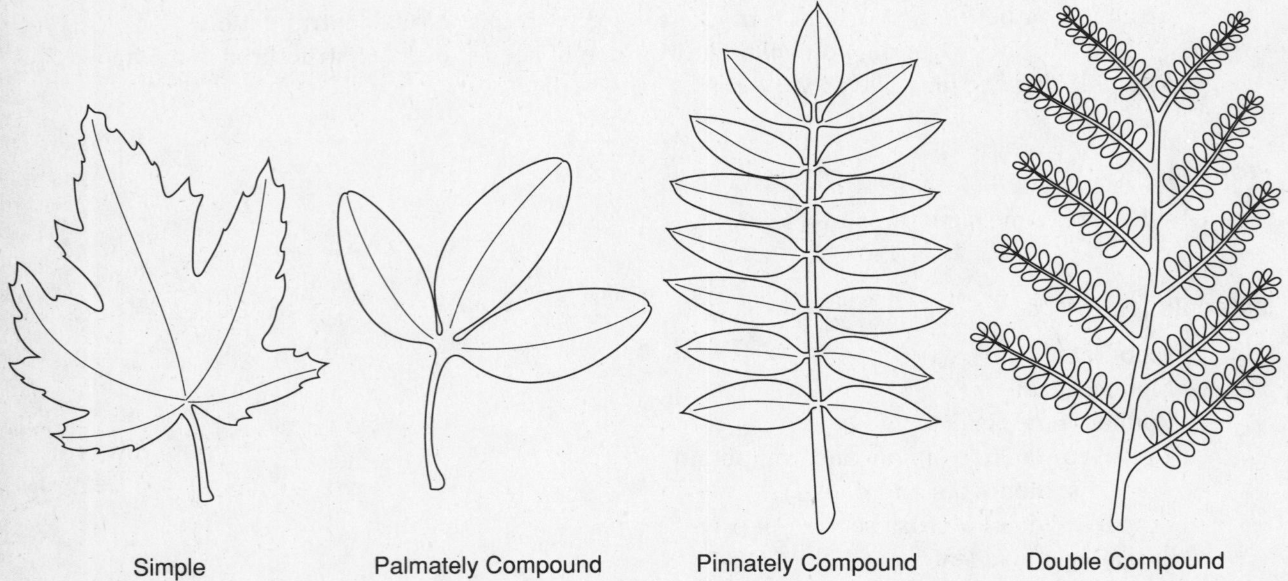

Figure 1 **Leaf Types**

3. Observe the venation, or vein pattern, of each leaf. In parallel venation, several main veins run the length of the leaf. Most monocots have parallel venation. In net venation, a netlike pattern of veins spreads throughout the leaf. Most dicots have net venation. To identify the vein pattern, hold each leaf specimen up to the light and use a hand lens to see the veins. Record the type of venation for each leaf in the Data Table.
4. Observe the pattern of leaf margin for each leaf. Use Figure 2 to identify types of leaf margins. Record the type of leaf margin in the Data Table.

Name ______________________ Class __________ Date __________

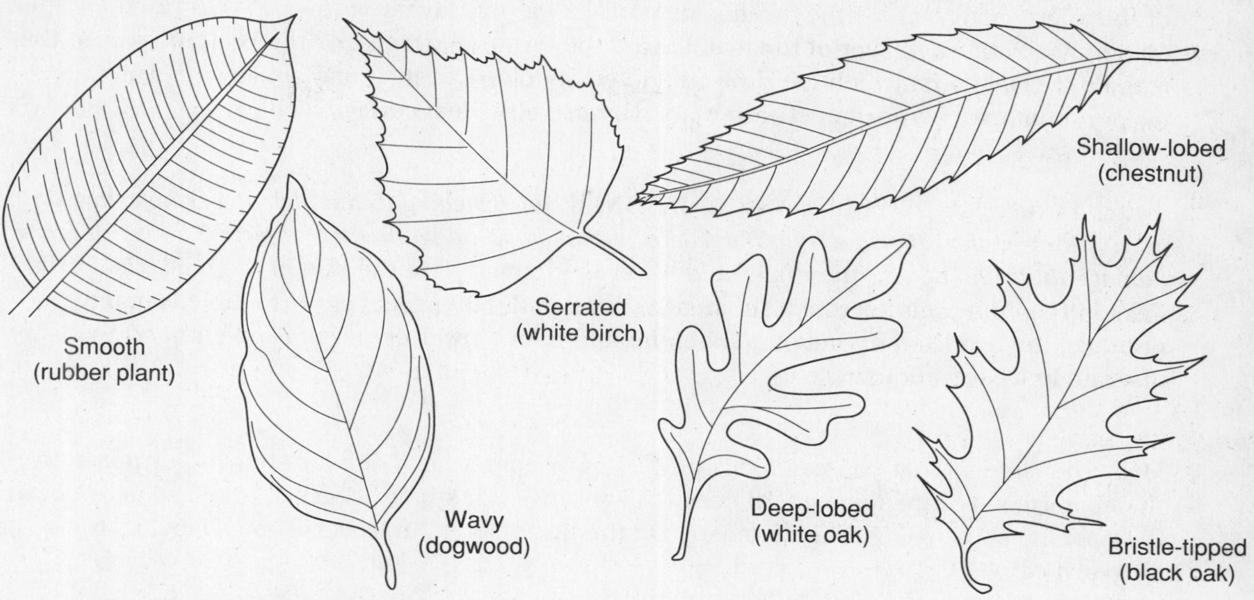

Figure 2 **Leaf Margin Types**

5. Observe the arrangement of the leaves or leaflets on the stem of each leaf. In an opposite arrangement, the leaves are arranged opposite each other on the stem. In an alternate arrangement, the leaves are attached at alternate positions on the stem. In a whorled arrangement, several leaves arise together around one part of the stem. Use Figure 3 to identify leaf or leaflet arrangements. Record the leaf arrangement for each leaf in the Data Table.

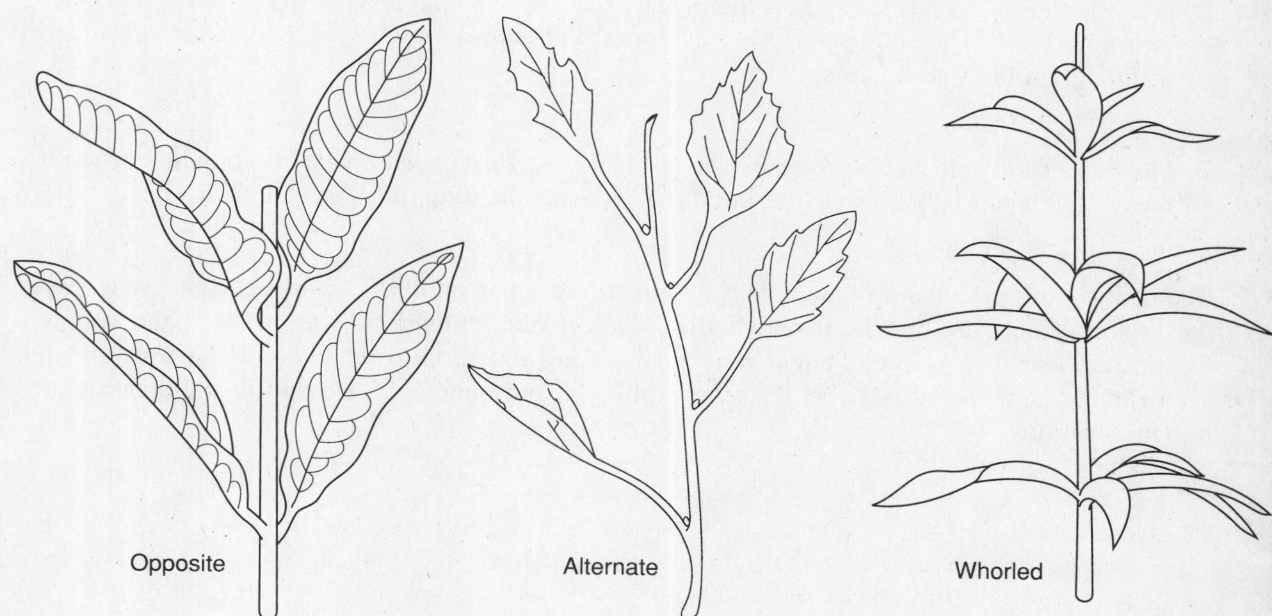

Figure 3 **Leaf Arrangement**

Part C. Examining a Cross Section of a Dicot Leaf

1. Examine a prepared slide of a privet or lilac leaf cross section under the low-power objective of the microscope. Notice the arrangement of the various layers of the leaf. Identify the upper epidermis, or the top layer of the leaf. Locate the lower epidermis, or the bottom layer of the leaf. Epidermal hairs, which cut down the intensity of bright light and slow down water evaporation, may be present. These hairs may make the leaf surface feel fuzzy.

2. Switch to the high-power objective. **CAUTION:** *When switching to the high-power objective, always look at the objective from the side of the microscope so that the objective does not hit or damage the slide.* Locate the upper and lower epidermis. Note the size of the epidermal cells. Cells in the epidermis secrete a substance called cutin, which covers the whole leaf and prevents the leaf from drying out. Label the upper and lower epidermis and cutin on the leaf diagram in Observations.

3. Move the slide so that you can focus on the lower epidermis. Look for stomata, or pores, in the epidermis. Locate some guard cells that border each stoma. These specialized cells control the opening and closing of the stomata. On the leaf diagram in Observations, label a stoma and its guard cells.

4. Locate a layer of elongated rectangular cells just below the upper epidermis. These cells are called palisade parenchyma cells. They contain many chloroplasts, which appear as tiny green bodies in the cells. Label a palisade parenchyma cell and a chloroplast on the leaf diagram in Observations.

5. Observe the irregularly shaped spongy parenchyma cells just below the palisade cells. Spongy parenchyma cells contain fewer chloroplasts than the palisade parenchyma cells. The spongy cells form a honeycomb pattern with intercellular air spaces, which connect with the stomata in the lower epidermis. Together, the palisade and spongy parenchyma cells make up the mesophyll layer. Label a spongy parenchyma cell, the mesophyll layer, and an air space on the leaf diagram in Observations.

6. Move the slide sideways and locate the circular structure called a vascular bundle in the mesophyll layer. Label a vascular bundle on the leaf diagram in Observations.

7. Locate some xylem and phloem cells within the vascular bundles. Xylem cells generally have thick walls and are bunched near the top of a leaf vein. Phloem cells are thinner and are located near the bottom of a leaf vein. Surrounding and supporting a vascular bundle are cells forming bundle sheaths. Label the xylem, phloem, and bundle sheath on the leaf diagram in Observations.

Name ______________________ Class __________ Date __________

Observations

Dicot Leaf

Monocot Leaf

Conifer Leaf

Data Table

Leaf Name	Leaf Type	Type of Venation	Leaf Margin	Leaf Arrangement

Leaf Cross Section

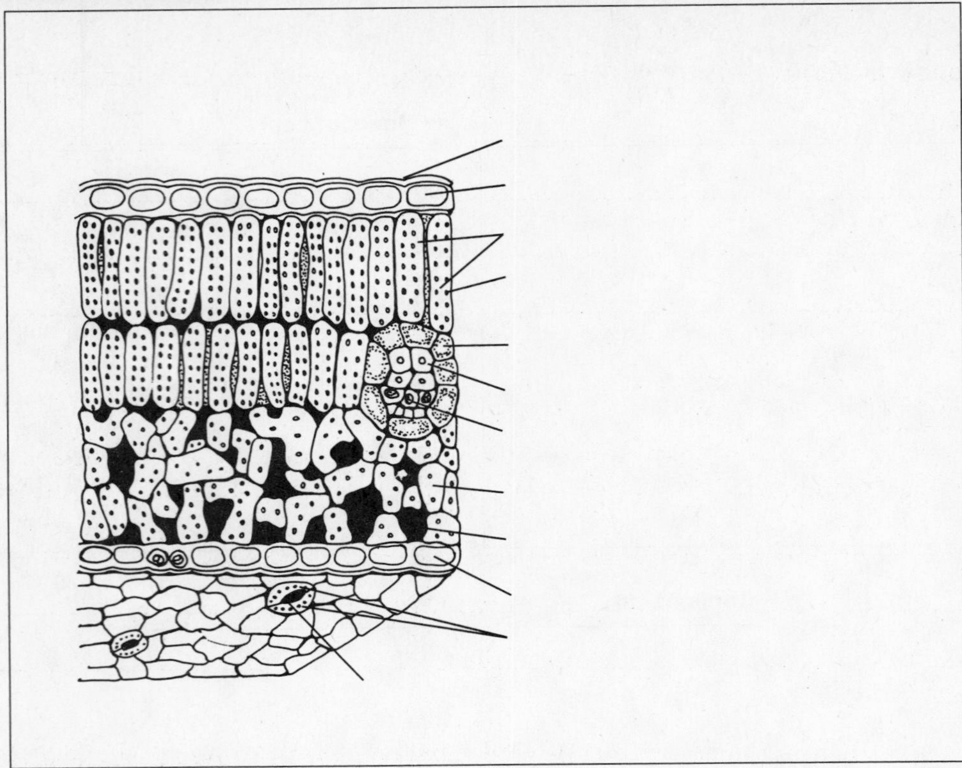

Analysis and Conclusions

1. Which surface of the leaf, upper or lower, contains the most chlorophyll? Explain why.

2. How is the structure that attaches the leaf to the stem different in monocots as compared to dicots?

3. How are the veins important to the leaf's function?

4. How does the conifer (pine) leaf differ from the dicot leaf?

5. Consult your Data Table. How many dicot leaves did you examine?

Name ______________________ Class ____________ Date ____________

6. How do the stomata and air spaces in the spongy parenchyma work together to help the leaf survive? ______________________

Critical Thinking and Application

1. A cross section of a pine leaf would show a thickened cutin and a thickened epidermis. How does this design help the pine leaf survive in the northern parts of the United States?

2. Some people mist their house plants with water. Do you think the plant can absorb the water through its leaves? Explain your answer. ______________________

3. The gypsy moth, *Lymantria dispar,* is an insect that is responsible for damaging millions of kilometers of trees in the northeastern United States. The larval stage of this insect is a caterpillar that feeds on the leaves of trees, particularly oak and apple. How do you account for the fact that after a few years of repeated defoliation by these insects, trees often die?

4. In order to keep your house plants healthy, why should you periodically remove dust from their leaves? ______________________

Going Further

Examine some examples of leaf adaptations. Leaf adaptations occur when the leaf changes its form for protection, storage, support, or some other need. Look for the following specially adapted leaves: an onion or flower bulb, a spine or prickle from a cactus, a tendril on a climbing plant, and a leaf from a succulent plant such as a jade. Also visit a greenhouse or a florist for an example of an insect-eating plant such as a Venus' flytrap. Use reference materials in the library to find photographs of plants that are carnivorous. Try to determine what function each of these different leaf adaptations serves for the plant.

Name ______________________ Class __________ Date __________

Estimating the Number of Stomata in a Leaf

Pre-Lab Discussion

The leaf is the main organ through which water is given off by a plant. As a plant absorbs water through its roots, the water is carried through the stem to the leaves. Some of the water is used in the leaves during photosynthesis. However, more water is taken in and transported to the leaves than can be used during photosynthesis. The excess water evaporates into the atmosphere through the *stomata,* or pores, in a process called *transpiration*.

Stomata are openings generally found on the undersides of leaves. They are formed by two specialized epidermal cells called *guard cells*. Guard cells regulate the passage of water vapor, oxygen, and carbon dioxide into and out of the leaf.

In this investigation, you will observe guard cells in a fresh lettuce leaf. You will also estimate the total number of guard cells in a lettuce leaf.

Problem

How can the number of stomata in a leaf be estimated?

Materials *(per group)*

Small piece of crisp lettuce in cold water
Small piece of wilted lettuce
Microscope
2 glass slides
2 coverslips
Medicine dropper
Forceps
Calculator (optional)
Transparent plastic metric ruler
Probe

Safety

Always handle the microscope with extreme care. You are responsible for its proper care and use. Use caution when handling glass slides as they can break easily and cut you. Be careful when handling sharp instruments. Note all safety alert symbols next to the steps in the Procedure and review the meanings of each symbol by referring to the symbol guide on page 10.

Procedure

Part A. Observing Guard Cells and Stomata in a Lettuce Leaf

1. Place a drop of water in the center of a glass slide.
2. Locate a large rib in the crisp lettuce leaf. As shown in Figure 1, bend the lettuce leaf against the curve until it snaps.

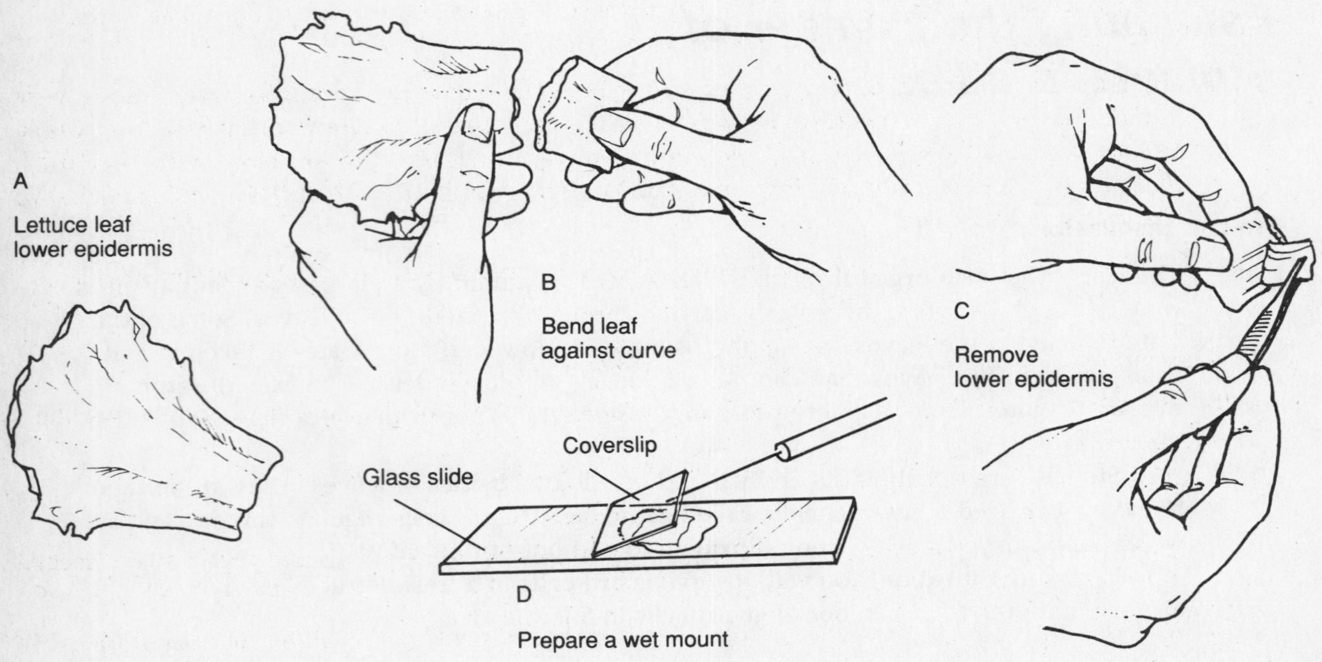

Figure 1

3. With the forceps, carefully remove the thin layer of epidermis from the piece of lettuce. Spread the epidermis out as smoothly as possible in the drop of water on the slide. **Note:** *If the epidermis becomes folded on the slide, use a probe to gently unfold and flatten it.* Add a coverslip.
4. Observe the lettuce epidermis under the low-power objective of the microscope. **Note**: *It may be necessary to adjust the diaphragm so that there is sufficient light passing through the cells.* Notice the irregular shapes of the epidermal cells. Notice the guard cells, which are pairs of kidney-shaped cells scattered throughout the epidermis. The spaces between the guard cells are the stomata.
5. Center a single stoma-guard cell unit in the center of the field of view under low power. Switch to high power and focus on the single stoma-guard cell unit. **CAUTION:** *When switching to the high-power objective, always look at the objective from the side of the microscope so that the objective does not hit or damage the slide.*
6. In the appropriate place in Observations, sketch a single stoma-guard cell unit as seen under high power of the microscope. Label the guard cells, stoma, and chloroplasts. Record the magnification of the microscope.
7. Set the wet-mount slide of the lettuce epidermis aside for now. It will be used in Part B of the investigation.
8. Repeat steps 1 through 6 using the wilted lettuce leaf.

Name ______________________ Class __________ Date __________

Part B. Estimating the Number of Stomata

1. Observe the slide of the crisp lettuce leaf epidermis under low power of the microscope. Count the number of stoma-guard cell units you see in the low-power field of view. Record the number in Data Table 1.
2. Carefully switch to high power. Count the number of stoma-guard cell units you see in the high-power field of view. Record the number in Data Table 1.
3. Using the transparent plastic metric ruler, measure the diameter of the low-power field of view in millimeters. **Note:** *See Laboratory Investigation 4 for step-by-step directions on how to measure the diameter of a field of view.* Calculate the diameter of the high-power field of view. Record the radii of the low-power and high-power fields of view in Data Table 2. The radius is equal to one half of the diameter.
4. Calculate the area of the low-power and high-power fields of view using the formula:

$$\text{area} = \pi r^2$$

Record this information in the appropriate place in Data Table 2.

5. To determine the number of stomata in a square centimeter of lettuce leaf epidermis, use the following formula:

$$\text{stomata/cm}^2 = \frac{\text{total number of stomata}}{\text{area (cm}^2\text{)}}$$

Calculate the number of stomata in a square centimeter for low power and high power. Record this information in the appropriate place in Data Table 2.

6. With the metric ruler, measure the size of the entire lettuce leaf. Calculate the area of the lettuce leaf using the following formula:

$$\text{area (cm}^2\text{)} = \text{width (cm)} \times \text{length (cm)}$$

Estimate the number of stomata in the lettuce leaf. Record this information in the appropriate place in Data Table 2.

Observations

Magnification __________

Crisp Lettuce Leaf

Magnification __________

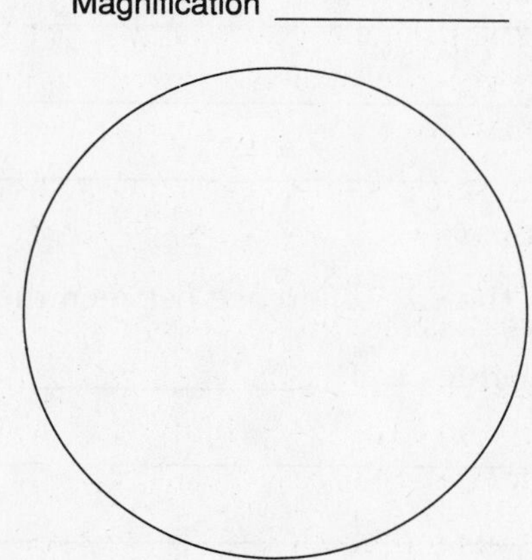

Wilted Lettuce Leaf

Data Table 1

	Low Power	High Power
Number of Stomata		

Data Table 2

Measurement	Low Power	High Power
Radius (mm)		
Radius^2 (mm^2)		
$\pi \times r^2$	3.14 x	3.14 x
Area (mm^2)		
Area in cm^2 (Area in $cm^2 = \frac{\text{area in } mm^2}{100}$)		
Stomata / cm^2 = $\frac{\text{Total number of stomata}}{\text{area in } cm^2}$		
Area of lettuce leaf (cm^2) = width (cm) x length (cm)		
Number of stomata/leaf = area of lettuce leaf x stomata/cm^2		

Analysis and Conclusions

1. Compare the sizes of the stomata in the crisp and wilted lettuce leaves.

____

____

____

2. Explain the size difference between the stomata in the crisp lettuce leaf and the stomata in the wilted lettuce leaf. ____

____

____

____

Name ______________________ Class __________ Date __________

3. What is the function of the guard cells? ______________________

4. Is the low-power estimated number of stomata close to the high-power estimated number? Why might the numbers be different? ______________________

Critical Thinking and Application

1. Where would you expect to find stomata on the leaves of a floating water lily plant? Explain your answer. ______________________

2. Why do you think the stomata of a plant are open during the day and closed at night?

3. What would happen to a plant if its leaves were completely coated with petroleum jelly? Explain your answer. ______________________

4. Would you expect the leaf of a desert cactus or the leaf of a water lily to have more stomata? Explain your answer. ______________________

Going Further

1. Design and conduct an experiment to show the effects of temperature, time of day, light, and soil moisture on stomata opening and closing. How will you know whether the stomata are open or closed?
2. Outline a leaf, such as that of a bryophyllum or African violet, on a sheet of graph paper. Determine the area of the leaf by counting the squares inside the outline. Using the procedure described in the investigation, count the number of stomata in one square centimeter. Calculate the total number of stomata on the leaf surface.

46

Plant Hormones

Pre-Lab Discussion

Both plants and animals produce hormones, one type of which regulates growth. Plant hormones are often called phytohormones. Gibberellins are a group of phytohormones that affect plant growth. The effects of these hormones were first observed in the 1920s in rice plants that were attacked by a fungus. The rice plants grew unusually tall and spindly. Japanese scientists discovered that the plants' extraordinary growth was caused by a substance produced by the fungus *Gibberella fujikuroi*. The substance, later called gibberellin, or gibberellic acid, was named after the fungus that produced it.

In 1956, gibberellins were also found in the bean plant *Phaseolus vulgaris*. Now scientists believe that gibberellins are produced by all higher plants. Some plants have a mutation that causes dwarfism. This mutation is thought to interfere with a plant's ability to produce gibberellic acid.

In this investigation, you will observe the effects of gibberellic acid on the growth of normal and dwarf pea plants.

Problem

What is the effect of gibberellic acid on the growth of normal and dwarf pea plants?

Materials *(per group)*

4 large flower pots
Small stones
Soil-and-sand mixture
4 plastic bags
4 twist ties
Scissors
25 toothpicks
Metric ruler
Wooden stakes
Sheet of unlined white paper
Cotton swabs
Masking tape
Red and black pens or pencils
16 Little Marvel pea seeds
16 Alaska pea seeds
Gibberellic acid solution

Safety

Put on a laboratory apron if one is available. Put on safety goggles. Handle all glassware carefully. Always use special caution when working with laboratory chemicals, as they may irritate the skin or cause staining of the skin or clothing. Never touch or taste any chemical unless instructed to do so. Be careful when handling sharp instruments. Note all safety alert symbols next to the steps in the Procedure and review the meanings of each symbol by referring to the symbol guide on page 10.

Procedure

1. Using strips of masking tape, label the four large flower pots as shown in Figure 1. Also include the name of one person in your group and the date on each flower-pot label.

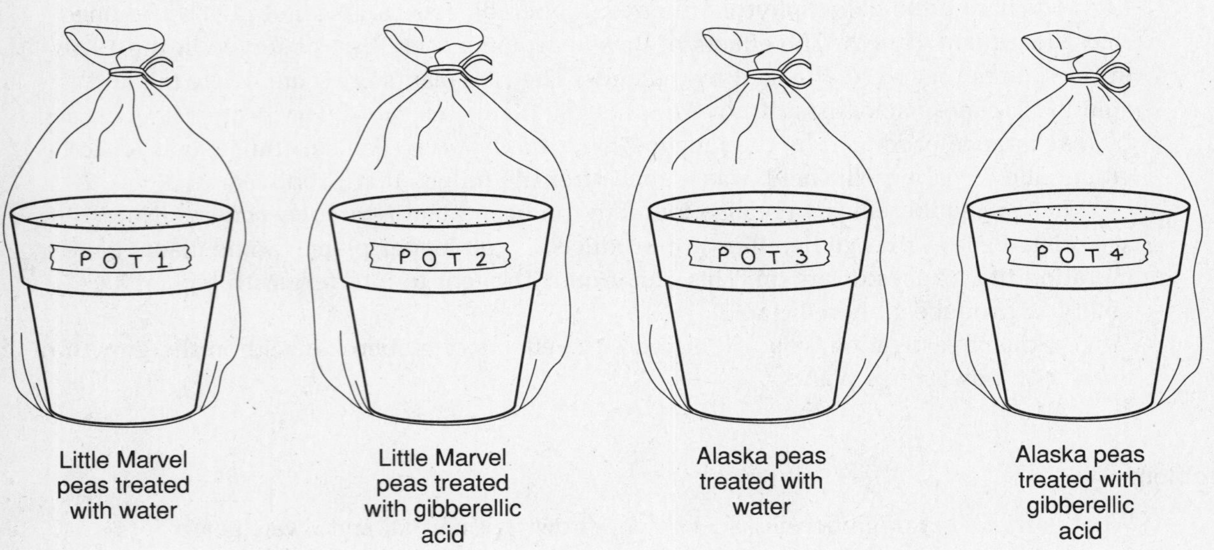

Figure 1

2. For drainage, place stones in the bottom of each flower pot. Then fill the flower pots two-thirds full with a mixture of soil and sand. Moisten the soil-and-sand mixture.
3. In each of the two pots labeled "Little Marvel peas," scatter eight Little Marvel pea seeds along the surface of the soil. Cover the seeds with a thin layer of soil. Moisten the seeds.
4. In each of the pots labeled "Alaska Peas," scatter eight Alaska pea seeds along the surface of the soil. Cover the seeds with a thin layer of soil. Moisten the seeds.
5. Place each pot in a plastic bag and seal the bag with a twist tie. Store the pots in a warm, dark place where they will remain undisturbed for four days.
6. When the seedlings begin to emerge from the soil, open the plastic bags and place the pots in a warm, sunny place. Add water as needed to keep the soil moist but not wet.
7. When the seedlings are about 3 cm tall (after about eight days), select five of the healthiest plants. With your thumb and index finger, pinch off the remaining seedlings at the soil line.

Name ______________________ Class ____________ Date ____________

8. To identify each seedling, cut out small "flags" from the unlined white paper. Number the flags 1 through 5. Attach these flags to the ends of toothpicks as shown in Figure 2. Place a toothpick in the soil next to each seedling.

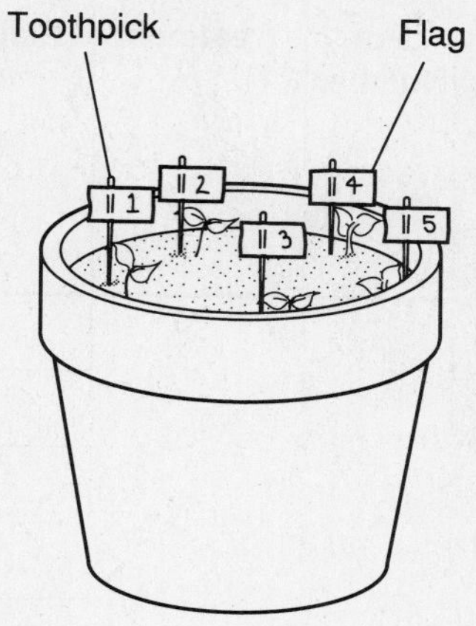

Figure 2

9. Using a metric ruler, measure the initial stem length of each seedling in millimeters. As shown in Figure 3, measure the stem length as the distance from the first pair of true leaves to the tallest part of the plant. Record these measurements in the appropriate places in the Data Table.

10. Using a cotton swab, apply water to the leaves and shoot tips of the seedlings in pots 1 and 3.

11. Using a clean cotton swab, carefully apply gibberellic acid solution to the leaves and shoot tips of the plants in pots 2 and 4. **CAUTION:** *Handle gibberellic acid with care to avoid spills.*

12. Arrange the pots so that each one receives the same quantity of light. Give each pot the same amount of water. As the plants grow taller, use wooden stakes to support them.

13. Each day for the next eight school days, measure the stem length of each plant to the nearest millimeter. Record the date and the measurements in the appropriate places in the Data Table. Find the average stem length of the five seedlings in each pot and record the average in the appropriate place in the Data Table.

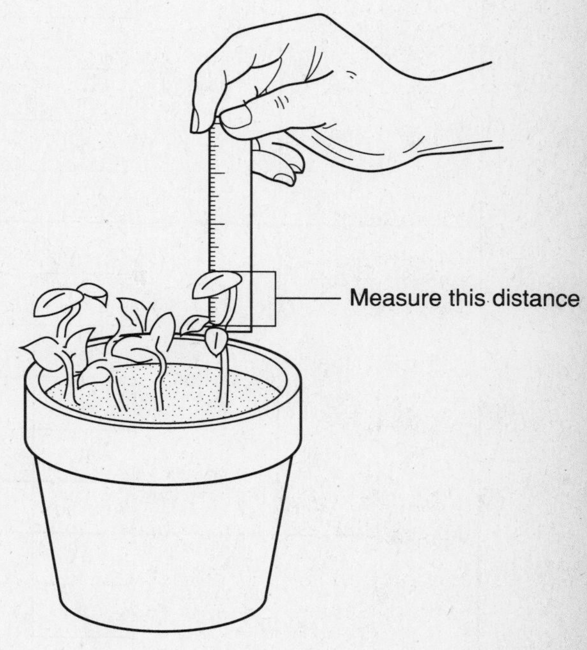

Figure 3

14. Using the averages in the Data Table, plot stem length versus number of days of growth on the graph in Observations. **Note:** *Because you have not recorded eight consecutive days of growth (you did not record on the weekend), you must account for these gaps on your graph. Leave space on the graph for the days when measurements could not be taken.*

Observations

Data Table

Plant Variety	Pot Number	Treatment	Plant Number	Stem Length (mm)							
				Day 1	Day 2	Day 3	Day 4	Day 5	Day 6	Day 7	Day 8
				Date:	Date:	Date:	Date:	Date:	Date:	Date:	Date:
Little Marvel	1	Water	1								
			2								
			3								
			4								
			5								
			Average								
	2	Gibberellic acid	1								
			2								
			3								
			4								
			5								
			Average								
Alaska	3	Water	1								
			2								
			3								
			4								
			5								
			Average								
	4	Gibberellic acid	1								
			2								
			3								
			4								
			5								
			Average								

Name ______________________ Class ____________ Date ____________

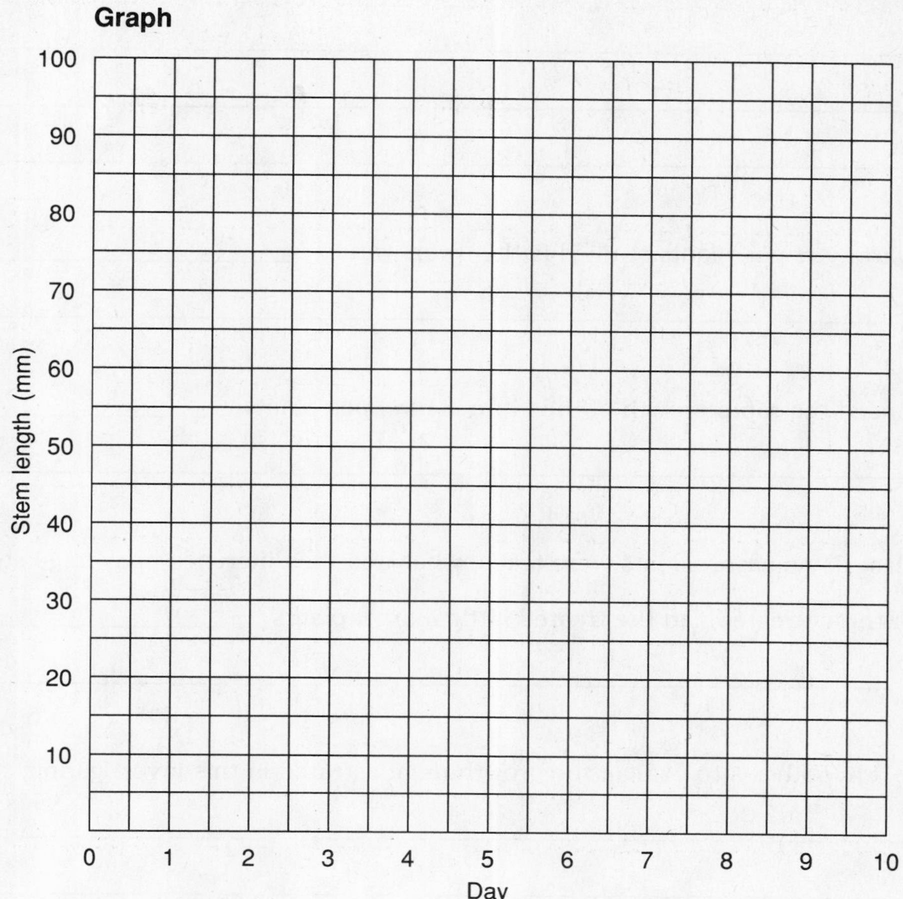

Analysis and Conclusions

1. Why was water applied to the leaves and stem tips of the plants in pots 1 and 3?

____

____

2. Based on the graph you constructed, explain which group of plants showed the fastest rate of growth. ____

____

3. Compare the growth of Little Marvel pea plants and Alaska pea plants with and without treatment with gibberellic acid. ______

______

______

______

4. Which variety of pea plants studied is the dwarf variety? ______

______

5. What caused the rapid growth of the Little Marvel pea plants? ______

______

6. On the last day measurements were taken, what was the difference in average stem length between the untreated and the treated Little Marvel plants? ______

______

7. Develop a hypothesis to explain the growth demonstrated in this investigation.

______

______

______

Critical Thinking and Application

1. Suggest how a commercial plant grower might apply knowledge of the effects of gibberellic acid. ______

______

______

2. A head of lettuce is actually a rosette of leaves on a very short stem. Explain what might happen if a lettuce head were treated with gibberellic acid. ______

______

______

Name _______________ Class _______ Date _______

3. The word hormone comes from a Greek word meaning "to arouse." How is this name suitable for the function of hormones? _______________

4. How might a knowledge of plant hormones be useful to a commercial gardener who supplies florists with fresh flowers for holidays even though some of these holidays do not coincide with the natural growing period of the plants in demand? _______________

Going Further

1. In the investigation you just completed, the gibberellic acid solution was applied directly to the leaves and stem tips of the plants. Design an experiment in which the gibberellic acid enters the plant through a route other than the leaves and stem tips. Perform the experiment and compare the results with those obtained in this investigation.
2. Gardeners often pinch off the stems of plants to enhance the plants' growth. This pinching off is related to the production of certain hormones. Determine which hormones are involved. Find out the mechanism of action of these hormones. Then design an experiment that shows the effect of pinching back the stem tips. Use four healthy plants — two experimental plants and two control plants.

Name ______________________ Class ____________ Date ____________

Germination Inhibitors

Pre-Lab Discussion

As a seed germinates, new cells are produced. As they grow larger, they differentiate into plant organs, such as roots, stems, leaves, and flowers. Eventually, a mature plant results.

Tomato seeds will usually germinate when exposed to the proper amounts of moisture and oxygen and a fairly warm temperature. Yet inside the tomato, where these conditions are also met, tomato seeds do not germinate. Why? Plants contain certain hormones that control their metabolism and growth processes.

In this investigation, you will observe the effects of a hormone called a germination inhibitor.

Problem

What are the effects of a germination inhibitor on tomato seeds? What effect does a germination inhibitor from one kind of tomato have on other kinds of tomatoes?

Materials *(per group)*

Tomatoes (2 different varieties)
Filter paper
Strainer
Mortar and pestle (or bowl and spoon)
Funnel
Beaker
4 petri dishes
Plastic wrap
Glass-marking pencil

Safety

Put on a laboratory apron if one is available. Handle all glassware carefully. Note all safety alert symbols next to the steps in the Procedure and review the meanings of each symbol by referring to the symbol guide on page 10.

Procedure

Part A. Observing the Effects of a Germination Inhibitor

1. With the glass-marking pencil, label the petri dishes 1 through 4.
2. Using a mortar and pestle, crush one whole tomato (Variety A). Strain the crushed tomato and collect the extract in the beaker. With a glass-marking pencil, label the beaker "Extract A" and set it aside for now.
3. From the tomato pulp, remove about 20 seeds and wash them.
4. Line petri dishes 1 and 2 with filter paper. Place 10 seeds in each dish.
5. Moisten petri dish 1 with water and petri dish 2 with the tomato extract. Cover the petri dishes with plastic wrap. Observe the seeds for several days, adding more water or tomato extract as needed to keep the filter paper moist.
6. Record the total number of seeds that germinate daily in each petri dish in the appropriate place in Data Table 1.

Part B. Comparing the Effects of Germination Inhibitors

1. Line petri dishes 3 and 4 with filter paper.
2. Using a mortar and pestle, crush a different kind of tomato (Variety B) and obtain 20 seeds from the tomato pulp. Wash the seeds and place 10 seeds in each petri dish
3. Moisten petri dish 3 with water. Moisten petri dish 4 with the extract in the beaker labeled Extract A, which you collected during Part A.
4. Cover the petri dishes with plastic wrap. Observe the seeds for several days. Record the total number of seeds that germinate daily in each petri dish in the appropriate place in Data Table 2.
5. Using the graph in Observations, construct four different line graphs representing the number of seeds that germinated in each of the four petri dishes. Label each of the line graphs.

Observations

Data Table 1

Petri Dish	Day 1	Day 2	Day 3	Day 4	Day 5	Day 6	Day 7	Day 8
1 (Variety A with water)								
2 (Variety A with extract A)								

Data Table 2

Petri Dish	Day 1	Day 2	Day 3	Day 4	Day 5	Day 6	Day 7	Day 8
3 (Variety B with water)								
4 (Variety B with extract A)								

Name ______________________ Class ____________ Date ____________

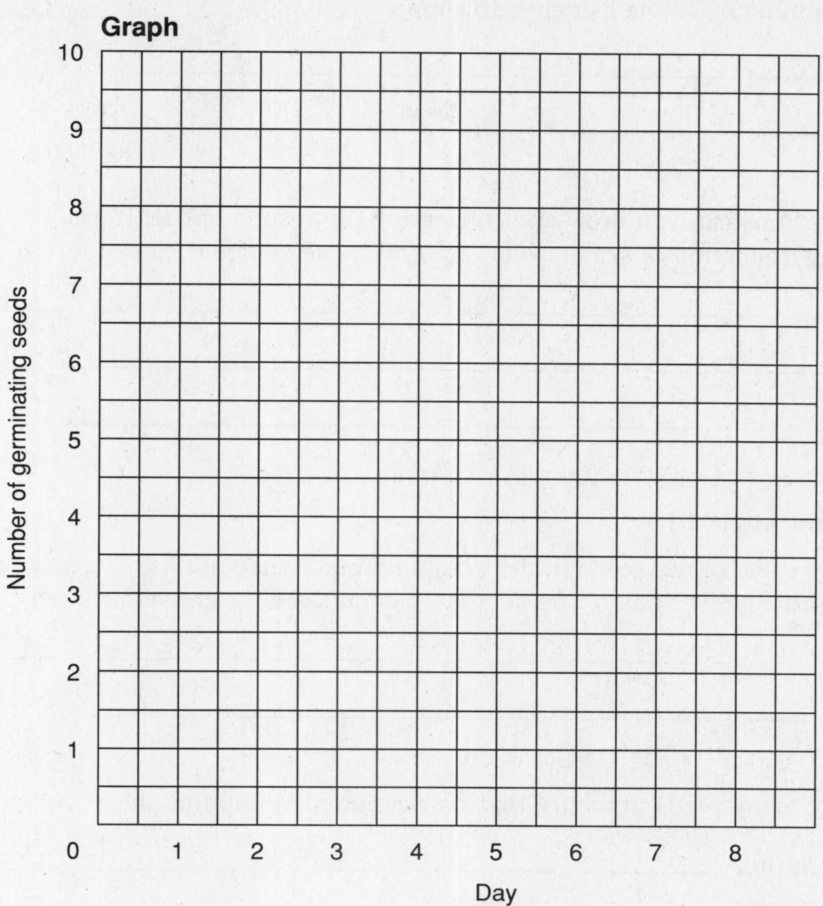

Analysis and Conclusions

1. What is the purpose of petri dishes 1 and 3? ______________________

____

2. What is seed germination? ______________________

____

____

3. Explain how you can determine whether or not a seed has germinated.

____

____

4. What conclusion can you draw after observing the results of Part A? Do tomatoes contain a germination inhibitor? Give evidence to support your answer. ______________________________

____

____

5. What conclusions can you draw after observing the results of Part B? Is Extract A effective in preventing germination of seeds from Variety B? Give evidence to support your answer.

____

____

____

Critical Thinking and Application

1. Explain why some of the seeds in petri dishes 1 and 3 may not have germinated, even though they were provided with all of the necessary conditions for growth.

____

____

2. Explain why many seeds in nature that do not contain inhibitors might begin to germinate during late spring. ______________________________

____

____

3. In addition to tomatoes, what are some fruits and/or vegetables that you think may contain germination inhibitors? ______________________________

____

____

Going Further

1. Test the effectiveness of the germination inhibitor on additional varieties of tomatoes as well as on other kinds of plants. You might want to try flax or tobacco seeds because they are in the same family as the tomato.
2. Is the germination inhibitor a protein? If it is, it will lose its effectiveness when boiled. Repeat Part A of this investigation using boiled tomato extract. What happens?

Name ______________________ Class ____________ Date ____________

48

Observing Plant Responses

Pre-Lab Discussion

The movement of a part of a plant in response to environmental stimuli such as light, gravity, water, or touch is called a *tropism.* The growth of a plant stem toward a light stimulus is called *phototropism.* The growth of a root downward toward the center of the Earth (gravity) is called *gravitropism.* The growth of a plant toward water is called *hydrotropism.* A plant's growth movement in response to touch is called *thigmotropism.*

When the plant part responds to a stimulus by growing toward the stimulus, the response is known as a *positive tropism.* When the plant part responds by growing away from the stimulus, the response is known as a *negative tropism.* Thus, stems show a positive tropism to light and a negative tropism to gravity. Roots, on the other hand, show a negative tropism to light and a positive tropism to gravity. In many tropisms, certain plant hormones called *auxins* regulate plant growth and help a plant respond to some stimuli.

Plant movements that are independent of the direction of the stimulus are not classified as tropisms. Instead, they are called *nastic movements.* While tropisms take a fairly long time to occur and are not easily reversed, nastic movements occur rapidly and are quickly reversible.

In this investigation, you will observe the processes of gravitropism and phototropism. You will also investigate the nastic movements of the *Mimosa* plant.

Problem

How do plants respond to external stimuli?

Materials *(per group)*

- 4 corn seeds
- 4 bean seeds
- Petri dish
- 2 paper cups
- Paper towels
- Masking tape
- Glass-marking pencil
- Clay
- Scissors
- Potted *Mimosa pudica* plant
- Toothpick
- Stopwatch or clock with second hand
- Metric ruler
- Large cardboard box with slit (one box for the entire class)
- Soil
- Protractor

Safety

Put on a laboratory apron if one is available. Handle all glassware carefully. Be careful when handling sharp instruments. Note all safety alert symbols next to the steps in the Procedure and review the meanings of each symbol by referring to the symbol guide on page 10.

Procedure

Part A. Observing Gravitropism

1. Arrange the four soaked corn seeds in a petri dish as shown in A of Figure 1. The pointed end of each seed should face the center of the petri dish. One seed should be placed at the 12 o'clock position, and the other seeds should be at the 3, 6, and 9 o'clock positions, respectively.
2. Cut a circle of paper towel the size of the petri dish and place it over the seeds. Then carefully pack the petri dish with enough pieces of paper towel so that the seeds are held firmly in place when the other half of the petri dish is put on. See B of Figure 1. **CAUTION:** *Be careful when using scissors.*

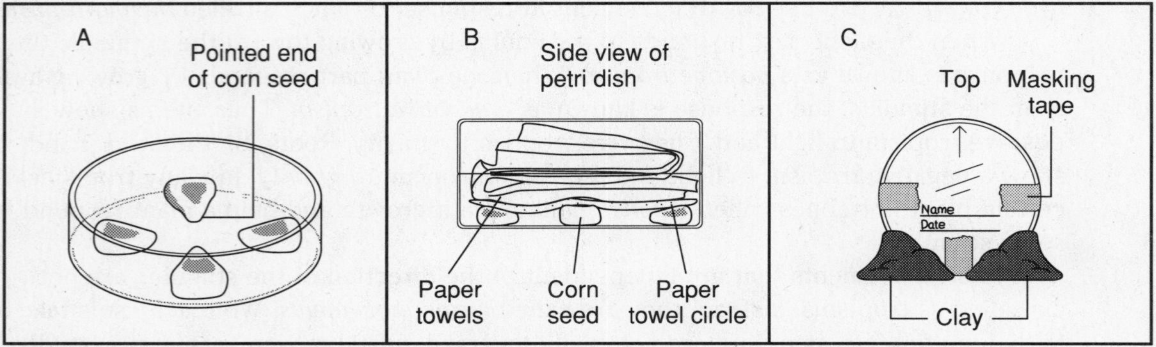

Figure 1

3. Moisten the paper towels with water. Cover the petri dish and seal the two halves together with strips of masking tape.
4. With a glass-marking pencil, draw an arrow on the lid pointing toward 12 o'clock. Label the lid with the name of one member of your group and the date.
5. As shown in C of Figure 1, prop the petri dish up with pieces of clay so that the arrow is pointing upward. Place the petri dish in a dark place and allow it to remain undisturbed for about 5 days.
6. After 5 days, remove the petri dish from the dark place. In the appropriate place in Observations, make a sketch of the seeds in the petri dish.

Part B. Observing Phototropism

1. Punch four holes in the bottom of each paper cup. Label the cups A and B and put the name of one member of your group on each cup.
2. Fill the cups nearly full with soil. Then plant two bean seeds about 4 cm apart in each cup and cover with about 1 cm of soil. Water the seeds and allow them to grow until they are about 5 cm high.
3. After the seedlings have reached a height of about 5 cm, place the cup labeled A in an area of the room that gets the same amount of light from all sides.
4. Place the cup labeled B in the cardboard box that your teacher prepared. Note that a small slit has been cut out on one side of the box. Place the cup in the box as shown in Figure 2.

Name ______________________ Class ____________ Date ____________

Figure 2

5. After each group has placed their cups in the cardboard box, it will be closed so that light enters only through the slit. Allow the cups to remain undisturbed in the box for 72 hours.
6. After 72 hours, observe the plants in in cups A and B. Use the protractor to measure the angle at which the plant bends in relation to the soil. See Figure 3. Record the information in the appropriate places in the Data Table. Also in the Data Table, record any observations on the height, color, stem and leaf shape, and leaf arrangement of each plant.

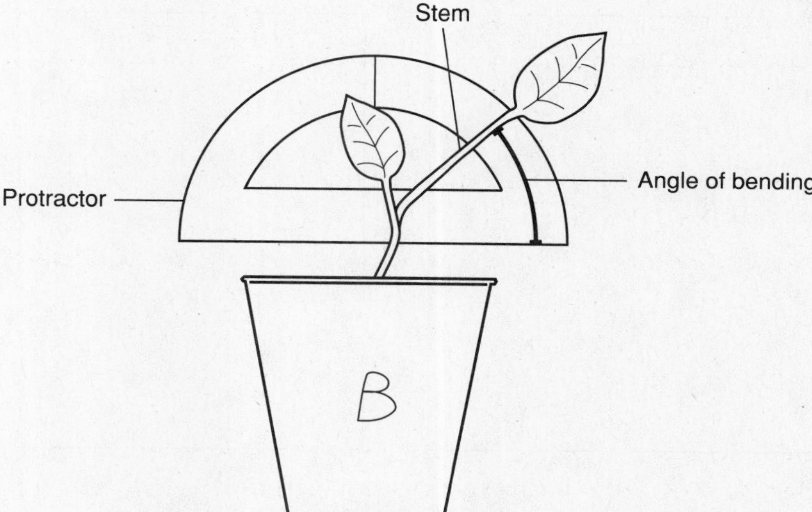

Figure 3

Part C. Observing Nastic Movements

1. Without touching them, observe the petiole and leaflets of the *Mimosa* plant. In the appropriate place in Observations, sketch these structures.
2. Position the toothpick between your index finger and thumb. Then as shown in Figure 4, allow the toothpick to spring forward so that it touches a leaflet. Observe the response of the leaflet to touch. Have one group member determine how long it takes the leaflet to respond to the stimulus and how long it takes the leaflet to return to its original position.

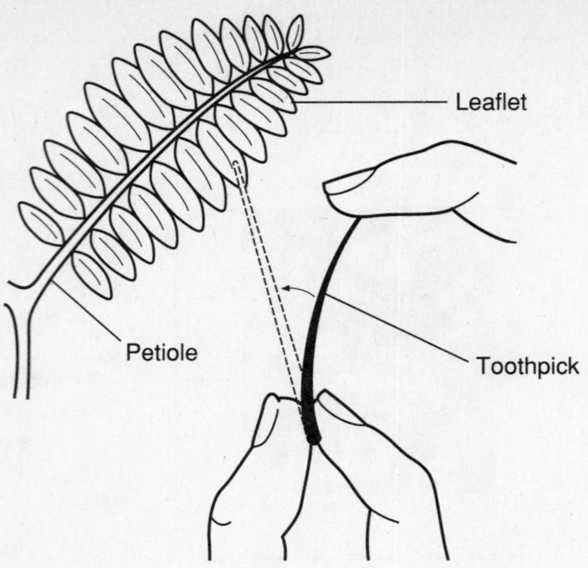

Figure 4

3. Repeat step 2 touching the petiole instead of the leaflet. Determine how long it takes the petiole to respond to the stimulus and how long it takes the petiole to return to its original position.

Observations

Germinating Corn Seeds

***Mimosa* Leaf**

Data Table

Cup	Plant	Angle of Bending	Observations
A	1		
	2		
B	1		
	2		

Name ______________________ Class __________ Date __________

1. How did the leaflet respond when you touched it with the toothpick?

____

2. How long did it take for the leaflet to respond to the stimulus and then to return to its original position? ____

____

3. How did the petiole respond when you touched it with the toothpick?

____

4. How long did it take for the petiole to respond to the stimulus and then to return to its original position? ____

____

Analysis and Conclusions

1. Which part(s) of the germinating corn seeds showed positive gravitropism? Which part(s) showed negative gravitropism? ____

____

____

2. Why were the corn seeds placed in the dark instead of near a light source?

____

3. Why was it important that the petri dish in Part A be kept in an upright position throughout the investigation? ____

____

____

4. Do plant leaves and stems show a positive or a negative phototropism? Give evidence to support your answer. ____

____

____

5. Why are *Mimosa* plants said to have nastic movements instead of negative or positive thigmotropism? ______________________________

____

____

Critical Thinking and Application

1. If an insect landed on a *Mimosa* plant, how do you think the leaflets would respond? How might this response be helpful to the plant? ______________________________

____

____

2. Why do many people regularly turn or rotate their houseplants? ______________________________

____

____

3. Certain types of vines display a positive thigmotropism by climbing on or clinging to objects such as stakes or fences. How is this an advantage in the process of photosynthesis?

____

____

4. How is a plant root's positive gravitropism an advantage for the survival of the plant?

____

____

____

Going Further

1. Design an experiment to observe another type of tropism such as hydrotropism or chemotropism.
2. Are bean seedlings responsive to white light only? Design and conduct an experiment to find out.

Name ______________________ Class ____________ Date ______________

49

Reproductive Structures in a Flower

Pre-Lab Discussion

Flowers are the reproductive organs of angiosperms, or flowering plants. Flowers vary in size, color, and arrangement. All flowers, regardless of variety, have the function of seed formation and the production of more plants. Flowers contain both nonreproductive and reproductive structures.

In this investigation, you will examine a flower and identify its nonreproductive and reproductive parts. You will also observe the male and female sex cells produced by a flower.

Problem

How is a flower adapted for sexual reproduction?

Materials *(per group)*

Variety of fresh flowers
Stamens and pistil from a dissected flower
Hand lens
Forceps
Scalpel or single-edged razor blade
Dissecting tray
Dissecting needle
Scissors
Microscope
2 glass slides
2 coverslips
Medicine dropper
Sheet of unlined white paper

Safety

Put on a laboratory apron if one is available. Handle all glassware carefully. Be careful when handling sharp instruments. Always handle the microscope with extreme care. You are responsible for its proper care and use. Use caution when handling glass slides as they can break easily and cut you. Note all safety alert symbols next to the steps in the Procedure and review the meanings of each symbol by referring to the symbol guide on page 10.

Procedure

Part A. Observing the External Anatomy of a Flower

1. Refer to Figure 1 as you examine each flower. Identify the receptacle, or structure to which all other flower parts are attached. Note the sepals, which are small, leaflike structures above the receptacle. All of the sepals together are called the calyx. Observe the brightly colored petals inside the calyx. All of the petals together are called the corolla. These structures make up the nonreproductive structures of a flower.
2. As you observe each flower, note the number and color of both the sepals and the petals. Record this information in the appropriate place in the Data Table.

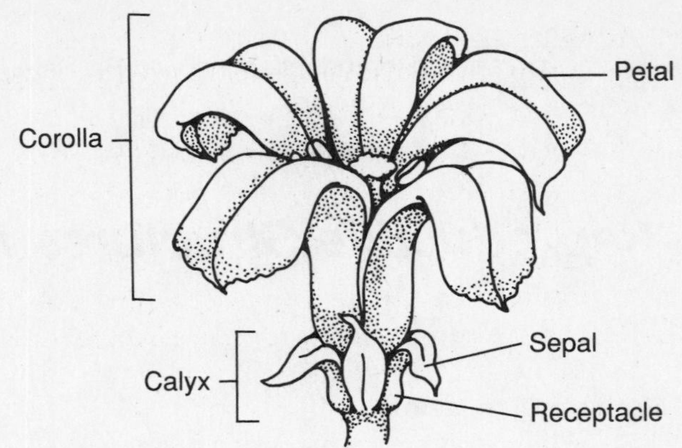

Figure 1

Part B. Reproductive Structures of a Flower

1. Choose one flower. Using Figure 2, identify the male reproductive structures, or stamens. Each stamen consists of an anther and a filament. Identify the female reproductive structure, or pistil. Each pistil consists of an ovary, a stigma, and a style.
2. In the appropriate place in Observations, sketch the flower you observed. On the sketch, label the stamen, anther, filament, pistil, ovary, stigma, and style.
3. With a scissors or forceps, remove the petals from the flower. **CAUTION:** *Be careful when handling sharp instruments.* Be very careful not to destroy any of the reproductive structures of the flower.

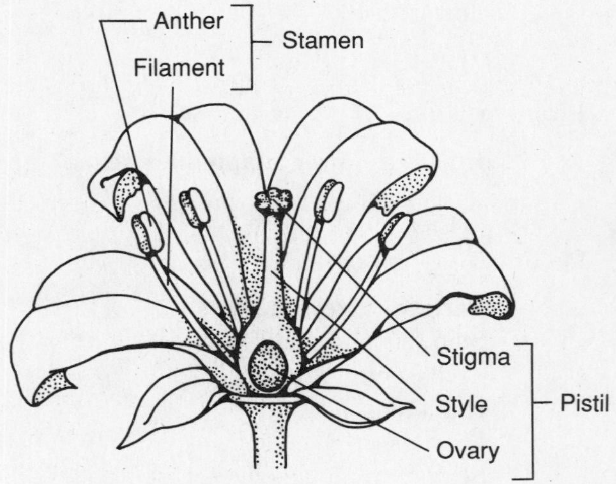

Figure 2

4. With a forceps, remove the stamens and place them on a sheet of unlined white paper. Set the stamens aside for use in Part C. In the appropriate place in Observations, sketch a stamen that you removed from the flower. Label the anther and the filament.

5. Using a scalpel or single-edged razor blade, carefully remove the pistil. Cut the ovary in half at its widest part as shown in Figure 3. **CAUTION:** *Be careful when handling sharp instruments. Always cut in a direction away from your hands and body.* Observe the inside of the ovary. Note the number and arrangement of the ovary chambers. The white, oval objects inside the chambers are the ovules, or eggs.

6. In the appropriate place in Observations, sketch a cross section of the ovary showing the number and arrangement of the chambers. Label the ovules, ovary walls, and chambers. Set the pistil aside for use in Part C.

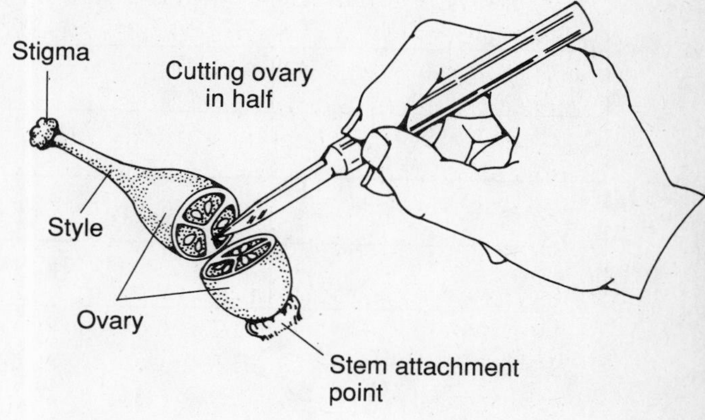

Figure 3

Part C. The Gametes of a Flower

1. With a forceps, remove an ovule from the dissected ovary. Place the ovule in a drop of water in the center of a clean glass slide and add a coverslip. Observe the ovule under the low-power objective of a microscope. In the appropriate place in Observations, sketch the ovule as seen through the microscope. Record the magnification of the microscope.

2. Gently brush an anther from one of the stamens across the stigma of the pistil. With a hand lens, look for pollen grains that may have stuck to the stigma.

3. Gently touch the point of a dissecting needle to one of the anthers. Transfer some pollen grains to a drop of water in the center of a clean glass slide. Add a coverslip. Examine the pollen grains under the high-power objective of a microscope. **CAUTION:** *When switching to the high-power objective, always look at the objective from the side of the microscope so that the objective does not hit or damage the slide.* The pollen grains may appear to be moving. This is due to the movement of water under the coverslip.

4. In the appropriate place in Observations, sketch several pollen grains as seen through the microscope. Record the magnification of the microscope.

Observations

Data Table

Flower Name	Number of Sepals	Color of Sepals	Number of Petals	Color of Petals

Reproductive Structures of a Flower

Flower Stamen

Flower Ovary Cross Section

Name ______________________ Class __________ Date __________

Magnification __________

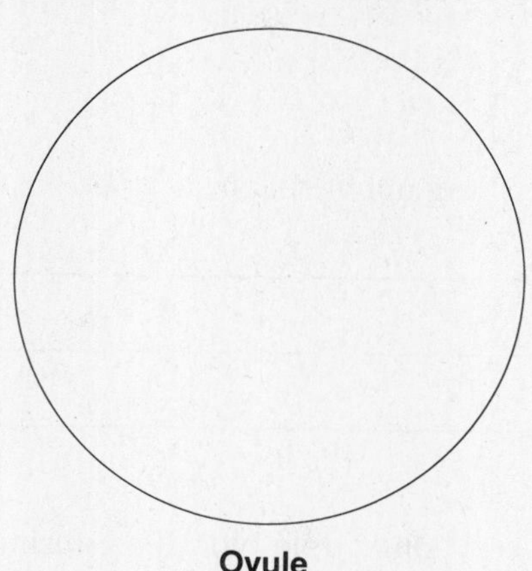

Ovule

Magnification __________

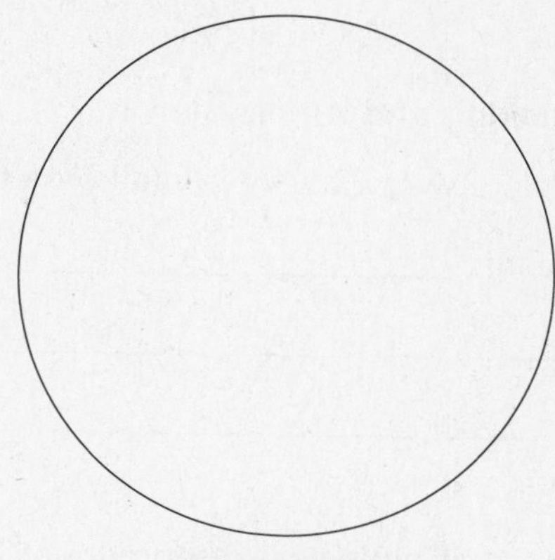

Pollen Grains

1. How many stamens are in the flower you dissected? ______________________

2. How many pistils are in the flower you dissected? ______________________

3. Approximately how many ovules do you think are in the ovary? (Make an educated guess.)

____

Analysis and Conclusions

1. The flower parts of monocot flowers usually occur in multiples of three. The flower parts of dicot flowers usually occur in multiples of four or five. From this information, identify the flower you dissected as a monocot or a dicot. ______________________

2. List the function of each of the following flower parts and note whether the parts comprise the pistil or the stamen.

a. anther ______________________

b. filament ______________________

c. ovary ______________________

d. ovule ______________________

e. pollen ______________________

f. stigma ______________________

g. style ______________________

3. Why does a flower have more pollen grains than ovules? ______________________________

____

____

Critical Thinking and Application

1. Explain why a heavy rainfall on a fruit orchard in the spring might result in a poor fruit yield in the fall. ____

____

____

2. For what biological reason do farmers object to people picking apple blossoms during the apple blossom season? ____

____

____

3. Flowers that are pollinated by the wind have smaller petals and sepals than flowers that are pollinated by insects or animals. Why are small petals and sepals an advantage to these flowers? ____

____

____

4. Why do flowers that are pollinated by insects have large, brightly colored petals?

____

____

____

Going Further

Germinate pollen tubes from the pollen of a mature stamen in a 10% sucrose solution. Prepare a wet mount of the pollen solution in a depression slide and allow it to set for a few days in a warm place. Examine the pollen under low power and high power of a microscope. Look for pollen tubes and sets of three nuclei in the pollen tubes.

Name ____________________ Class __________ Date __________

 50

Fruits and Seeds

Pre-Lab Discussion

The ripened ovary, with its enclosed seeds, and other associated parts form the *fruit*. Associated parts include sepals, petals, and the receptacle. The term "fruit" not only refers to the apples, pears, and peaches that are commonly found in the supermarket, but it also includes such things as string beans, peanuts, tomatoes, chestnuts, and cucumbers.

In the life cycle of a flowering plant, the fruit is essential to the dispersal of seeds. For example, animals may eat the fruit and thereby carry the seeds far from the parent plant. The seeds often pass unchanged through the animal's body. When the seeds pass out of the animal's body, they are dropped in places were they may germinate. Some fruits burst open, thus releasing their enclosed seeds and dispersing them some distance from the parent plant. Other fruits float in water and can be carried hundreds of kilometers from where they originate. Still other fruits are adapted to being carried by the wind.

A *seed* is made up of a seed coat, stored food, and an embryo. Each seed contains one or more *cotyledons,* or seed leaves, that store and digest foods. A plant that has two cotyledons is called a *dicot,* whereas a plant that has one cotyledon is called a *monocot.* Some monocot seeds have an *endosperm,* which makes up a large portion of the seed. The endosperm stores food for the embryo plant.

In this investigation, you will examine some of the major structures of some common fruits and seeds.

Problem

What is the structure of some common fruits and seeds?

Materials *(per group)*

Scapel
Dissecting tray
Peach
Apple
Orange
Kidney bean seed
Corn seed
Single-edged razor blade
Hand lens

Safety 

Put on a laboratory apron if one is available. Handle all glassware carefully. Be careful when handling sharp instruments. Note all safety alert symbols next to the steps in the Procedure and review the meanings of each symbol by referring to the symbol guide on page 10.

Part A. Examining the Parts of the Pericarp

1. Using a scalpel, carefully cut the peach lengthwise. Do not cut into the pit. **CAUTION:** *Be careful when handling sharp instruments. Always cut in a direction away from your hands and body.*

2. Observe the ripened ovary wall, which is called the pericarp. The pericarp may be fleshy or dry, hard or soft. A pericarp is made up of three layers: the exocarp, or outer layer; the mesocarp, or middle layer; and the endocarp, or inner layer. Because the thickness and texture of these layers vary in different types of fruits, the variations are used to classify the different types of fruits. Examine Figure 1, which shows a stone fruit and its pericarp layers. A stone fruit is a type of fruit in which the endocarp is stony or hard. Find the parts labeled in Figure 1 in the peach. In Data Table 1, describe the three different layers of the pericarp.

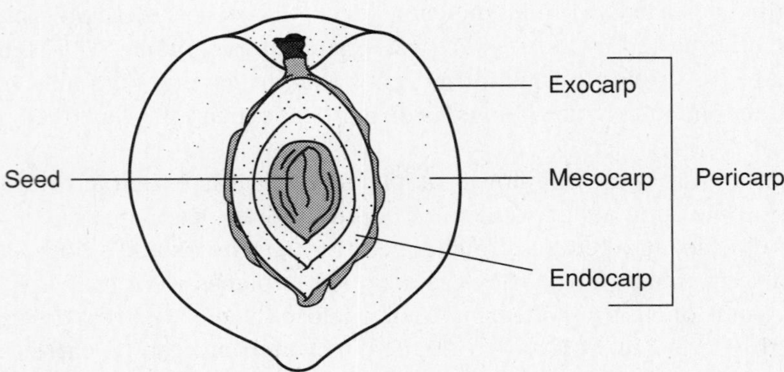

Figure 1

Part B. Examining Some Simple Fruits

1. Examine the peach that you cut open in Part A. Because the peach develops from a single pistil, it is known as a simple fruit. A simple fruit in which all or most of the pericarp is fleshy is called a simple fleshy fruit.

2. Notice that the pericarp of the peach has three distinctly different layers. A simple fleshy fruit with a pericarp divided into three distinct layers is called a drupe.

3. Place the orange in the dissecting tray. Using a scalpel, cut the orange in half so that you can see it in cross section. **CAUTION:** *Be careful when handling sharp instruments. Always cut in a direction away from your hands and body.* Look for small oil glands just below the surface of the skin, which together with the rind forms the pericarp. The large fleshy portion of the fruit, which contains the liquid, is called the juice sac. The orange, which has a pericarp that is entirely fleshy, is called a berry. In the appropriate place in Observations, sketch the cross section of the orange and label the pericarp, oil gland, juice sac, and seed.

4. Using a scalpel, cut through the center of the apple. The apple is an example of a pome. A pome is a simple fleshy fruit whose pericarp is surrounded by a fleshy, tasty floral tube. The floral tube is formed from the fused bases of the petals, sepals, and stamens.

5. Study Figure 2. In the appropriate place in Observations sketch the cross section of the apple and label the floral tube, exocarp, mesocarp, and endocarp.

Name ______________________ Class ____________ Date ____________

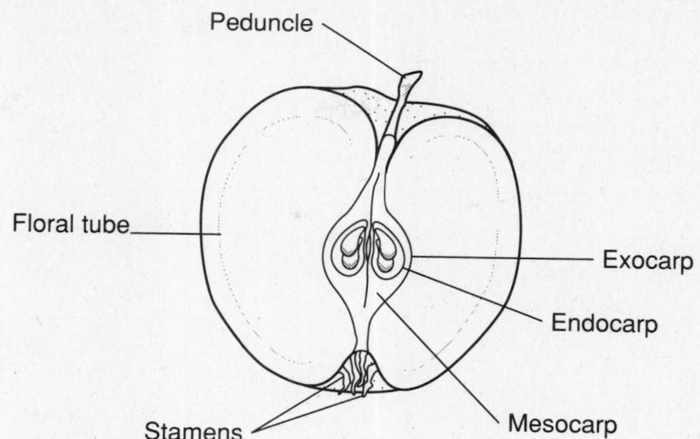

Figure 2

6. Complete Data Table 2 based on your observations.

Part C. Structure of Seeds

1. Locate the hilum, or oval scar, on the side of the kidney bean seed. The hilum is the point of attachment of the ovule. Look for a small dot called the micropyle directly above or below the hilum. The micropyle is the opening through which the pollen tube enters the ovule. Examine the seed coat, or thick outer covering. In the appropriate place in Observations, label the hilum, micropyle, and seed coat.
2. Carefully remove the seed coat from the bean seed with a single-edged razor blade. **CAUTION:** *Be careful when handling a sharp instrument. Always cut in a direction away from your hands and body.*
3. Separate the seed into two equal halves. Each of these halves is a cotyledon, or seed leaf, which stores food for the developing embryo plant.
4. Use a hand lens to observe the stemlike structure located near the edge of one of the cotyledons. The upper portion of this structure is the hypocotyl, which forms the stem of the plant. Observe the lower portion of the hypocotyl, which is called the radicle and forms the primary root. Locate the epicotyl, or small leaflike structure connected to the hypocotyl. The epicotyl forms the first leaves of the plant. In the appropriate place in Observations, label the cotyledons, hypocotyl, radicle, and epicotyl.
5. Examine the seed coat of the corn seed and identify the silk scar, which is a tiny pointed structure at the top of the kernel. In the appropriate place in Observations, label the seed coat and silk scar.
6. With the single-edged razor blade, carefully cut the corn seed in half from top to bottom. **CAUTION:** *Be careful when handling a sharp instrument. Always cut in a direction away from your hands and body.* The cotyledon is the whitish area in which the embryo is embedded. The epicotyl is the part of the embryo formed by the two leaves rolled into a spear. The hypocotyl is the small shaft of tissue below the epicotyl. The lower tip of the hypocotyl is the radicle. The endosperm occupies the remainder of the kernel outside of the cotyledon. In the appropriate place in Observations, label the hypocotyl, epicotyl, radicle, cotyledon, and endosperm.

Observations

Data Table 1

Pericarp Layer	Description
Exocarp	
Mesocarp	
Endocarp	

Orange Cross Section

Apple Cross Section

Data Table 2

Type of Fruit	Description	Examples
Berry		
Drupe		
Pome		

Name ______________________ Class __________ Date __________

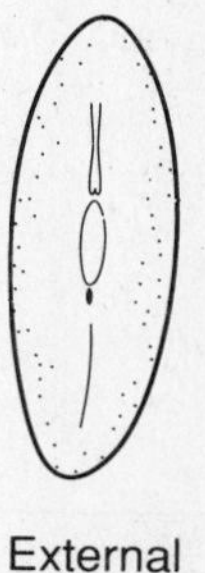

External

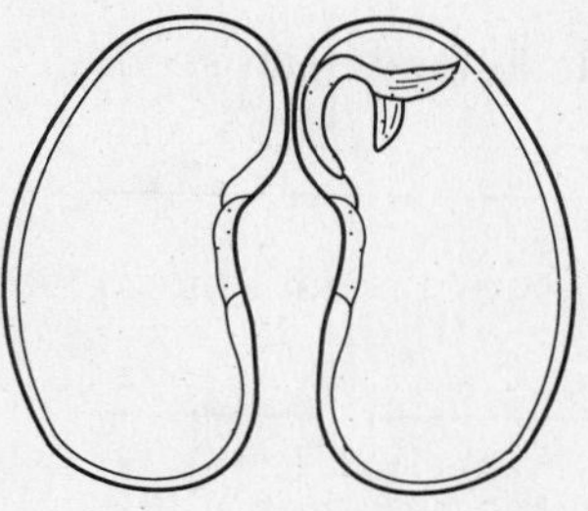

Internal

Kidney Bean Seed

Figure 4

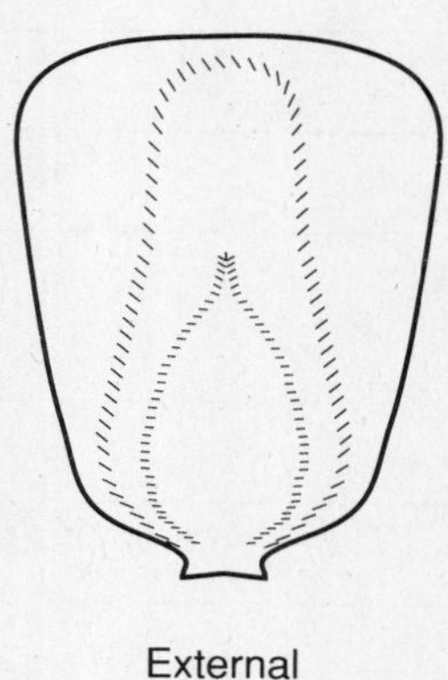

External

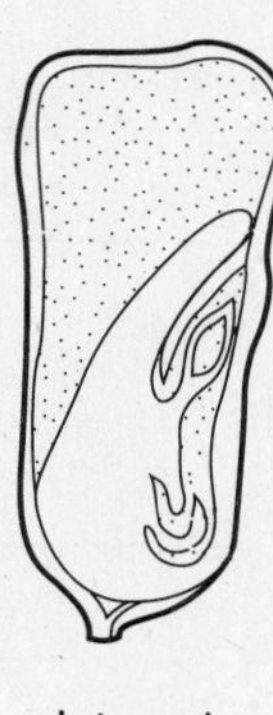

Internal

Corn Seed

Figure 5

Analysis and Conclusions

1. Why do flower structures dry up as fruits develop? ______________________

____

2. A fleshy fruit with a rind is called a hesperidium. Other than the orange name a common hesperidium. ______________________

3. From your observations, what is one difference between a drupe and a berry?

____

____

4. From your observations, what is one difference between a drupe and a pome?

5. a. Is the kidney bean seed a monocot or a dicot? Explain your answer.

   b. Is the corn seed a monocot or a dicot? Explain your answer.

6. What is the function of the cotyledons?

Critical Thinking and Application

1. People sometimes get into a dispute about whether a tomato is a fruit or a vegetable. As a botanist, how could you settle this argument?

2. How are seeds structurally adapted to life on land?

3. Why is seed dispersal essential to the survival of a plant species?

4. List two examples of fruits in which the seed is eaten and the fruit is discarded.

Going Further

1. Fruits that develop from several flowers whose ovaries ripen and develop as one mass are known as multiple fruits. Find out which fruits are multiple fruits. Cut some multiple fruits lengthwise so that you can see the internal organization of the fruit. Make labeled sketches of what you see.
2. Seeds have various structures that assure their dispersal to environments that are suitable for their germination and growth. Using reference materials, make a chart of the different dispersal methods. Include drawings or photographs showing the types of seed dispersal.

Name ______________________ Class __________ Date __________

51

Germination and Seedling Development

Pre-Lab Discussion

When conditions are suitable, a seed undergoes *germination,* or the development of an embryo into a seedling. For germination to occur, water, warmth, and oxygen must be available in the proper amounts. The amounts vary from species to species.

Germination can occur only in *viable seeds,* or seeds in which the embryo is alive. Not all viable seeds will germinate, even when given the proper amounts of water, warmth, and oxygen. Many seeds must go through a period of *dormancy,* or a period during which the embryo is alive but not growing. Dormancy is an adaptation that prevents germination of the seed until conditions are suitable. Once dormancy is complete and all the needed environmental conditions are met, germination begins.

In this investigation, you will observe some of the processes associated with seed germination and seedling development.

Problem

What changes occur in a seed during germination and seedling development?

Materials *(per group)*

10 *Brassica rapa* seeds
Petri dish
Forceps
Hand lens
Filter paper
Base of a 2-L soft-drink bottle
Metric ruler
Pencil
2 colored pencils
Fluorescent plant light (if available)

Safety

Put on a laboratory apron if one is available. Handle all glassware carefully. Observe proper laboratory procedures when using electrical appliances. Note all safety alert symbols next to the steps in the Procedure and review the meanings of each symbol by referring to the symbol guide on page 10.

Procedure

1. As shown in Figure 1, use a metric ruler and pencil to draw a line across the filter paper about 3 cm from the top edge. Label the bottom edge of the filter paper with the seed type, date, and name of one member of your group. **Note:** *Be sure to use pencil to label the filter paper because ink will smear when water is added.*

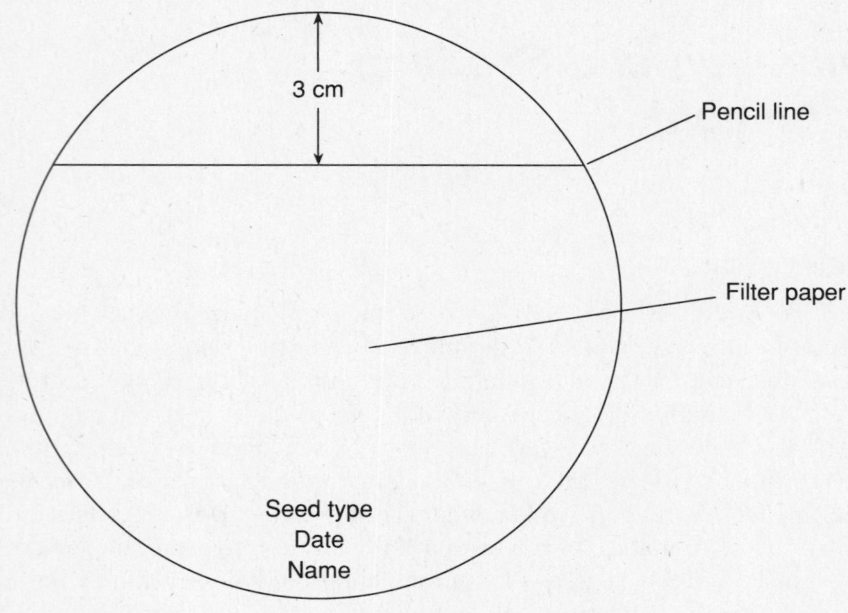

Figure 1

2. Place the filter paper in the top of a petri dish. Thoroughly wet the filter paper.
3. With a forceps place 10 *Brassica rapa* seeds on the line you drew on the filter paper. Space the seeds out evenly across the line as shown in Figure 2.

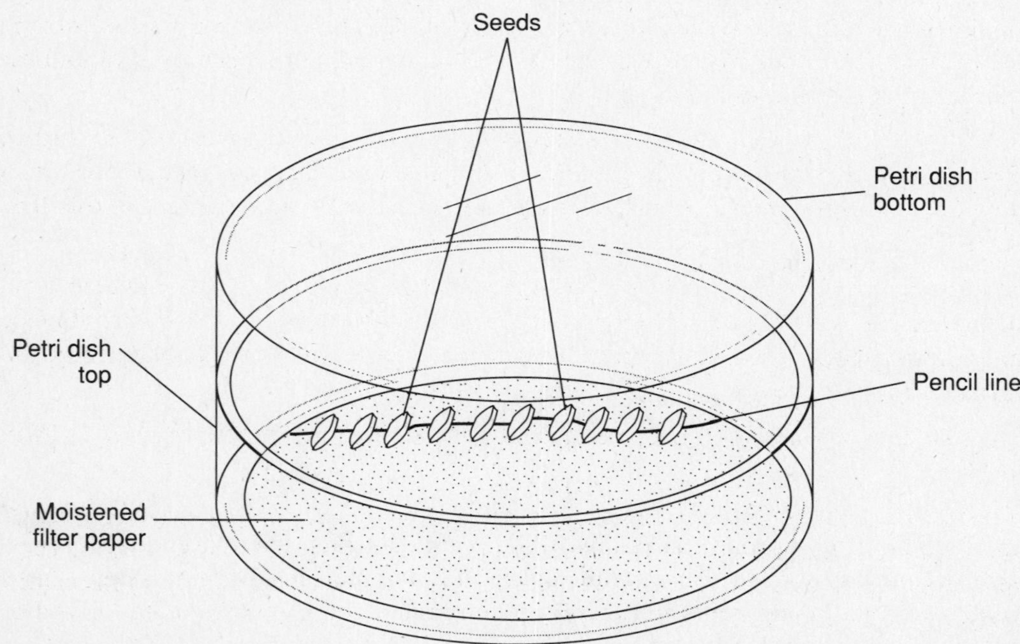

Figure 2

4. Cover the seeds with the bottom half of the petri dish.
5. Carefully place the petri dish in the base of the 2-liter soft-drink bottle so that the seeds are positioned at the top and the petri dish is angled slightly. See Figure 3. Make sure that none of the seeds have fallen from their original positions. Slowly add water to the soft-drink-bottle base from the side until the water reaches a depth of about 2 cm.

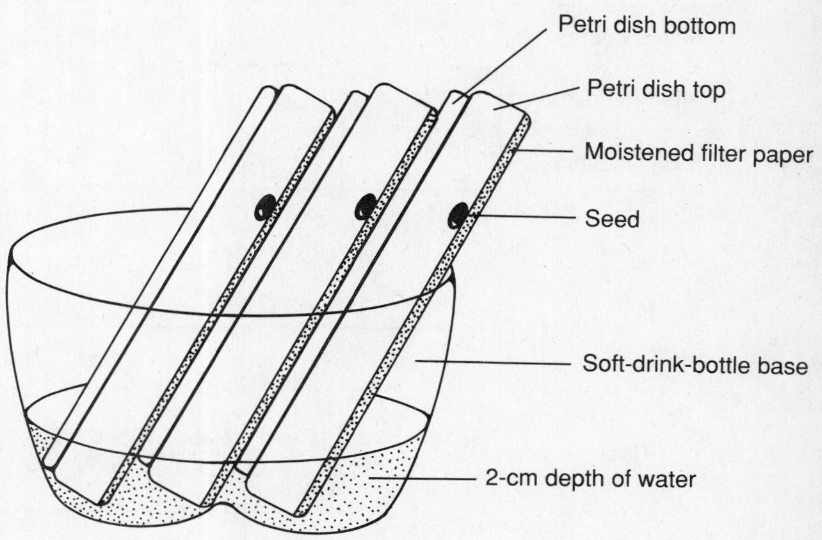

Figure 3

6. Place the soft-drink-bottle base under a fluorescent light. If a fluorescent light is not available, place the base near the best source of light in the room.
7. After 24 hours, observe the germination of the seeds. If necessary, use the hand lens to observe the germinating seeds. Note the number of seeds that have a split seed coat, an emerging radicle, hypocotyl, or epicotyl, or the appearance of a primary root and root hairs. Record the information in Data Table 1.
8. Measure the primary root length of each of the 10 seeds in millimeters and record this information in Data Table 2. If no primary root has emerged from a seed, record its length as 0 mm. Calculate the average root length for each of the 10 seeds and record this information in Data Table 2.
9. Measure the hypocotyl length of each of the 10 seeds in millimeters and record this information in Data Table 3. If no hypocotyl has emerged from a seed, record its length as 0 mm. Calculate the average hypocotyl length for each of the 10 seeds and record this information in Data Table 3.
10. Repeat steps 8 and 9 after two more 24-hour periods. Record the information in Data Tables 1, 2, and 3.
11. On the graph in Observations, construct a line graph showing the average root length over the 72-hour observation period. On the same graph, construct a line graph showing the average hypocotyl length over the 72-hour observation period. Use pencils of different colors to construct the two line graphs. Label each line graph.
12. In the appropriate place in Observations, sketch one of your 72-hour-old seedlings. Label the hypocotyl, primary root, root hairs, and cotyledons.

Observations

Total number of seeds in petri dish = ________

Data Table 1

Time	Number of Seeds					
	Split Seed Coat	Radicle	Primary Root	Root Hairs	Hypocotyl	Epicotyl
After 24 hours						
After 48 hours						
After 72 hours						

Data Table 2

Time	Root Length (mm)										
	Seed 1	Seed 2	Seed 3	Seed 4	Seed 5	Seed 6	Seed 7	Seed 8	Seed 9	Seed 10	Average
After 24 hours											
After 48 hours											
After 72 hours											

Data Table 3

Time	Hypocotyl Length (mm)										
	Seed 1	Seed 2	Seed 3	Seed 4	Seed 5	Seed 6	Seed 7	Seed 8	Seed 9	Seed 10	Average
After 24 hours											
After 48 hours											
After 72 hours											

Name ______________________ Class __________ Date __________

Graph

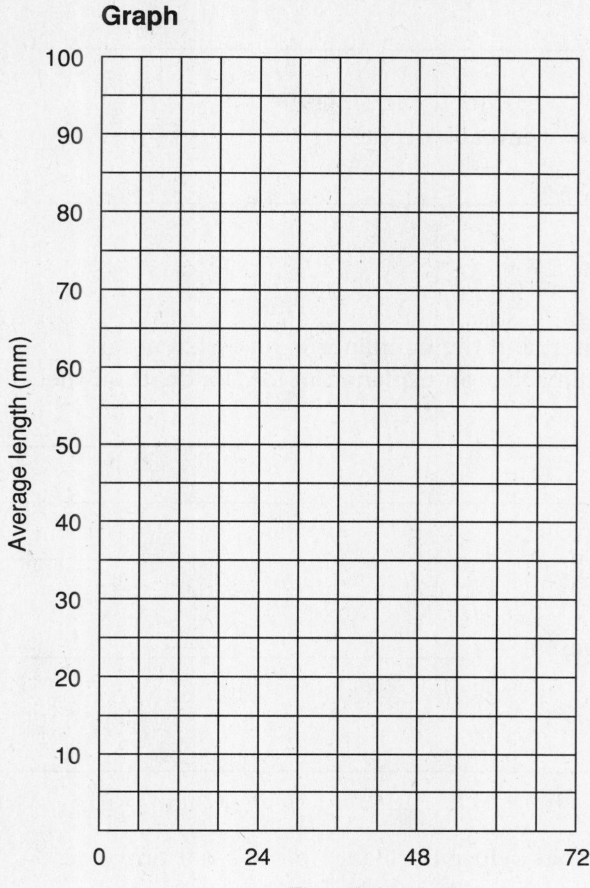

***Brassica* Seedling**

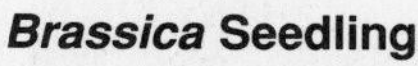

Analysis and Conclusions

1. Why might some of the *Brassica* seeds not germinate? ______________________

____

____

2. Why is it important for the dry seed to absorb water before it begins germination?

____

____

____

3. What is the first structure to emerge from inside the seed? What is the function of this structure? ______________________

____

____

4. What is the function of the cotyledons? ______________________________

5. In which part of the seedling will photosynthesis eventually occur?

Critical Thinking and Application

1. Several seeds are put in a moist, airtight container, and the container is placed in a dark closet. The seeds begin to germinate and then die. Offer an explanation for the death of the seeds. ______________________________

2. How is dormancy an adaptation for plant survival? ______________________________

3. If certain seeds require six weeks of prolonged cold before dormancy ends, what prevents them from germinating in the middle of January after a particularly cold November and December? ______________________________

Going Further

1. To observe the effect of light on seed germination and seedling development, prepare two petri dishes using the procedure described in this investigation. Completely cover one of the petri dishes with aluminum foil and allow the other petri dish to remain uncovered. Compare the rate of germination and the lengths of the primary roots and hypocotyls of the two sets of seeds. Construct data tables and line graphs to record your observations.
2. To observe the effect of temperature on seed germination and seedling development, prepare three petri dishes using the procedure described in this investigation. Keep one petri dish at room temperature, place another in an area that is warmer than room temperature, and put the remaining one in the refrigerator. The three dishes need to receive equal amounts of light; if it is not possible to provide light for the seeds in the warm and cold environments, cover all of the petri dishes so that none of the seeds receive any light. Compare the germination rates and lengths of the primary roots and hypocotyls of the three sets of seeds. Construct data tables and line graphs to record your observations. Are the seeds that were kept in the cold environment dead or merely dormant? Design and conduct an experiment to find out.

Name ______________________ Class ____________ Date ____________

52

Sponges and Hydras

Pre-Lab Discussion

Sponges and cnidarians are simple invertebrates. Sponges make up the phylum Porifera, which means "pore-bearing." The phylum is named for the many pores that cover the body of a sponge. Cnidarians are members of the phylum Cnidaria. Hydras, jellyfishes, and sea corals are types of cnidarians.

Sponges have no true tissues or organs, no digestive tract, and no nervous system. Their bodies are loosely organized into two cell layers. Support comes from hard structures called *spicules* or from flexible material called *spongin*. Some sponges have both spicules and spongin.

Cnidarians have true tissues. Their bodies consist of two layers—an outer epidermis and an inner gastrodermis. Each cnidarian also has a body cavity where food is digested. Cnidarians have a simple nervous system that allows them to respond to stimuli.

In this investigation, you will examine the characteristics of sponges and cnidarians.

Problem

What are some characteristics of a sponge and a hydra?

Materials *(per group)*

Preserved whole specimen of *Grantia*
Prepared slide of *Grantia,* longitudinal section
Hand lens
Microscope
Scalpel or single-edged razor blade
Glass slide
Coverslip
Chlorine bleach solution
Toothpick
2 medicine droppers
Prepared slide of hydra, whole mount
Prepared slide of hydra, longitudinal section

Safety

Put on a laboratory apron if one is available. Put on safety goggles. Handle all glassware carefully. Always use special caution when working with laboratory chemicals, as they may irritate the skin or cause staining of the skin or clothing. Never touch or taste any chemical unless instructed to do so. Always handle the microscope with extreme care. You are responsible for its proper care and use. Use caution when handling glass slides as they can break easily and cut you. Be careful when handling sharp instruments. Note all safety alert symbols next to the steps in the Procedure and review the meanings of each symbol by referring to the symbol guide on page 10.

Procedure

Part A. Examining the Anatomy of a Sponge

1. With a hand lens, examine the external structure of the simple marine sponge *Grantia.* Find the osculum, which is a large opening at the top through which water flows out of the sponge. Locate several ostia (singular, ostium), or pore cells. Water flows into the sponge through the ostia. Note the long, straight spicules that encircle the osculum and project through the outer surface of the sponge. In the appropriate place in Observations, sketch what you see. Label the osculum, ostia, and spicules.

2 Under the low-power objective of a microscope, examine a prepared slide of a longitudinal section of *Grantia.* Note the two cell layers. The outer cell layer is called the ectoderm and the inner cell layer is called the endoderm. Look for the flagellated cells called collar cells in the endoderm. Collar cells collect food particles from the water that passes through the sponge's body cavity, or spongocoel. Between the ectoderm and endoderm is mesenchyme, or a jellylike material containing some cells. Observe the spicules, hard structures made of calcium carbonate. In the appropriate place in Observations, sketch the sponge you have observed under low power. Label ectoderm, mesenchyme, collar cell, flagellum, spicules, and spongocoel. Record the magnification of the microscope.

3. Using a scalpel or a single-edged razor blade, cut a small piece from the *Grantia* specimen. **CAUTION:** *Be very careful when handling sharp instruments. Always cut in a direction away from your hands and body.*

4. Put on safety goggles. Place the piece of *Grantia* on a glass slide. With a medicine dropper, add two drops of chlorine bleach solution to the piece of sponge. **CAUTION:** *Be very careful when using chlorine bleach. It may burn your skin or clothing.* Using a toothpick, gently stir the sponge and chlorine bleach solution.

5. Using another medicine dropper, add a drop of water to the slide. Then cover with a coverslip. Observe the *Grantia* spicules under the low-power objective of the microscope. In the appropriate place in Observations, sketch several spicules. Record the magnification of the microscope.

Name ______________________ Class __________ Date __________

Part B. Examining the Anatomy of a Hydra

1. Under low power of the microscope, examine a prepared, whole mount slide of a hydra. Locate the basal disk at the posterior end of the body. The basal disk is the part with which the hydra attaches itself to surfaces. At the anterior end is the mouth. Look for several long tentacles. In the appropriate place in Observations, sketch the hydra under low power. Label the mouth, tentacle, body, and basal disk.

2. Under low power of the microscope, examine a prepared slide of a longitudinal section of a hydra. Locate the three layers: the outer epidermis, the inner gastrodermis, and the mesoglea, or thin layer of jellylike material between the epidermis and gastrodermis. Note the body cavity called the gastrovascular cavity.

3. Locate one of the hydra's tentacles. Notice the small bumps on the tentacle. They contain stinging structures called nematocysts.

4. Most species of hydra have separate male and female animals. However, some species contain both male and female reproductive structures. Look for the testes, or cone-shaped structures found on the upper half of the body, and the ovary, found on the lower half of the body. Examine the hydra to see if it has testes or an ovary (or both). In the appropriate place in Observations, sketch the hydra under low power. Label the mouth, tentacle, nematocyst, testis (if present), ovary (if present), bud (if present), epidermis, mesoglea, gastrodermis, and gastrovascular cavity.

Observations

Grantia

Magnification ______________

Grantia, **Longitudinal Section**

Magnification ______________

Grantia **Spicules**

Magnification ______________

Hydra, Whole Mount

Magnification ______________

Hydra, Longitudinal Section

Name ______________________ Class __________ Date __________

Analysis and Conclusions

1. How does a sponge get oxygen and food? ______________________

2. What is the function of the spicules of a sponge? ______________________

3. What was the purpose of the chlorine bleach solution in Part A of the investigation?

4. What body structures make the sponge well adapted for living in water?

5. What is the function of the gastrovascular cavity of a hydra? ______________________

6. Compare the sponge and the hydra in terms of body symmetry and tissue structure.

Critical Thinking and Application

1. How do sponges differ from flagellate protists? ______________________

2. Why do you think a *Grantia* sponge should not be used to wash cars or porcelain sinks?

3. How did the discovery and production of artificial sponges change the population of natural sponges? ____

____

____

4. "Cnid" is the Greek word for nettle, or stinging hair. Is the phylum name Cnidaria appropriate to the group of organisms such as hydra and jellyfishes? Explain your answer.

____

____

____

Going Further

1. Obtain and examine several different types of natural sponges, as well as a commonly used synthetic sponge. Note any similarities and differences between the sponges.
2. To examine the feeding habits of the hydra, place several drops of a hydra culture that has remained unfed for 24 hours in a watch glass. Add a drop of *Daphnia* culture to the watch glass. Place the watch glass under a dissecting microscope for observation. Describe what you see.

Name ______________________ Class ____________ Date ____________

53

Flatworms and Roundworms

Pre-Lab Discussion

Flatworms are classified in the phylum Platyhelminthes. The phylum name is derived from the Greek words *platy,* meaning flat, and *helminth,* meaning worm. The worms in this phylum have flat bodies and no body cavity. Most flatworms are parasites, or organisms that live on or in other living things. Some flatworms, however, are free-living organisms. Although most free-living flatworms live in the oceans, some live in fresh water or in soil.

Roundworms are classified in the phylum Nematoda. The worms in this phylum have long, slender bodies that taper to points at both ends. There are more species of roundworms than any other kind of worm. Roundworms are found in nearly every kind of environment, including deserts, arctic regions, hot springs, and oceans. Like flatworms, some roundworms are parasitic, whereas others are free-living.

In this investigation, you will observe some of the characteristics of parasitic and free-living flatworms and roundworms.

Problem

What are some characteristics of flatworms and roundworms?

Materials *(per group)*

Microscope
Prepared whole-mount slides of:
- Planarian *(Dugesia)*
- Pork tapeworm *(Taenia)*, with view of scolex and proglottids
- Vinegar eel *(Turbatrix aceti)*
- Trichina worm *(Trichinella),* encysted in a muscle

Safety

Handle all glassware carefully. Always handle the microscope with extreme care. You are responsible for its proper care and use. Use caution when handling glass slides as they can break easily and cut you. Note all safety alert symbols next to the steps in the Procedure and review the meanings of each symbol by referring to the symbol guide on page 10.

Procedure

Part A. Observing Flatworms

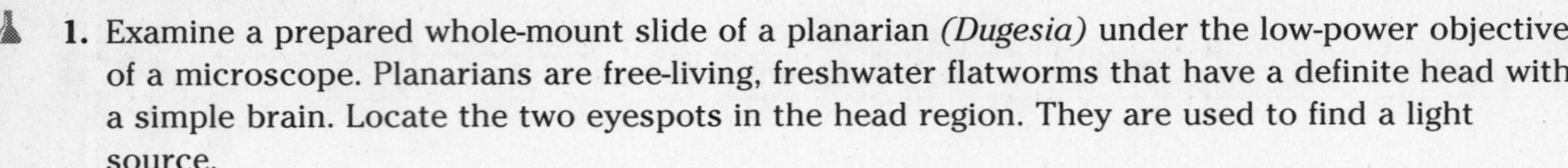

1. Examine a prepared whole-mount slide of a planarian *(Dugesia)* under the low-power objective of a microscope. Planarians are free-living, freshwater flatworms that have a definite head with a simple brain. Locate the two eyespots in the head region. They are used to find a light source.
2. Locate the mouth in the middle of the body on the ventral, or lower, surface. Note a long tube, called the pharynx, that draws food from the mouth into the gastrovascular cavity. In the appropriate place in Observations, sketch a planarian under low power. Label the eyespots, mouth, pharynx, and gastrovascular cavity. Record the magnification of the microscope.
3. Examine a prepared whole-mount slide of a tapeworm *(Taenia)* under low power. Locate the head, called the scolex. Note the presence of several suckers and a ring of hooks.
4. Behind the scolex is a narrow neck. The remainder of the tapeworm's body is made up of nearly square sections called proglottids, which grow from the neck. Thus the youngest proglottids are closest to the neck. Proglottids contain the male and female reproductive organs. In the appropriate place in Observations, sketch the tapeworm under low power. Label the scolex, suckers, hooks, young proglottids, and older proglottids. Record the magnification of the microscope.

Part B. Observing Roundworms

1. Examine the prepared whole-mount slide of a vinegar eel *(Turbatrix aceti)* under low power. This roundworm is named for the liquid in which it is normally found. Locate the mouth at its anterior, or front, end and the anus at its posterior, or tail, end. Note the bulblike pharynx and long intestine.
2. If the vinegar eel is a female, eggs will be found lined up in the uterus. If it is a male, it will contain a single testis. In the appropriate place in Observations, sketch the vinegar eel under low power. Label the mouth, pharynx, intestine, anus, eggs (if female), and testis (if male). Record the magnification of the microscope.
3. Examine a prepared whole-mount slide of a trichina worm *(Trichinella)* larva encysted in muscle tissue under low power. In the appropriate place in Observations, sketch what you see. Label the trichina worm larva, cyst, and muscle tissue. Record the magnification of the microscope.

Observations

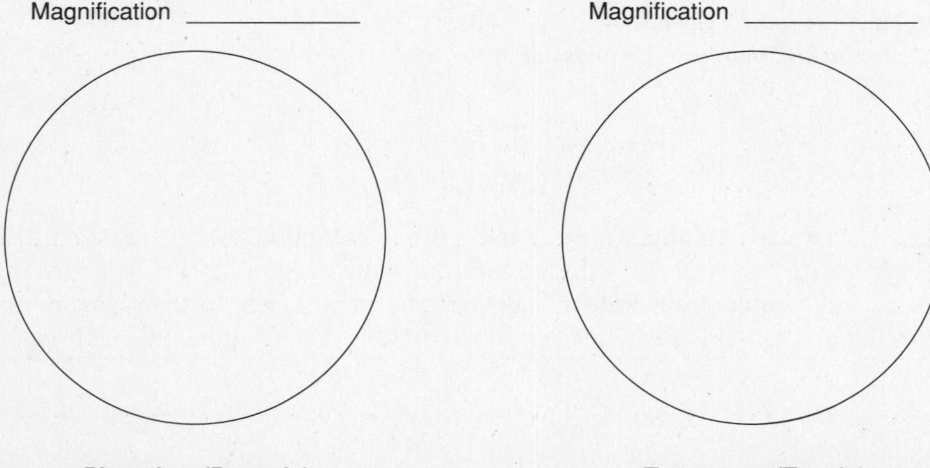

Planarian *(Dugesia)* | Tapeworm *(Taenia)*

Name ______________________ Class ____________ Date ____________

Magnification ____________

Vinegar eel ***(Turbatrix aceti)***

Magnification ____________

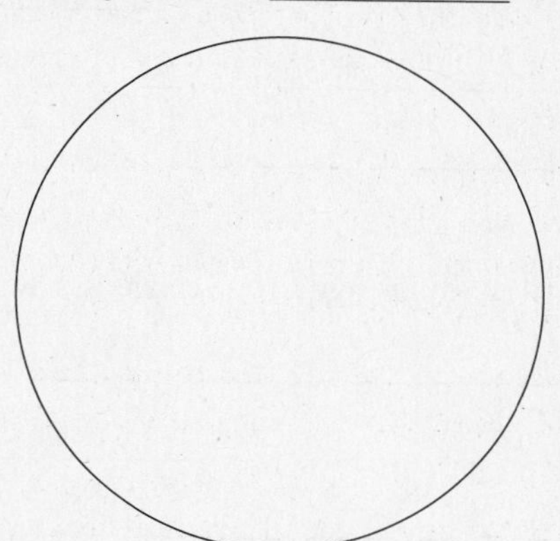

Trichinella **encysted in muscle tissue**

Analysis and Conclusions

1. How do flatworms and roundworms differ in body structure? ______________________

____

____

2. How are flatworms similar to roundworms in body structure? ______________________

____

____

3. Why are structures for locomotion lacking in parasitic flatworms and roundworms?

____

____

4. Would the nervous and digestive systems be more complex in the free-living forms or the parasitic forms of the flatworms and roundworms? Explain your answer.

____

____

____

____

Critical Thinking and Application

1. Why is the term "eyespot" instead of "eye" used when refering to the structure at the anterior end of the planarian body? ______________________________

2. List two advantages that parasitic worms have over free-living worms.

3. Why is it rare that an individual parasite will kill its host? ______________________________

4. You have taken a job as a product inspector in a large meat-packing company. As your first assignment, you must inspect pork for the presence of trichina worms. What should you look for? ______________________________

5. A person infected with a tapeworm will display symptoms such as weight loss, weakness, and a low blood count. How can a doctor be certain that a patient has a tapeworm and not some other illness with similar symptoms? ______________________________

Going Further

1. To collect planarians, tie a piece of string around a piece of liver and lower it into a pond or lake. After a few hours, raise the string. Why does the liver attract planarians?
2. Fill a clear glass petri dish with pond water. Cover half of the petri dish with cardboard, so that it will be in darkness. Place planarians in the exposed side of the petri dish. After an hour, determine in which side of the petri dish the planarians are found. Do your observations show planarians to have a positive or a negative response to light?

Name ______________________ Class ____________ Date ____________

Examining a Clam

Pre-Lab Discussion

The phylum Mollusca has many species, including clams, snails, slugs, octopi, and squids. Mollusks are soft-bodied invertebrates that have a muscular foot and a mantle. In most mollusks, the mantle secretes a hard shell. The many different types of mollusks may swim freely, float, or burrow into the sand or mud. Although most mollusks are marine, or sea-living, some mollusks live in fresh water, and others live on land. For thousands of years, people have eaten many types of mollusks, such as clams, scallops, oysters, and squids. Other kinds of mollusks have been used to make buttons, ornaments, tools, money, and dyes.

One representative mollusk is the clam. Clams are pelecypods, or bivalves, and have a two-part hinged shell. They are found in fresh water in streams, ponds, and lakes. They are also commonly found burrowed into the mud of ocean mud flats.

In this investigation, you will observe the behavior of a clam. You will also examine the internal and external structures of a clam.

Problem

What are some structures of a clam?

Materials *(per group)*

Live freshwater clam
Fresh or preserved clam
One valve of a clam shell
Large beaker containing water and sand
Carmine solution in dropper bottle
Glass stirring rod
10% hydrochloric acid in dropper bottle
Dissecting tray
Scalpel
Paper towels

Safety

Put on a laboratory apron if one is available. Put on safety goggles. Handle all glassware carefully. Always use special caution when working with laboratory chemicals, as they may irritate the skin or cause staining of the skin or clothing. Never touch or taste any chemical unless instructed to do so. Be careful when handling sharp instruments. Follow your teacher's directions and all appropriate safety procedures when handling live animals. Note all safety alert symbols next to the steps in the Procedure and review the meanings of each symbol by referring to the symbol guide on page 10.

Procedure

Part A. Observing a Live Clam

1. Place a live clam on its side in a large beaker of water (or an aquarium) with at least 4 cm of sand on the bottom. Observe what happens. Record your observations in the Data Table. Continue to observe the clam for several minutes without disturbing it. Answer question 1 in Observations.

2. Locate the incurrent and excurrent siphons on the live clam. These are the areas where water enters and exits the body of the clam. Obtain a dropper bottle of carmine solution. Place one or two drops of carmine solution near the siphons. **CAUTION:** *Handle the carmine solution with care because it stains the skin and clothing.* Observe what happens to the carmine solution. Record your observations in the Data Table.

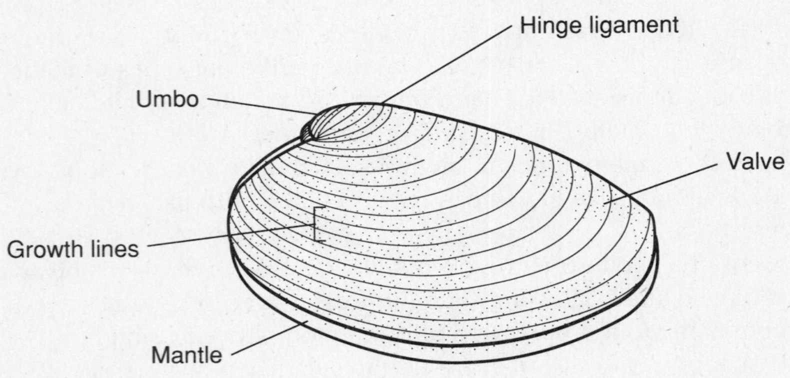

Figure 1

3. Locate the mantle on the live clam. See Figure 1. With a glass stirring rod, carefully and gently touch the clam on different parts of the mantle. Record your observations in the Data Table.

4. Return the live clam in the beaker to your teacher when you have completed your observations.

Part B. External Anatomy of the Clam

1. Thoroughly rinse a preserved clam to remove excess preservative. **CAUTION:** *The preservative used on the clam can irritate your skin. Avoid touching your eyes while working with the preserved clam.* Place the clam in a dissecting tray. Observe the bivalve shell. Notice the hinge ligament. Answer question 2 in Observations.

2. Locate the small, pointed area near the hinge ligament. This is the umbo, which is the oldest part of the clam. Note that the umbo is on the dorsal side toward the anterior end of the clam. Notice the concentric growth lines. They represent alternating periods of slow and rapid growth. Answer question 3 in Observations. Locate the posterior, anterior, dorsal, and ventral surfaces of the clam shell. Hold the clam shell with the anterior end up and the hinge facing toward you. Locate the hinge, right valve, and left valve of the clam shell. In the appropriate place in Observations, label the surfaces of the clam shell as they appear. Also label the hinge and right and left valves.

3. Hold the clam in the dissecting tray as shown in A of Figure 2. With a scalpel, carefully expose the middle layer by scraping away some of the horny outer layer of the shell. **CAUTION:** *Be careful when handling sharp instruments. Scrape in a direction away from your hand to avoid cutting yourself.*

Name ______________________ Class ____________ Date ____________

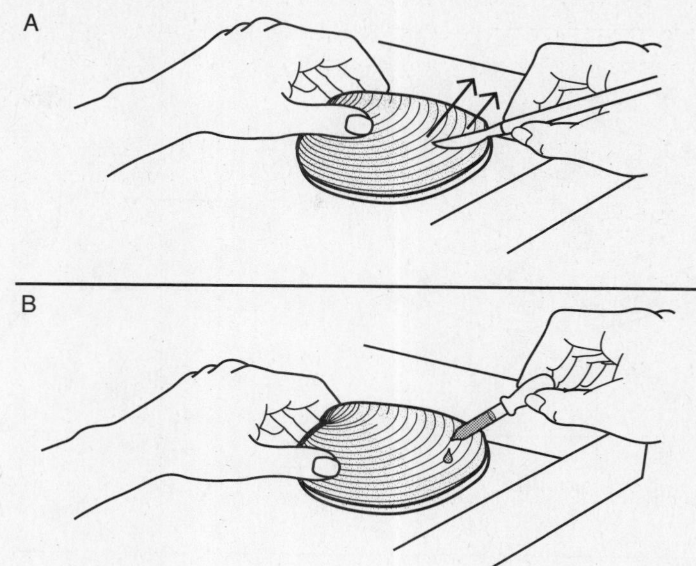

Figure 2

4. The shell of a clam is made up of three layers: the horny outer layer; the thick, middle layer called the prismatic layer; and the innermost layer called the pearly layer. Place one drop of hydrocholoric acid on the exposed prismatic layer, as shown in B of Figure 2. **CAUTION:** *Be careful when handling acids as they can burn the skin.* The bubbling of the acid indicates that calcium carbonate is present in the shell. Answer question 4 in Observations.
5. Follow your teacher's directions for storing the clam for further use or properly disposing of the clam.
6. Observe a valve of an opened clam shell. Note the inner surface of the valve and the pearly layer. Answer question 5 in Observations. Locate the "scars" left by the anterior and posterior adductor muscles on the inner surface of the valve. In the appropriate place in Observations, label the anterior adductor muscle, posterior adductor muscle, mantle, and umbo.

Observations

Data Table

	Observations
Clam placed on side	
Movement of carmine solution	
Touching mantle with glass rod	

Clam Shell

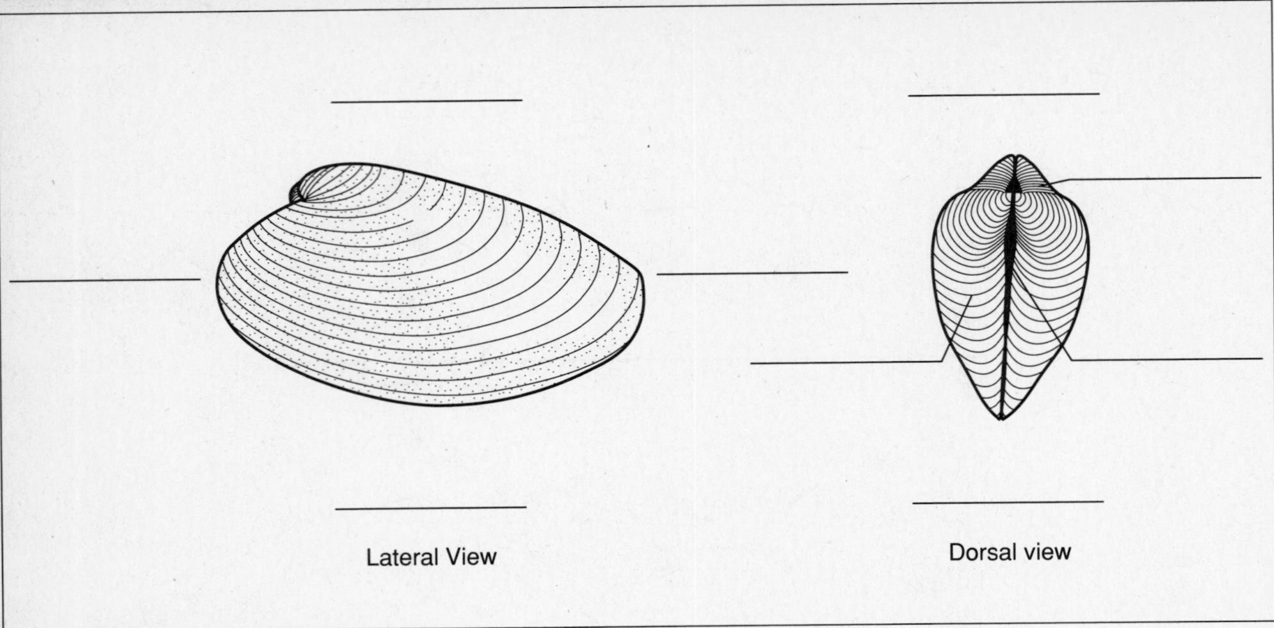

Clam Shell Valve

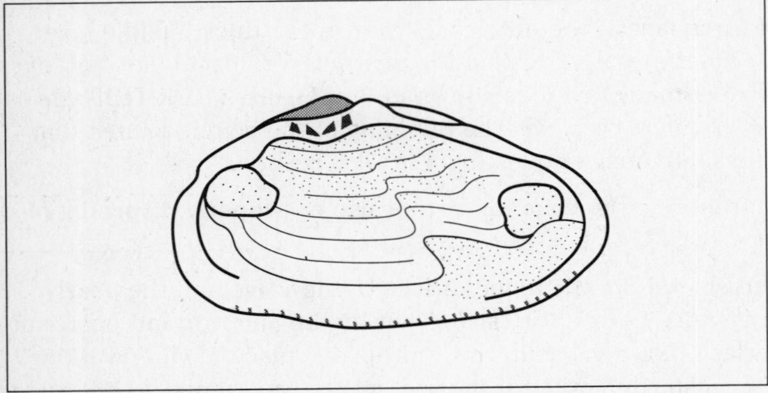

1. Describe how a clam moves and buries itself in the sand. ______________________

____

____

2. Describe the location of the hinge ligament in relation to the two halves of the shell.

____

____

3. How many growth lines did you count on your clam shell? ______________________

Name ______________________ Class ____________ Date ____________

4. Describe what happened when acid was placed on the exposed prismatic layer of the clam shell. ______________________

____

5. Describe the inner layer of the clam shell. ______________________

____

Analysis and Conclusions

1. Why is it important for a clam to take in water through the incurrent siphon?

____

____

2. Why is it important for a clam to expel water through the excurrent siphon?

____

____

3. What is the function of the hinge ligament? ______________________

____

4. Why would the number of growth lines vary among clams? ______________________

____

5. Of what is the prismatic layer of a clam's shell composed? ______________________

____

6. In addition to the valve you observed, where else would adductor muscles be found?

____

____

Critical Thinking and Application

1. The nervous system of a clam is very simple. How can a clam function with such a simple nervous system? ______

2. List one advantage and one disadvantage of the shell surrounding a clam.

3. Why is it important for the hinge ligament and the adductor muscles of the clam to be very strong? ______

4. Clams reproduce by releasing eggs and sperm into the water. Would you expect clams to release large or small numbers of eggs and sperm? Explain your answer.

5. Why are fossils of mollusks more abundant than those of the worm phyla?

Going Further

1. Examine the gills from a preserved clam under a dissecting microscope. Remove some of the gill tissue and prepare a wet mount of it for examination under the microscope. Sketch what you see.
2. Shell collecting is an interesting hobby. Many books on shell collecting and identification are available in bookstores and libraries. If you live near the ocean, or if you vacation at the seashore, you may wish to start a shell collection.

Name ______________________ Class ____________ Date ____________

55

The Earthworm

Pre-Lab Discussion

The earthworm, *Limbricus terrestris*, is a member of the phylum Annelida. An earthworm is quite well-adapted to a life of burrowing through the soil. Its streamlined shape helps it move through the soil. A coating of mucus secreted by the skin lubricates the earthworm as it passes through the soil. The mucus coating also helps oxygen pass through the earthworm's skin while it is in the air or under the water. The earthworm moves through the soil by sucking the soil in its path into its mouth with the aid of its muscular pharynx. As material passes through the tubelike digestive system, sand grains in the gizzard help grind the food, which is then digested and absorbed in the intestine.

In this investigation, you will examine the external and internal structure of the earthworm.

Problem

How is the earthworm's body adapted to its environment?

Materials *(per group)*

Preserved earthworm
Dissecting tray
Scalpel
Scissors
Medicine dropper
Probe
Dissecting pins
Forceps
Hand lens
Resealable plastic bag
Paper towels

Safety

Put on a laboratory apron if one is available. Handle all glassware carefully. Be careful when handling sharp instruments. Note all safety alert symbols next to the steps in the Procedure and review the meanings of each symbol by referring to the symbol guide on page 10.

Procedure

Part A. The Technique of Animal Dissection

1. Dissection is the technique of exposing the internal structures of an organism for observation. Dissection is commonly used in the study of large and complex plants and animals. The opportunity to dissect an animal should be thought of as a unique opportunity to gain firsthand knowledge of an animal you know little about. As you dissect an animal, you should think in terms of structure related to function.

2. Obtain the following tools and instruments needed for dissection: dissecting tray, scalpel, scissors, probe, dissecting pins, and forceps. **CAUTION:** *The scalpel, scissors, and probe are sharp. Use extreme caution when handling these instruments to avoid cuts. Always cut in a direction away from your hands and body.* Become familiar with these tools and instruments used in dissection.

3. Most cutting in dissection is done with scissors rather than a scalpel. Most of the actual dissection involves the forceps, probe, and fingers. These instruments are used to tear, separate, and move or lift parts instead of cutting them.

4. Read the following rules for dissection.
   - Before beginning a dissection, identify all external parts.
   - Determine the proper order in which internal structures are to be exposed.
   - Identify which structures could be easily damaged if dissection is not done properly.
   - Do not completely remove any body part unless instructed to do so. If a body part is to be removed, leave a small portion of it attached as a reference point.
   - When making the first cut, insert the point of the scissors just below the skin. Cut with short, clipping motions. Keep the lower blade of the scissors pointing upward, away from the internal structure of the animal being dissected.

Part B. External Anatomy of the Earthworm

1. Rinse a preserved earthworm thoroughly with water to remove excess preservative. Place the earthworm in a dissecting tray lined with moist paper towels. Turn the earthworm over and observe the difference between the darker-colored dorsal side and the lighter-colored ventral side of the cuticle (outer layer). With your fingers, feel the shape of the dorsal and ventral surfaces. **CAUTION:** *The preservative used on the earthworm can irritate your skin. Avoid touching your eyes while working with the preserved earthworm.* Answer question 1 in Observations.

2. Locate the slightly pointed anterior end and the blunt posterior end of the earthworm's body. Notice the somites, or segments, of the body. Count the total number of segments in the earthworm. The segments of an earthworm are counted from anterior to posterior. In the anterior portion of the earthworm identify the clitellum, the light brown enlarged band around the body. The clitellum functions in reproduction by secreting the cocoon into which eggs are deposited. Answer question 2 in Observations.

3. Locate the mouth on the ventral side of the anterior end of the earthworm. The flap of skin overhanging the mouth is the prostomium. **Note:** *The prostomium is not counted as the first segment in the earthworm's body.* Locate the anus, the opening in the last segment of the posterior end. The mouth and anus are the two openings of the tubelike digestive system.

4. Slide your fingers along the ventral surface of the earthworm from anterior to posterior and feel the bristlelike setae. The setae anchor the earthworm to the ground and function in movement. Then slide your fingers along the dorsal surface of the worm from anterior to posterior. With a hand lens, observe the location and number of setae on each segment of the earthworm's body. Answer question 3 in Observations.

Name ______________________ Class __________ Date __________

5. Use a hand lens to find the openings of the seminal receptacles, which are located on the ventral surface in the grooves between segments 9 and 10 and segments 10 and 11. The seminal receptacles receive sperm from another earthworm during reproduction. On the sides of segment 14, locate the openings of the oviducts. These openings are pores through which eggs are released. On the side of segment 15, locate the openings of the sperm ducts. These openings are pores through which sperm are released and are surrounded by swollen "lips." You may be able to observe slight ridges that run posteriorly from the sperm duct openings to the clitellum. These ridges direct sperm into the clitellum.

6. In the appropriate place in Observations, label the following structures of the earthworm's external anatomy: anterior end, posterior end, mouth, prostomium, setae, segments, seminal receptacles, oviduct openings, sperm duct openings, clitellum, and anus. Give the number(s) of the segment(s) in which each structure appears.

Part C. Internal Anatomy of the Earthworm

1. Place the preserved earthworm in the dissecting tray with the dorsal surface up. Slightly stretch out the earthworm's body. With dissecting pins, pin the first and last segments of the earthworm to the bottom of the dissecting tray. Locate the dorsal blood vessel, a dark line that runs along the midline of the dorsal surface. The vessel runs from the anterior end to the posterior end. Insert the tip of a pair of scissors just to the right of the dorsal blood vessel and about 10 segments from the posterior end. Cut along the blood vessel to the anterior end as shown in Figure 1. Keep the scissors parallel to the bottom of the dissecting tray. **CAUTION:** *When using scissors, cut in a direction away from your hand and body to avoid cutting yourself.* Keep the cut as shallow as possible. The body wall of the earthworm is very thin, and the internal organs lie just inside. **Note:** *The major internal organs of the earthworm are anterior to the clitellum. Be very careful when cutting in this area.*

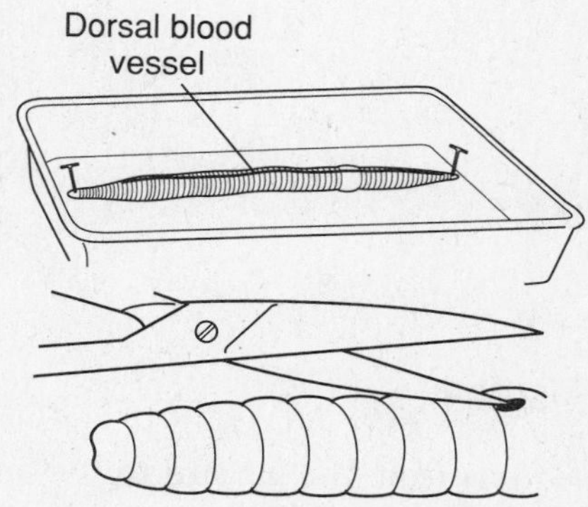

Figure 1

2. When the cut is complete, carefully open the body wall. Notice that the coelom is not one large cavity but is separated into segments by septa (singular, septum), or thin partitions of tissue. The septa are continuations of the external segments. Beginning at the anterior end, use a probe or dissecting pin to carefully tear the septa from the anterior to the posterior end. Place dissecting pins in segments 5, 15, 25, 35, and 45 to hold open the body wall. Place the pins at an angle as shown in Figure 2. Placing pins in every tenth segment will help you locate the major internal organs that are located in specific segments. Use a medicine dropper to place several drops of water on the exposed organs of the earthworm. **Note:** *Throughout the dissection, add water to the earthworm to keep it from drying out.*

3. Trace the digestive system of the earthworm from the mouth to the anus. The mouth is located in the first three segments. Locate a slight swelling, the muscular-walled pharynx, posterior to the mouth in segments 3 to 6. The slender esophagus, located in segments 6 to 14, empties into the thin-walled crop, located in segments 15 and 16. The crop temporarily stores food. Examine the gizzard, a grinding organ located in segments 17 and 18. The gizzard mixes food with sand from the soil and physically breaks the food into smaller pieces. Use a probe to feel the difference in the walls of the crop and gizzard. Locate the intestine, or straight tube

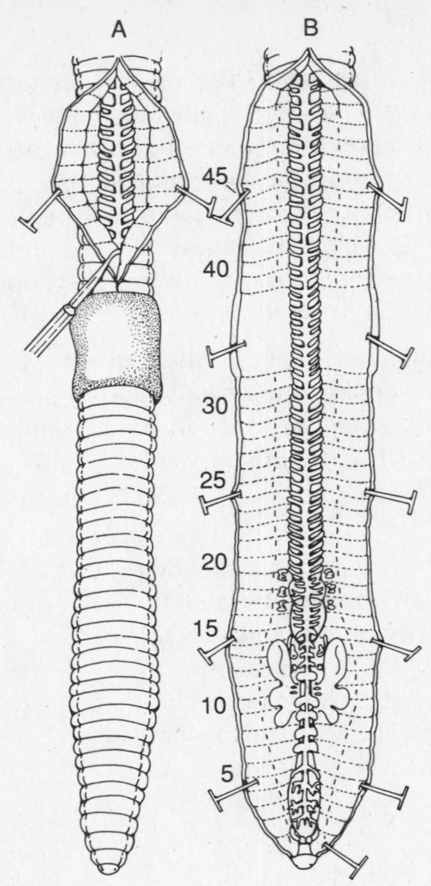

Figure 2

leading from the gizzard to the anus. Near the posterior end of the earthworm, cut out a 1-cm section of the intestine and observe it with a hand lens. Locate the dorsal fold of tissue in the inside of the intestine. This tissue fold is the typhlosole. Answer question 4 in Observations.

4. In the appropriate place in Observations, label the following parts of the earthworm's digestive system: mouth, pharynx, esophagus, crop, gizzard, and intestine.
5. Trace the earthworm's closed circulatory system by following the dorsal blood vessel anteriorly along the midline to the five aortic arches, located in segments 7 to 11. In the posterior section use a probe to carefully move aside the intestine. Locate the ventral blood vessel. The aortic arches connect the dorsal blood vessel with the ventral blood vessel.
6. In the appropriate place in Observations, label the following parts of the earthworm's circulatory system: aortic arches, dorsal blood vessel, and ventral blood vessel.
7. In the appropriate place in Observations, sketch the plan of the earthworm's circulatory system as seen from the side. Label the aortic arches, dorsal blood vessel, and ventral blood vessel. Include arrows that indicate the flow of blood in each vessel.
8. Locate the brain, which is a white mass of tissue found in the third segment anterior and dorsal to the pharynx. With a probe, gently push aside the pharynx and esophagus to expose the white ventral nerve cord. The ventral nerve cord runs along the inner ventral surface from segment 3 to the last segment. With a hand lens, you may be able to see the two branched nerves that run from the brain around the esophagus to the ventral nerve cord. In each segment, locate a mass of nerve cells, or ganglia, attached to the ventral nerve cord.
9. In the appropriate place in Observations, label the following parts of the earthworm's nervous system: brain, ventral nerve cord, and ganglion.

Name ______________________________ Class ______________ Date ______________

10. With a probe, push aside the intestine in a region just posterior to the clitellum. With a hand lens, look for the white, coiled nephridia (singular, nephridium). A pair of nephridia occur in every segment except the first three segments and the last segment. Each nephridium collects wastes from the coelom and carries them to the outside of the body through pores, or openings, in the body wall of each segment.

11. In the appropriate place in Observations, label the nephridia, or structures of the earthworm's excretory system.

12. An earthworm has both male and female reproductive organs and can produce both eggs and sperm. Use a hand lens during the remainder of the dissection. First locate the male reproductive organs. Find the two seminal vesicles in segments 9 to 13. The seminal vesicles are three-lobed, light-colored organs along the sides of the esophagus. Two small pairs of testes are located in segments 10 and 11. The female reproductive organs include a pair of small, spherical seminal receptacles, located in segments 9 and 10, and a pair of ovaries, located in the septum between segments 12 and 13. With the exception of the seminal vesicles, most of the reproductive organs are small and difficult to see in preserved specimens.

13. In the appropriate place in Observations, label the following parts of the earthworm's reproductive system: seminal vesicles, testes, seminal receptacles, and ovary.

14. Follow your teacher's instructions for storing the earthworm for further use or properly disposing of the earthworm and its parts. Thoroughly wash and dry your dissecting tray, scissors, pins, probe, and any other equipment you may have used. Wash your hands with soap and water.

Observations

External Anatomy of the Earthworm

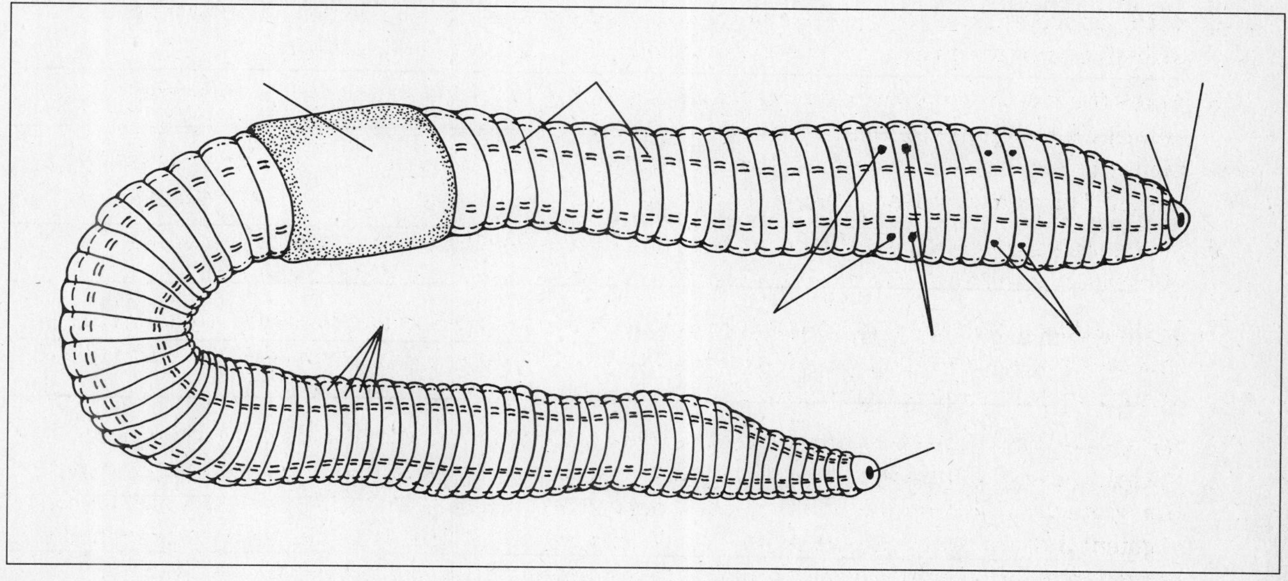

Circulatory System of the Earthworm

Internal Anatomy of the Earthworm

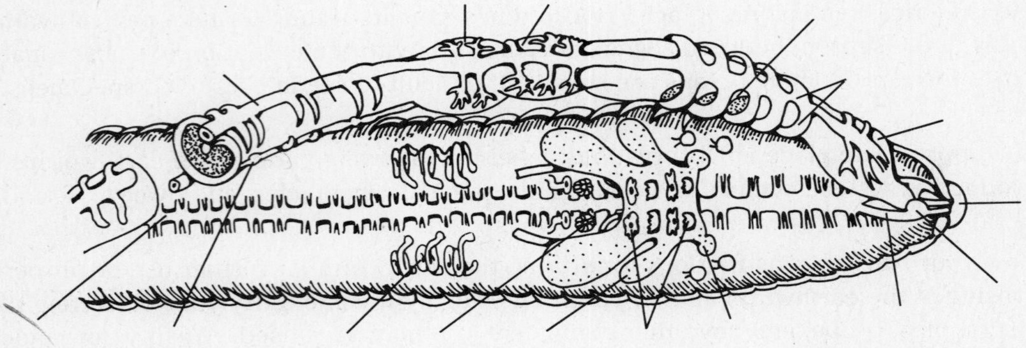

1. Describe the shape of the dorsal and ventral surfaces of the earthworm's body.

____

____

2. a. How many segments are in your earthworm? _______________________

____

b. In which segments is the clitellum located? _______________________

____

3. a. Where are setae located on the earthworm? _______________________

____

b. How many setae are on each segment? _______________________

____

Name ______________________ Class ____________ Date ____________

4. Compare the walls of the crop and the gizzard. ______________________

____

Analysis and Conclusions

1. Why is it important not to make a deep cut with the scissors when dissecting your earthworm specimen? ______________________

____

____

2. What do you think is the function of the typhlosole? ______________________

____

3. How does the earthworm's digestive system adapt it to filtering food out of the soil?

____

____

____

4. Describe two ways in which an earthworm's body is adapted to life in the soil.

____

____

____

Critical Thinking and Application

1. Imagine an assembly line for disassembling things instead of putting them together. Compare the earthworm's tubelike digestive system to such a "disassembly line." What is the advantage of a tubelike digestive system over a saclike digestive system?

____

____

____

____

____

____

2. Describe one way in which an earthworm is poorly adapted to life on land.

3. How might an earthworm's lack of appendages be an adaptation to burrowing?

4. Explain how an earthworm enriches and aerates the soil, thus improving it for plant growth.

Going Further

1. With a probe, carefully tear open each organ of the digestive system and observe its contents.
2. Observe a prepared transverse section of an earthworm through a dissecting microscope. Make a sketch of the section and label the following parts: dorsal blood vessel, ventral blood vessel, ventral nerve cord, intestine, typhlosole, nephridia, setae, circular muscle fibers, longitudinal muscle fibers, and cuticle.
3. Set up a terrarium with live earthworms and observe their burrowing, feeding, and mating behavior.
4. Obtain earthworm cocoons from a biological supply house or collect them outdoors. Place the cocoons on moist paper towels in a covered petri dish. Watch for 12 to 20 earthworms to hatch in 1 to 30 days. Describe what you observe.

Name ______________________ Class __________ Date __________

The Grasshopper

Pre-Lab Discussion

Grasshoppers are members of the phylum Arthropoda and the class Insecta. The class Insecta contains more than 900,000 species, which is about five times as many species as in all the other animal groups combined. Insects are mainly land animals and occupy almost every environmental habitat on land. The development of wings gave insects such distinct advantages over other land invertebrates as greater dispersal, access to more food supplies, and more efficient methods of escape from predators.

Insects have an exoskeleton composed of chitin. They have three pairs of jointed legs and three distinct body regions. These regions are the head, thorax, and abdomen. Usually two pairs of wings are attached to the thorax. As members of the class Insecta, grasshoppers have one pair of antennae and one pair of large compound eyes. They also have a tracheal tube-spiracle system for gas exchange. The efficient respiratory system of insects provides oxygen for the rapid movement of muscles. The high reproductive capacity and relatively short reproductive cycle of grasshoppers enables them to increase their populations at a rapid rate.

In this investigation, you will observe the behavior and movement of a live grasshopper. You also will examine the external features of the grasshopper and identify parts of its anatomy.

Problem

What are the parts of a grasshopper?

Materials *(per group)*

Live grasshopper in glass jar
Preserved grasshopper
Paper towels
Dissecting tray
Probe
Scalpel
Hand lens
Lettuce
Dissecting microscope
Glass slide
Medicine dropper

Safety

Put on a laboratory apron if one is available. Handle all glassware carefully. Always handle the microscope with extreme care. You are responsible for its proper care and use. Use caution when handling glass slides as they can break easily and cut you. Be careful when handling sharp instruments. Always use special caution when working with laboratory chemicals, as they may irritate the skin or cause staining of the skin or clothing. Never touch or taste any chemical unless instructed to do so. Follow your teacher's directions and all appropriate safety procedures when handling live animals. Note all safety alert symbols next to the steps in the Procedure and review the meanings of each symbol by referring to the symbol guide on page 10.

Procedure

Part A. Observing the Reactions of a Live Grasshopper

1. Referring to Figure 1, locate the head, thorax, abdomen, three pairs of legs, and two pairs of wings of a live grasshopper in a glass jar. Observe the grasshopper for several minutes without disturbing it. Notice how the grasshopper moves and note which legs it uses when walking. Gently tap the side of the jar to make the grasshopper jump. Observe which legs are used in jumping. Answer questions 1 and 2 in Observations.

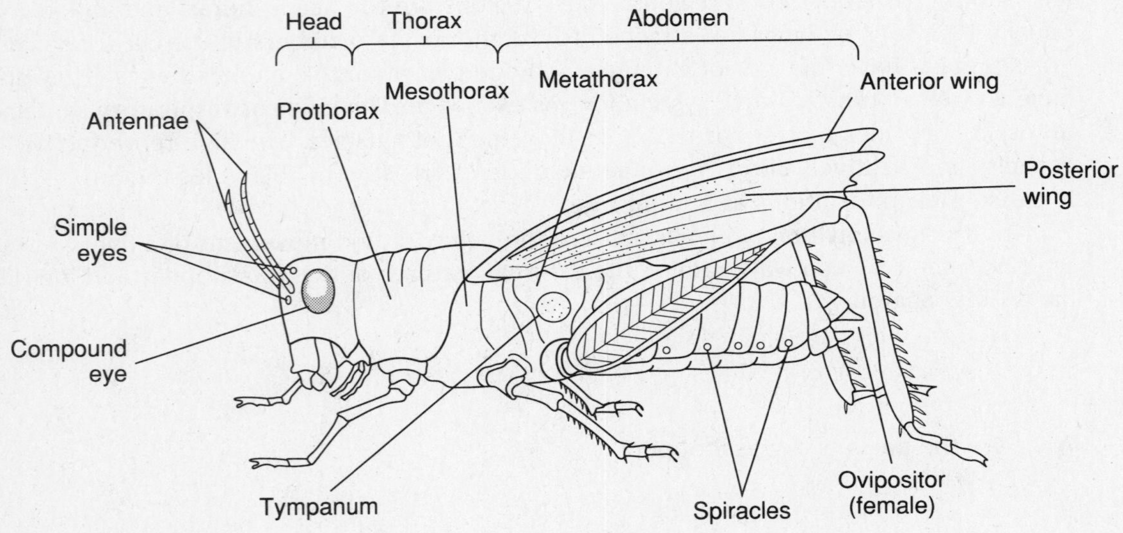

Figure 1

2. Observe the position and movements of the grasshopper when it is at rest. Use the hand lens to observe the movements of the abdomen that are associated with respiration. Answer question 3 in Observations.
3. Tear off a small piece of lettuce and offer it to the grasshopper. Notice how the animal eats. Closely observe the movement of the mouthparts. Answer question 4 in Observations.
4. Return the grasshopper to your teacher when you have completed your observations.

Part B. Observing the External Anatomy of the Grasshopper

1. Rinse the preserved grasshopper with water to remove excess preservative. Place the grasshopper in a dissecting tray. Touch the chitinous exoskeleton and apply gentle pressure. **CAUTION:** *The preservative used on the grasshopper can irritate your skin. Avoid touching your eyes while working with the preserved grasshopper.* Answer question 5 in Observations.

2. Locate the three body segments of the grasshopper: the head, the thorax, and the abdomen. Use the hand lens to examine the head. Notice two long antennae, the sensory organs for touch, located at the front top of the head. Closely examine the antennae with a hand lens. Three simple eyes, or ocelli, are located in the head—one at the base of each antenna and one in the center front of the head. The head also contains two compound eyes, one located on each side of the head. Simple and compound eyes are sensory organs for vision. Answer question 6 in Observations.

3. With a scalpel, carefully cut off a portion of the compound eye and place it on a clean glass slide with a drop of water. **CAUTION:** *When using a scalpel, cut in a direction away from your hand and body to avoid cutting yourself.* Observe the piece of the eye through low power of a dissecting microscope. The compound eye is made up of over 2000 facets, or surfaces, that allow the grasshopper to see almost all of its surroundings. In the appropriate place in Observations, sketch the compound eye as it appears through low power. Record the magnification of the microscope. Answer question 7 in Observations.

4. Closely examine the mouthparts with a hand lens. Identify the following parts as shown in Figure 2: the labrum (upper lip), the mandibles (jaws), the maxillae with sensory palps for tasting, the labium (lower lip) with sensory palps for tasting, and the hypopharynx, the tonguelike structure inside the mouth. Figure 2 also shows what each individual mouthpart looks like.

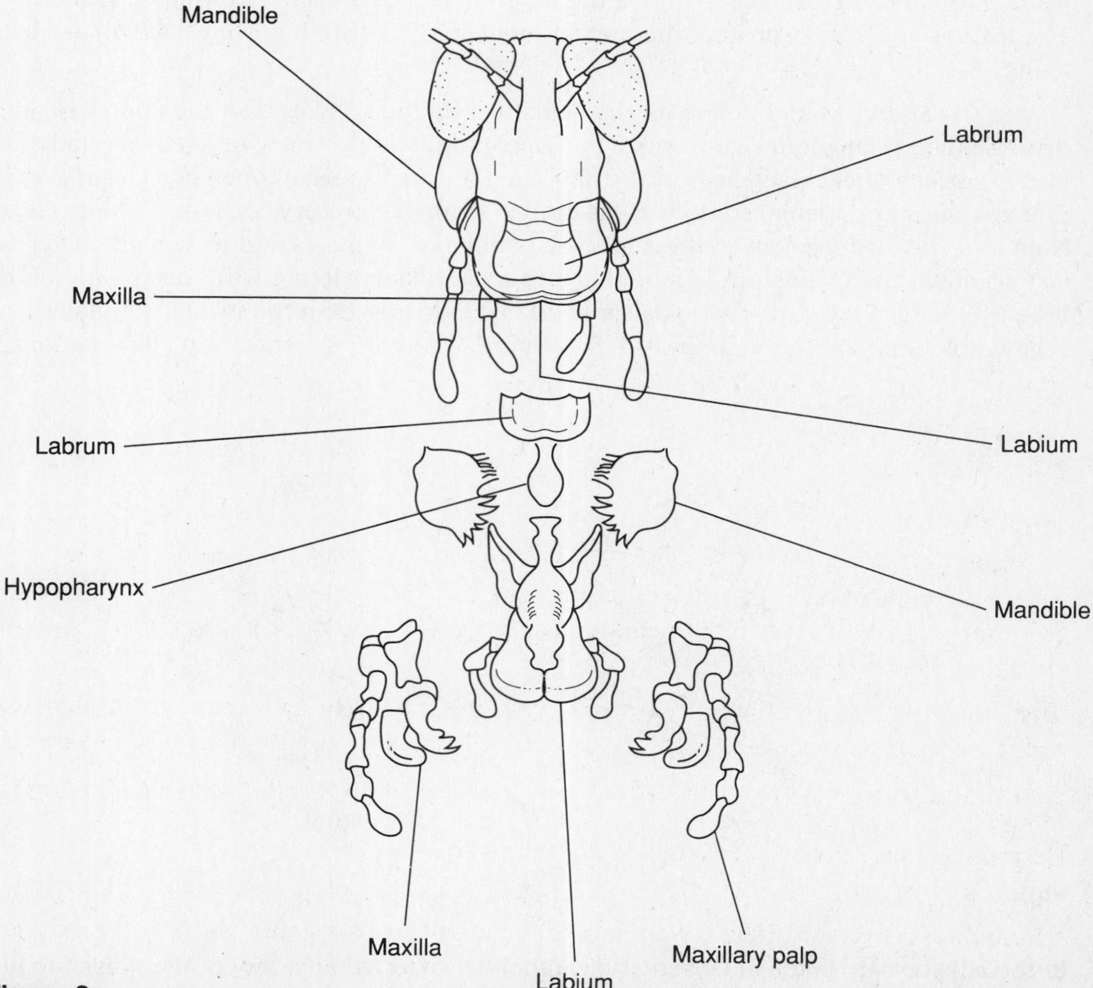

Figure 2

5. Locate the thorax. Notice that the thorax is divided into three segments. One pair of legs is attached to each segment. Observe the two front legs, or forelegs, and the third and largest pair of legs, the jumping legs, or hindlegs. Notice that each leg is composed of the coxa, trochanter, femur, tibia, and tarsus, as shown in Figure 3. Answer question 8 in Observations.

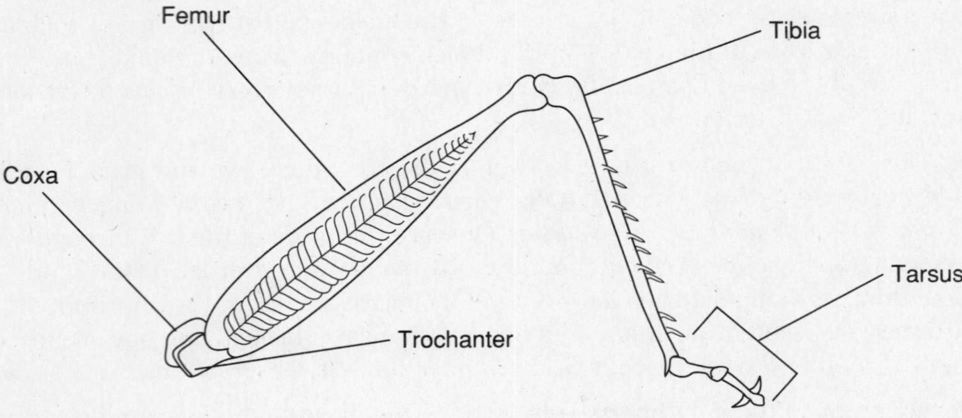

Figure 3

6. Observe the two pairs of wings, which also are located on the thorax. Use your fingers to gently spread open the wings. Notice the difference between the forewings and the hindwings. The leathery forewings protect the membranous and delicate hindwings, which are used for flying.

7. Locate the abdomen and notice its segments. On the first segment of the abdomen, locate the drum-shaped tympanum (eardrum). The tympanum is the sensory organ for sound. On the lateral surface of each segment of the abdomen, locate the small openings called spiracles. The spiracles are openings into the gas-exchange, or respiratory, system of the grasshopper. Note that the last segment of the abdomen is different in males and females. In females, the last segment, the ovipositor, is modified into a clawlike structure with four points used to dig a hole where eggs are laid. The female's abdomen is longer than the male's. In males, the last segment is blunt and curved upward. See Figure 4. Answer question 9 in Observations.

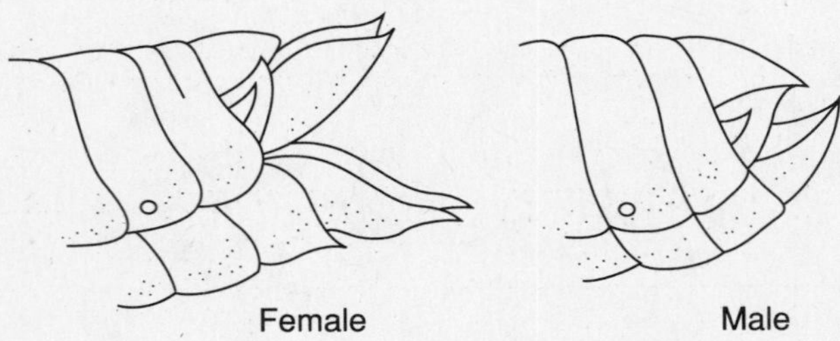

Figure 4

8. In the appropriate place in Observations label the external anatomy of the grasshopper.

Name ______________________ Class __________ Date __________

Observations

Grasshopper Compound Eye

Grasshopper

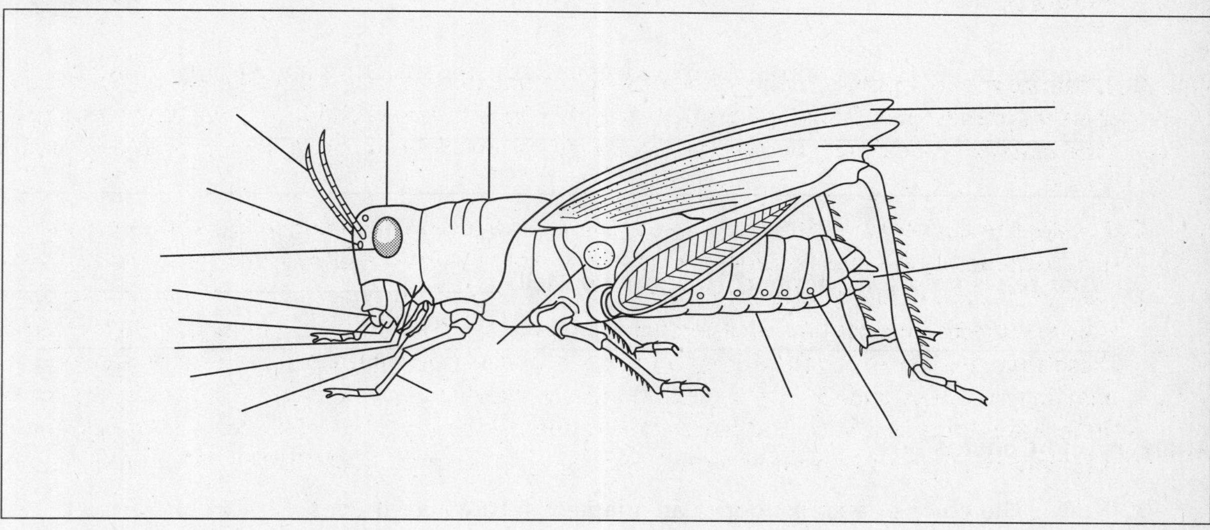

1. Describe the color of the grasshopper. ______________________

2. Which legs are used for walking or crawling? For jumping? ______________________

3. Describe the movements of the grasshopper's abdomen associated with respiration.

4. Compare the movement of the grasshopper's jaws to those of humans.

5. Describe the texture of the grasshopper's exoskeleton.

6. Describe the appearance of the antennae as seen through a hand lens.

7. How do simple and compound eyes differ?

8. Compare the forelegs and the jumping legs in terms of structure and size.

9. What sex is your grasshopper? How can you tell?

Analysis and Conclusions

1. How is the color of a grasshopper an adaptation to where it lives?

2. List two characteristics of chitin that make it a good material for the exoskeleton and wings of arthropods.

3. How is the grasshopper adapted to detect moving objects in the environment?

Name ______________________ Class ____________ Date ____________

4. What are the mouthparts of a grasshopper adapted to do? ______________________

5. Of the three body regions of the grasshopper, which one is specialized for locomotion?

6. What do you think is the function of the spinelike structures found on the tibia and tarsus?

Critical Thinking and Application

1. How can a grasshopper's exoskeleton be both an advantage and a disadvantage?

2. List three structures of the grasshopper that are adaptations for life on dry land.

3. Most insects are small as compared to most vertebrates. Do you think being small is more of an advantage or a disadvantage to insects? Give evidence to support your answer.

4. Can you drown a grasshopper by holding its head under water? Explain your answer.

____

____

____

Going Further

1. Study other behaviors of living grasshoppers by setting up experiments to determine whether grasshoppers can live in close quarters with other species of insects, what foods grasshoppers prefer, and whether grasshoppers react to certain stimuli, such as light.
2. Obtain preserved specimens from as many of the classes of arthropods as possible. Compare the characteristics of each, such as the number of body regions, number of legs, types of mouthparts, and types of respiratory organs. Record your observations in a data table.

Name ______________________ Class __________ Date __________

Isopod Environments

Pre-Lab Discussion

The phylum Arthropoda is made up of several different subphyla. The subphylum Crustacea contains over 35,000 species. Animals in the subphylum Crustacea include crabs, lobsters, crayfishes, and shrimps. The crustaceans are familiar as food sources for people. In general, crustaceans are characterized by a hard exoskeleton, two pairs of antennae, and mouthparts called mandibles. Although most are aquatic, a few crustaceans, such as the isopods, live on land. The pill bug and the sow bug are examples of isopods.

In this investigation, you will examine the external anatomy of a pill bug. You will also observe the behavior of pill bugs.

Problem

What are some characteristic structures of a pill bug? What type of environment is preferred by the pill bug?

Materials *(per group)*

6 live pill bugs in a glass jar
Paper towels
Sheet of black construction paper
Petri dish and cover
Masking or transparent tape
Pen or pencil
Scissors
2 meduim-sized paper cups
Ice cubes
Warm water
Hand lens

Safety

Put on a laboratory apron if one is available. Handle all glassware carefully. Be careful when handling sharp instruments. Follow your teacher's directions and all appropriate safety procedures when handling live animals. Note all safety alert symbols next to the steps in the Procedure and review the meanings of each symbol by referring to the symbol guide on page 10.

Procedure

Part A. Observing the Behavior of a Pill Bug

1. Referring to Figure 1, locate the head, antennae, chitinous exoskeleton, and seven pairs of legs of live pill bugs in a glass jar.
2. Open the jar and gently touch one of the pill bugs as it walks across the bottom of the jar. Answer questions 1 and 2 in Observations.
3. Use a hand lens to observe a pill bug as it walks across the jar. Answer question 3 in Observations.

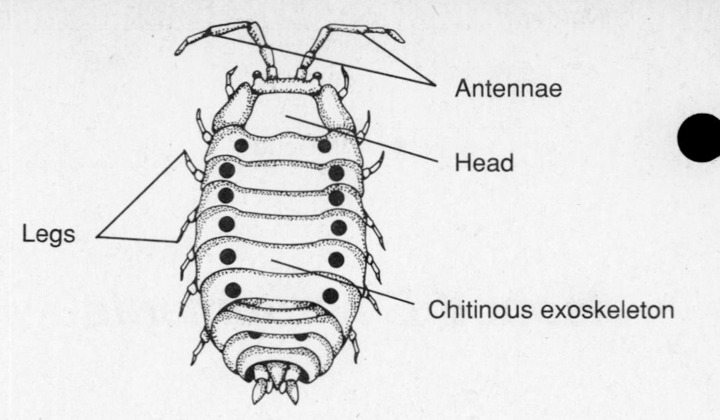

Figure 1

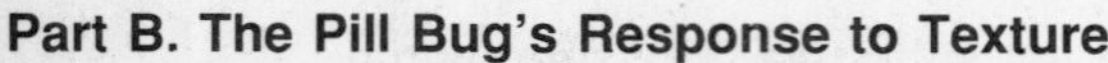

Part B. The Pill Bug's Response to Texture

1. Place the bottom of a petri dish on a paper towel. Using the pencil or pen, trace the bottom of the petri dish on the paper towel.
2. Use scissors to cut out the paper outline of the petri dish. Then fold the outline in half. **CAUTION:** *Be careful when handling sharp instruments.*
3. As shown in Figure 2, place the folded paper towel in the bottom of the petri dish. Tape the paper towel in place along all sides. **Note:** *Place the petri dish on top of a paper towel. This will provide a background color that is similar on both sides of the petri dish.*
4. Place 6 isopods from the glass jar in the center of the petri dish. Cover the petri dish and allow the pill bugs to remain there undisturbed for 5 minutes.

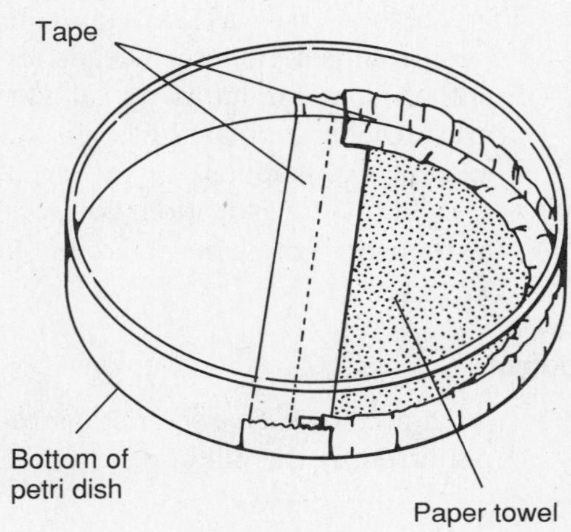

Figure 2

5. After 5 minutes, examine the petri dish. Count the number of pill bugs that moved onto the toweled side and the number of pill bugs that moved onto the untoweled side. Record this information in the appropriate place in the Data Table.
6. Return the pill bugs to the glass jar.

Part C. The Pill Bug's Response to Moisture

1. Repeat steps 1 and 2 of Part B. Place the folded paper towel on the other side of the bottom of the petri dish used in Part B. As shown in Figure 3, moisten one of the folded paper towels with water.
2. Place 6 pill bugs into the center of the petri dish. Cover the petri dish and allow the pill bugs to remain there undisturbed for 5 minutes.
3. After 5 minutes, examine the petri dish. Count the number of pill bugs that moved onto the moistened side of the petri dish and the number of pill bugs that moved onto the dry side of the petri dish. Record this information in the appropriate place in the Data Table.
4. Return the pill bugs to the glass jar and remove the paper towels and tape from the petri dish.

Name ______________________ Class ____________ Date ____________

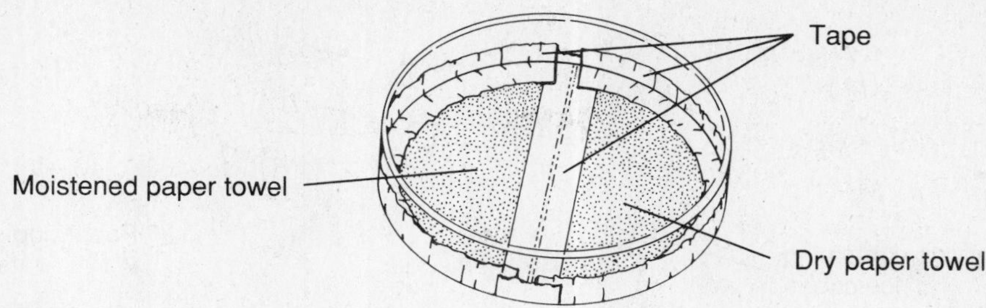

Figure 3

Part D. The Pill Bug's Response to Light

1. Repeat step 1 of Part B.
2. Use scissors to cut out the paper towel outline of the dish. **CAUTION:** *Handle scissors with care.*
3. Place the paper towel circle in the bottom of the petri dish and moisten the entire circle with water.
4. As shown in Figure 4, cover one-half of the top of the petri dish with black construction paper. Use tape to keep the construction paper in place.

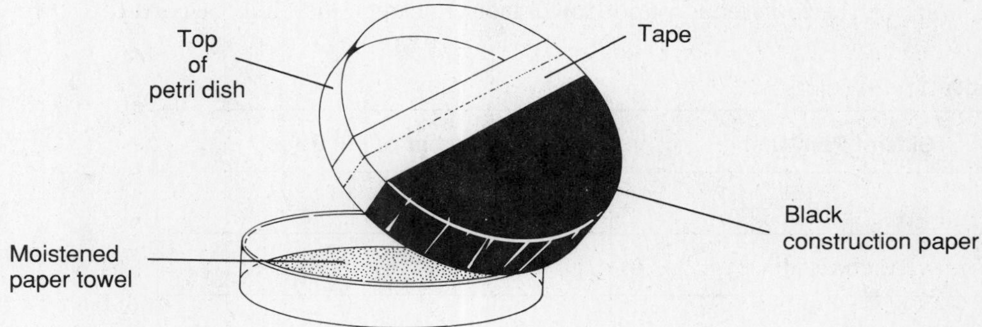

Figure 4

5. Position another paper towel under the petri dish. Add 6 pill bugs to the center of the bottom of the petri dish. Cover the petri dish and allow the pill bugs to remain there undisturbed for 5 minutes.
6. After 5 minutes, examine the petri dish. Count the number of pill bugs that moved onto the dark side of the petri dish and the number of pill bugs that moved onto the lighted side of the petri dish. Record this information in the appropriate place in the Data Table.
7. Return the pill bugs to the glass jar. Remove the black construction paper and tape from the top of the petri dish, but allow the moistened paper towel to remain in the bottom of the dish.

Part E. The Pill Bug's Response to Temperature

1. Fill one cup with ice cubes and fill the other cup with warm water.
2. Place the two cups side by side, as shown in Figure 5. Place the petri dish on top of the cups. Make sure that the center of the petri dish is over the area where the two cups touch each other.

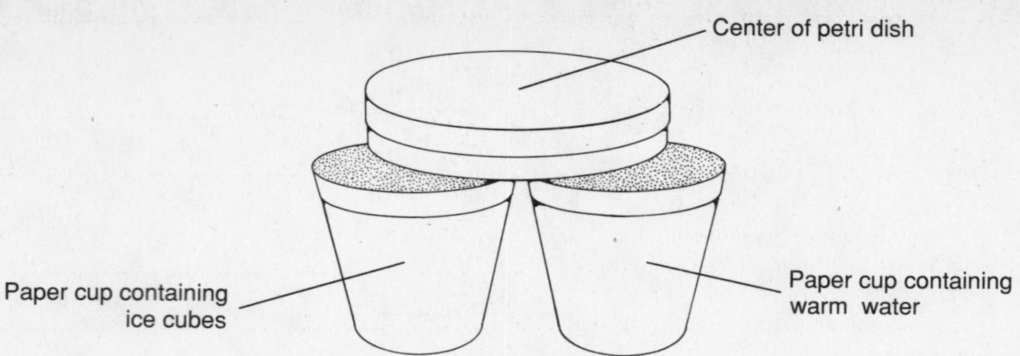

Figure 5

3. Place 6 pill bugs in the center of the petri dish. Cover the petri dish and allow the pill bugs to remain there undisturbed for 5 minutes.
4. After 5 minutes, examine the petri dish. Count the number of pill bugs found on the cold side of the petri dish and the number of pill bugs found on the warm side of the petri dish. Record this information in the appropriate place in the Data Table.
5. Return the pill bugs to the glass jar and follow your teacher's instructions for storing the pill bugs for further use.
6. Remove the moistened paper towel from the petri dish and discard the towel properly.

Data Table

Side of Petri Dish	Number of Pill Bugs
Toweled	
Untoweled	
Moist	
Dry	
Dark	
Light	
Cold	
Warm	

Observations

1. Describe the texture and appearance of the pill bug's exoskeleton.

____

____

Name ______________________ Class __________ Date __________

2. What response does the pill bug show when touched? ______________________

____

3. Describe the way in which the pill bug moves. ______________________

____

Analysis and Conclusions

1. What is the function of the pill bug's exoskeleton? ______________________

____

2. Of what advantage to the pill bug is its response to touch? ______________________

____

3. Does the pill bug prefer a rough-textured or a smooth-textured environment?

____

____

4. Does the pill bug prefer a dry or a moist environment? Why would a pill bug prefer this type of environment? ______________________

____

____

5. Does the pill bug prefer a light or a dark environment? Why would a pill bug prefer this type of environment? ______________________

____

____

6. Does the pill bug prefer a warm or a cold environment? ______________________

____

Critical Thinking and Application

1. Like other crustaceans, pill bugs use gills to breathe. Do the results of this investigation seem to support this information? Explain your answer. ______________________________

____

____

2. The word *isopod* comes from Greek words meaning "same feet." How is this name appropriate for the pill bug? ______________________________

____

____

3. How would a permeable exoskeleton on a land-living arthropod affect the type of environment in which you would find this arthropod? ______________________________

____

____

Going Further

To observe another crustacean, obtain a fiddler crab from a local pet store. Place the crab in the middle of a shoe box containing a layer of sand 5 cm deep and some small and medium-sized stones. Where does the crab go? Why is this reaction of value to the crab? Next, place the crab in a small aquarium filled with water to a depth of 10 cm. Gently touch the feelers of the crab with a pencil. Observe what happens. Use a net to remove the crab from the water. Compare the behavior of the crab in water with its behavior on land. Make an oral presentation to your class.

58

The Starfish

Pre-Lab Discussion

The starfish, or sea star, is a spiny-skinned marine invertebrate that is a member of the phylum Echinodermata. Echinoderms are animals whose bodies are often covered with hundreds of small spines. Brittle stars, sand dollars, sea cucumbers, and sea urchins are also types of echinoderms.

Starfish are the most familiar echinoderms. They are found in coastal waters and along rocky seashores. They feed on oysters, clams, snails, barnacles, and worms. Starfish are known for their characteristic star-shaped bodies. They typically have five rays, or arms, branching out from a central disk. Because of their division into five parts, starfish are said to have pentaradial symmetry. If a ray breaks off, the starfish is able to regenerate, or grow back, a new one.

In this investigation, you will examine the external structures of a starfish.

Problem

How is the anatomy of a starfish adapted to life in a marine environment?

Materials *(per group)*

Preserved starfish
Dissecting tray
Hand lens

Safety

Put on a laboratory apron if one is available. Handle all glassware carefully. Always use special caution when working with laboratory chemicals, as they may irritate the skin or cause staining of the skin and clothing. Never touch or taste any chemical unless instructed to do so. Note all safety alert symbols next to the steps in the Procedure and review the meanings of each symbol by referring to the symbol guide on page 10.

Procedure

1. Rinse the starfish thoroughly with water to remove any extra preservative. **CAUTION:** *The preservative used on the starfish can irritate your skin. Avoid touching your eyes while working with the preserved starfish.*

2. Place the starfish in the dissecting tray with its aboral, or top, surface facing up. Notice that the starfish's body plan consists of 5 rays radiating out from a central disk. Although most starfish have 5 rays, sun stars have 7 to 14 rays, and some sea stars have 15 to 24 rays.

3. Using a hand lens, examine the skin on the aboral surface. Notice the many coarse spines that cover the entire aboral surface. The epidermis is spiny and irregular because parts of the endoskeleton protrude through the skin. Around the base of the spines are pedicellariae, which are jawlike structures. They capture small animals and keep the epidermis free of foreign objects. The skin gills are soft projections from the aboral surface that are lined by tissue of the inner cavity. They provide a large, moist area across which oxygen can be removed from the water. The skin gills are protected by the pedicellariae.

4. Referring to Figure 1, use a hand lens to locate a spine and the pedicellariae around it. Locate skin gills near some pedicellariae. Answer question 1 in Observations.

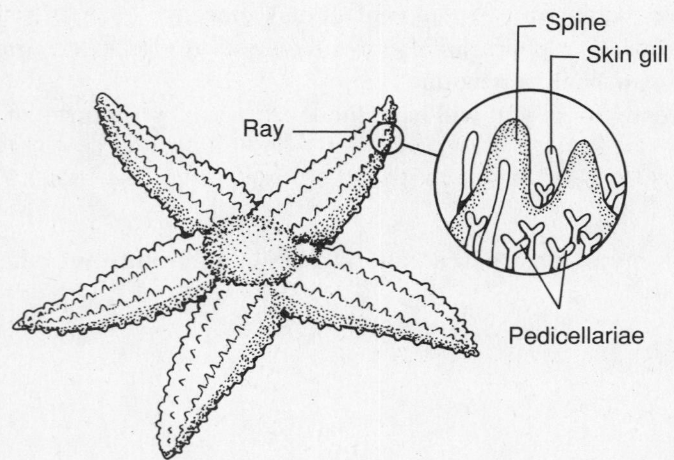

Aboral Surface of Starfish

Figure 1

5. Study the aboral surface of the central disk. Answer question 2 in Observations.

6. Locate a small red or yellow buttonlike structure on the central disk. This structure is the madreporite, or sieve plate. The madreporite contains many tiny pores through which water enters the water-vascular system. The water-vascular system is a system of water-filled canals and appendages that function primarily in locomotion and feeding.

7. Try to find the anus in the center of the central disk. The anus, which opens out from the intestine, is the opening through which solid wastes are eliminated from the body.

Name ______________________ Class __________ Date __________

8. In the appropriate place in Observations, label the following structures on the aboral side of the starfish: central disk, rays, spines, madreporite, and anus.

9. Turn the starfish over so that its oral, or bottom, surface is facing up.

10. With the hand lens, examine the mouth, an opening in the middle of the central disk through which food is taken in. Notice the small spines that surround the mouth. Many types of starfish feed by pushing part of the stomach out through the mouth. The stomach secretes enzymes that digest prey.

11. Find the ambulacral groove that begins at the mouth and extends down the center of each ray. Find the small tube feet that line the groove. The tube feet are part of the water-vascular system. The tube foot is a hollow, thin-walled cylinder with a bulblike structure called the ampulla at one end and a sucker at the tip. Fluid-pressure changes caused by muscle contractions in the ampulla and tube foot create suction in the tip of the tube foot. The suction enables the starfish to pull itself along a surface or to grasp prey. Answer question 3 in Observations.

12. Locate the eyespots, which are tiny red or pink dots that are found on the oral surface of a small tentacle at the tip of each ray. An eyespot is made up of 80 to 200 ocelli, which contain granules of red pigment. The eyespots are sensitive to light but are not capable of forming images.

13. In the appropriate place in Observations, label the following structures on the oral side of the starfish: ambulacral groove, mouth, tube feet, oral spines, and ray spines.

Observations

Aboral Side of Starfish

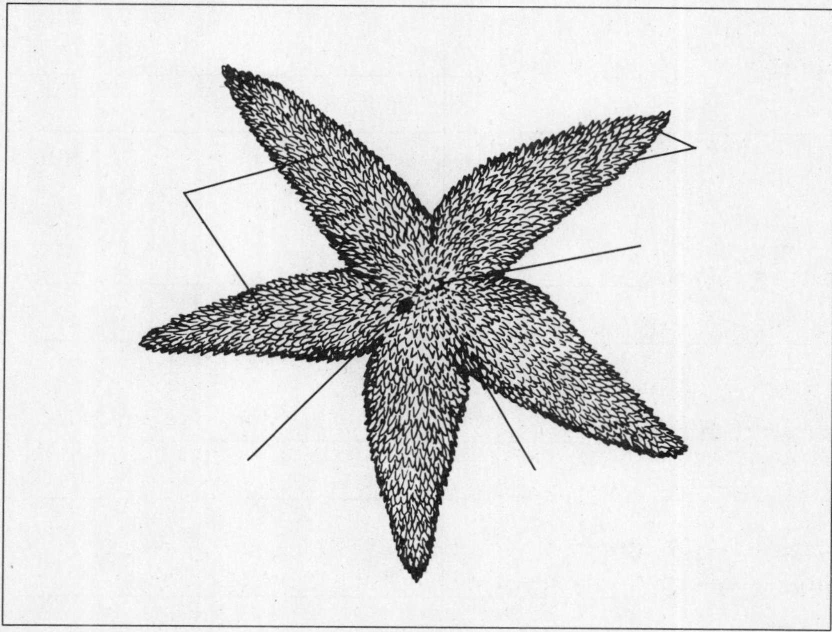

Oral Side of Starfish

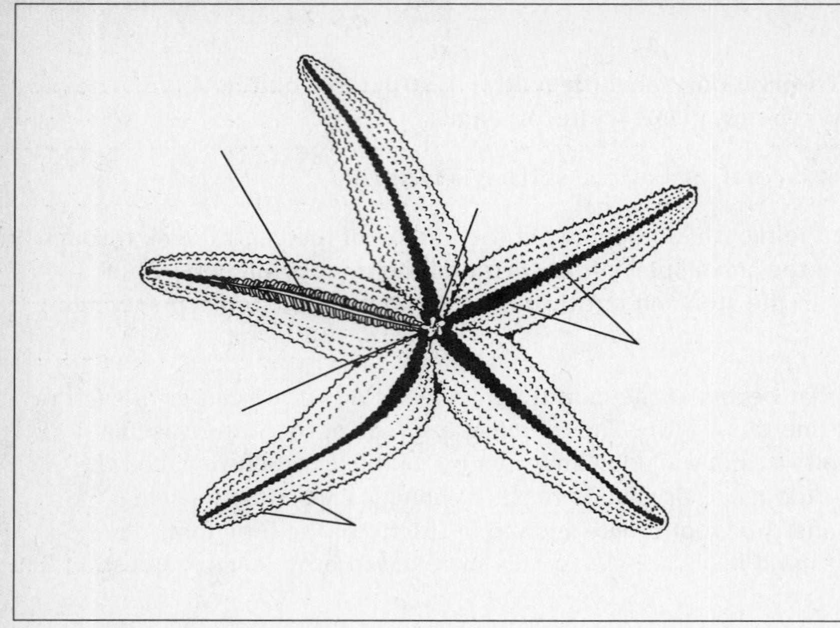

1. Describe the feel of a starfish's spines. ______________________________

____

2. How does the number of spines in the central disk compare to the number of spines in a ray?

____

____

3. How many rows of tube feet does your specimen have? ______________________________

____

Analysis and Conclusions

1. What is the function of a starfish's spines? ______________________________

____

2. What kind of body plan does a starfish have? ______________________________

____

3. What are two functions of the tube feet? ______________________________

____

Name ______________________ Class ____________ Date ____________

4. How does a starfish take in food? ______________________

5. How does respiration occur in the starfish? ______________________

6. List two adaptations of the starfish that make it well adapted to life in marine waters.

Critical Thinking and Application

1. Starfish produce large numbers of eggs and sperm. How is this production an adaptive advantage? ______________________

2. When a starfish pries open the shell of a clam or oyster, the mollusk resists. Even if the shell opens only slightly, the starfish will get its meal. How does this occur?

3. Because starfish were eating many clams and oysters, divers were hired to go out and chop the starfish into pieces. After this, fishermen found even more empty clam and oyster shells than before. Why did this occur? ______________________

4. Why can a starfish move equally well in any direction? ______________________

5. Many echinoderms, which are bottom-dwellers as adults, have free-swimming larvae. What advantage do free-swimming larvae provide for the echinoderms?

____

____

Going Further

1. Sand dollars, sea urchins, sea lilies, sea cucumbers, and brittle stars belong to the same phylum as starfish. Obtain preserved specimens of these animals. Compare their external structures with those of the starfish.
2. Set up a saltwater aquarium in your classroom. Obtain a live starfish and other types of echinoderms and observe their behavior in the aquarium.

Name ______________________ Class ____________ Date ____________

Comparing Invertebrate Body Plans

Pre-Lab Discussion

Invertebrates, like all other organisms, are divided into groups based on certain distinguishing characteristics. Two characteristics that are examined when grouping invertebrates are cell layers and body cavities.

The number of cell layers making up the body varies among invertebrates. Cnidarians—jellyfishes and sea anemones, for example—possess only two cell layers in their body wall: an inner gastroderm and an outer epidermis. More advanced invertebrates—worms, mollusks, arthropods, and echinoderms, to name a few—have three basic cell layers: an inner endoderm, a middle mesoderm, and an outer ectoderm.

The animals that possess three basic cell layers can be divided into groups based on the structure of their body cavity. The body cavity, if present, is a fluid-filled hollow in the body wall that is located between the endoderm and the ectoderm. Animals that lack a body cavity are called acoelomates. Animals that have a body cavity that is only partially lined with mesoderm are called pseudocoelomates. And animals that have a body cavity that is completely lined with mesoderm are called coelomates.

In this investigation, you will compare the body plans and structures of invertebrates from four different phyla.

Problem

What are the differences in body plans and structures of the cnidarians, flatworms, roundworms, and annelids?

Materials *(per group)*

Microscope
Prepared slides of cross sections of:
- Cnidarian *(Hydra)*
- Flatworm *(Dugesia)*
- Roundworm *(Ascaris lumbricoides)*
- Earthworm *(Lumbricus terrestris)*

Safety

Handle all glassware carefully. Always handle the microscope with extreme care. You are responsible for its proper care and use. Use caution when handling glass slides as they can break easily and cut you. Note all safety alert symbols next to the steps in the Procedure and review the meanings of each symbol by referring to the symbol guide on page 10.

Procedure

Part A. Examining the Body Plan and Structures of the Cnidarian

1. Study Figure 1, which shows the basic body plans of the four invertebrates you will be examining.

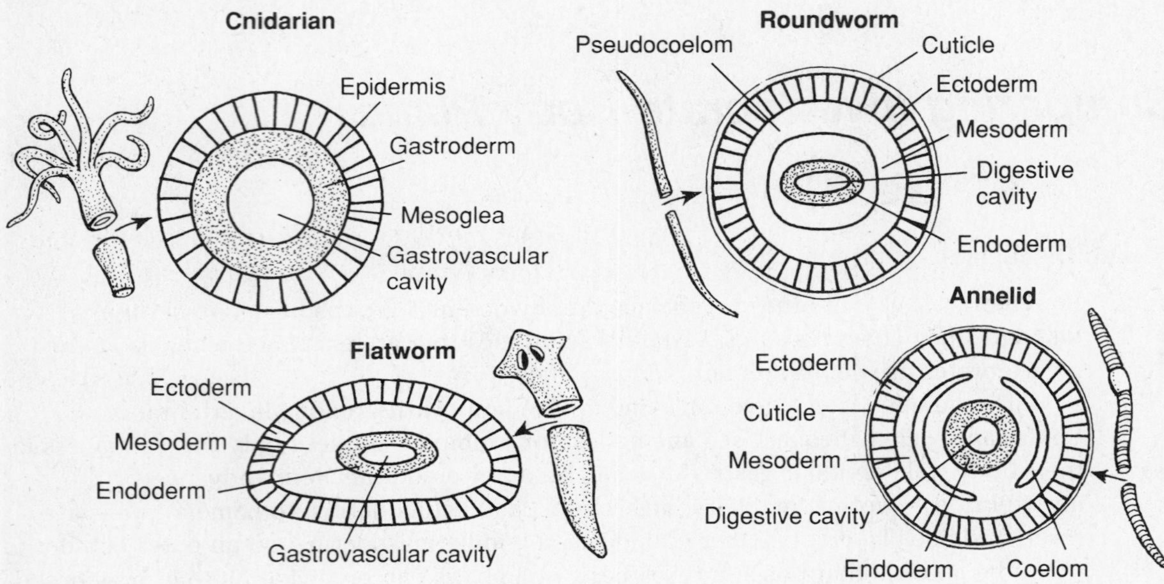

Figure 1

2. Examine a prepared slide of a cross section of a cnidarian under the low-power objective of a microscope. Look at the body layers. Switch to high power to see the specific structures in greater detail. **CAUTION**: *When switching to the high-power objective, always look at the objective from the side of the microscope so that the objective does not hit or damage the slide.*
3. Locate the epidermis, or outer layer of cells. Try to find some barbed cells called cnidocytes, each of which contains a stinging structure called a nematocyst.
4. Locate the gastroderm, or inner cell layer. Notice how some of the cells in the gastroderm have flagella. These flagellated cells help to circulate food and other materials within the gastrovascular cavity.
5. Locate the gastrovascular cavity in the center of the cross section of the cnidarian. Food digestion occurs within the gastrovascular cavity.
6. Locate the mesoglea, which is a noncellular jellylike material between the epidermis and the gastroderm.
7. In the appropriate place in Observations, label the following structures on the cross section of the cnidarian *(Hydra)*: epidermis, mesoglea, gastroderm, cnidocyte, nematocyst, and gastrovascular cavity.

Part B. Observing the Body Plan and Structures of the Flatworm

1. Examine a prepared slide of a cross section of a flatworm under low power. Locate the ectoderm, endoderm, and gastrovascular cavity.
2. Find the middle cell layer, or mesoderm, between the ectoderm and the endoderm. The mesoderm makes up most of the flatworm and consists of muscles, glands, organs, loose cells, and many other kinds of structures. Notice the absence of a body cavity.
3. In the appropriate place in Observations, label the following structures on the cross section of the flatworm: ectoderm, mesoderm, endoderm, and gastrovascular cavity.

Name ______________________ Class ____________ Date ______________

Part C. Examining the Body Plan and Structures of the Roundworm

1. Examine a prepared slide of a cross section of a roundworm under low power. Locate the thick outer covering called the cuticle. This tough coating keeps a parasitic roundworm from being digested by its host. Just inside of the cuticle is the ectoderm, a thin layer of cells. Locate the mesoderm, the fiberlike layer just inside the epidermis.
2. Toward the center of the roundworm, look for the endoderm. The space inside the endoderm is the inside of the roundworm's digestive tract. The space between the endoderm and the mesoderm is called a pseudocoelom.
3. In the appropriate place in Observations, label the following structures on the cross section of the roundworm: cuticle, ectoderm, mesoderm, endoderm, and pseudocoelom.

Part D. Examining the Body Plan and Structures of the Earthworm

1. Examine a prepared slide of a cross section of an earthworm. Find the thin protective outer layer called the cuticle. Just below the cuticle, look for the layer of cells of the ectoderm.
2. Inside the ectoderm are two muscle layers. The outer layer contains the circular muscles, which run circularly around the earthworm. The thick inner layer contains the longitudinal muscles, which run the length of the worm. These two layers of muscle make up the mesoderm.
3. Study the center of the slide. You will see a round or horseshoe-shaped space that is the inside of the digestive tract. The layer of cells that surrounds the space is the endoderm. Surrounding the endoderm is another set of circular and longitudinal muscles.
4. Notice a relatively open space between the muscles surrounding the digestive tract and the muscles just inside the epidermis. This body cavity is called a coelom.
5. Locate a long, fiberlike structure inside the coelom. This structure, called a nephridium, is involved in excretion. There should be one nephridium on each side of the earthworm's body.
6. On the outer surface of the earthworm, find some bristly structures called setae. The earthworm uses the setae to help move its body through the soil.
7. In the appropriate place in Observations, label the following structures on the cross section of the earthworm: cuticle, ectoderm, circular and longitudinal muscles (both sets), endoderm, coelom, nephridium, and setae.

Observations

Cross Section of Cnidarian *(Hydra)*

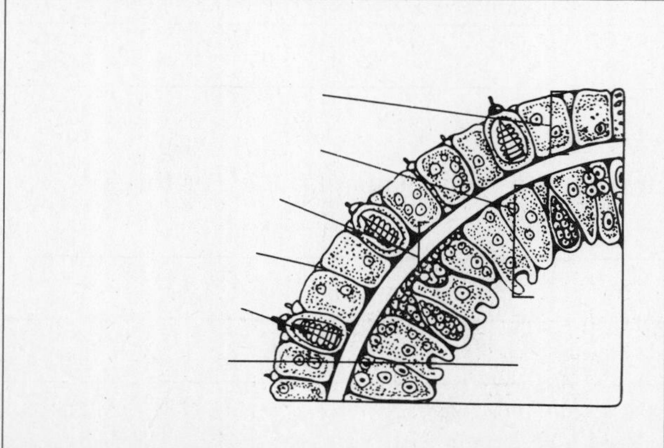

Cross Section of Flatworm *(Dugesia)*

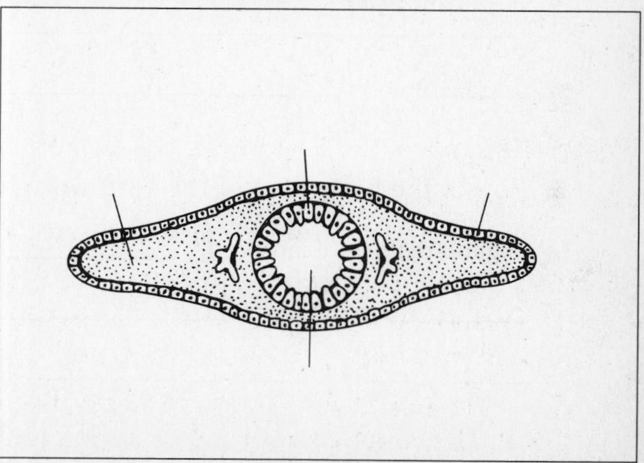

Cross Section of Roundworm *(Ascaris)*

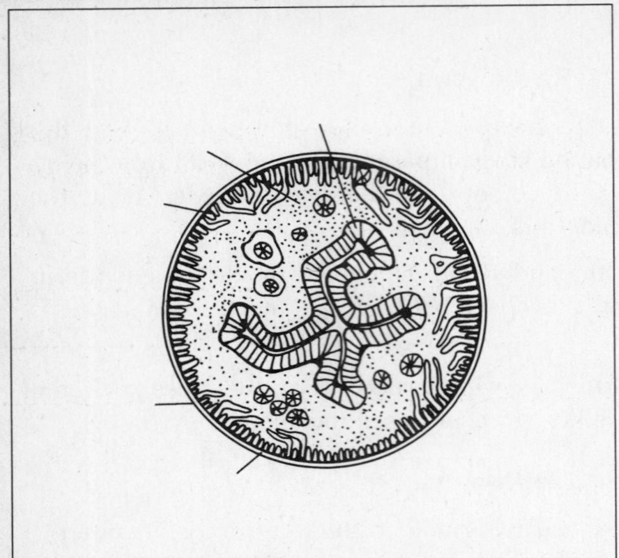

Cross Section of Earthworm *(Lumbricus)*

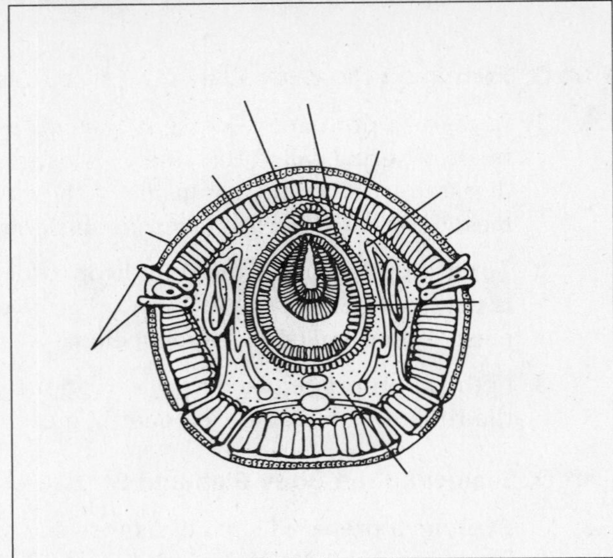

Analysis and Conclusions

1. What structures in the cnidarian correspond to the digestive cavity, ectoderm, and endoderm in the earthworm? ______________________________

____

2. How is the body plan of the cnidarian different from the body plan of the flatworm?

____

____

3. How is the body plan of the flatworm similar to the body plan of the roundworm?

____

____

4. How is the body plan of the flatworm different from the body plan of the roundworm?

____

____

5. How is the body plan of the earthworm different from the body plans of the other three organisms you examined? ______________________________

____

____

Name ______________________ Class ____________ Date ____________

Critical Thinking and Application

1. Suppose a newly discovered organism is found to have three body layers. Scientists think that it is a worm but disagree as to the group in which it should be placed. How could they assign it to the proper group based on body structure? ____________________

2. What are two advantages of the presence of a coelom in an organism?

3. Suppose a new phylum of invertebrates is discovered. These invertebrates have a gastrovascular cavity and a very thin layer of mesodermal cells but show no organ development. Where would you place this new phylum with respect to the ones studied in this investigation? Explain your placement. ____________________

Going Further

Construct or draw a new form of invertebrate having characteristics of one existing phylum or intermediate characteristics of two related phyla. From your completed model or drawing, your classmates should be able to identify the animal's symmetry, locomotion, nervous system, method of circulation, excretory system, and type of digestive system and, based upon its gas-exchange structure, whether your new "species" is terrestrial or aquatic.

Name ______________________ Class __________ Date __________

Identifying Invertebrates

Pre-Lab Discussion

Invertebrates are an extremely diverse group of organisms. For example, invertebrates perform the function of movement in many different ways. Hawk moths zip and hover agilely in the air; snails creep slowly over the ground; soft-bodied earthworms burrow in the dark, damp soil; brittle stars thrash their arms to move across the ocean floor; jellyfishes pulse through the water; sponges do not move at all. The ways in which invertebrates carry out their other life functions are also quite varied.

Because invertebrates are so different from one another, it is necessary to classify them, or place them into groups that share certain key characteristics. This makes it easier to study them and to organize information about them in a coherent way.

In this investigation, you will identify the phylum to which each of a number of unknown invertebrate specimens belongs.

Problem

What external characteristics can be used to identify unknown invertebrates?

Materials *(per group)*

Unidentified preserved invertebrate specimens
Dissecting tray
Hand lens

Safety

Put on a laboratory apron if one is available. Handle all glassware carefully. Always use special caution when working with laboratory chemicals, as they may irritate the skin or cause staining of the skin or clothing. Never touch or taste any chemical unless instructed to do so. Note all safety alert symbols next to the steps in the Procedure and review the meanings of each symbol by referring to the symbol guide on page 10.

Procedure

1. Rinse an unidentified invertebrate specimen thoroughly with water in order to remove any excess preservative. **CAUTION:** *The preservative used on the specimen can irritate your skin. Avoid touching your eyes while working with the preserved specimen.*

2. Throughout this investigation, you will be examining a number of invertebrate specimens for the presence or absence of certain characteristics and structures. Use the hand lens whenever you need to see a characteristic more clearly. Record all of your observations in the appropriate place in the Data Table in Observations. If an organism does not display a particular characteristic, write the word "none" in the appropriate place in the Data Table.

3. Examine your specimen carefully.
   - Does the animal have no symmetry, radial symmetry, or bilateral symmetry?
   - Does the animal have a hard body or a soft body?
   - Does the animal have a hard exoskeleton?
   - Does the animal's body show segmentation?
   - Is the animal motile, or free to move around, or is it sessile, or not free to move around?
   - If the animal is motile, what is its method of locomotion?

   Record the observations in the appropriate places in the Data Table.

4. Observe the external structures of the organism's digestive system.
   - Does the animal have a mouth?
   - Does the animal have an anus?

   Record the observations in the appropriate place in the Data Table.

5. Observe the external structures of the organism's respiratory system.
   - Does the animal have gills?
   - Does respiration occur through the external surface of the organism?

   Record the observations in the appropriate place in the Data Table.

6. Observe any external evidence of reproductive structures on the organism.
   - Is the animal a hermaphrodite, having the ability to produce both male and female sex cells, or are the sexes separate?

   Record the observations in the appropriate place in the Data Table.

7. Are there any other external characteristics that might help you to identify this animal? Record the observations in the appropriate place in the Data Table.

8. Review all your observations. Based on your observations, write the name of the phylum to which this organism belongs in the last column of the Data Table.

9. Repeat steps 3 through 8 for all of the other specimens provided by your teacher.

Name ______________________ Class ____________ Date ____________

Observations

Data Table

Specimen	Type of Symmetry	Hard or Soft Body	Hard Exoskeleton	Segmentation	Motile or Sessile	Method of Locomotion	Mouth	Anus	Type of Respiratory System	Hermaphrodite or Single Sex	Other Characteristics	Phylum
1												
2												
3												
4												
5												
6												
7												
8												

Analysis and Conclusions

1. Why should you examine the same characteristics in each animal that you are trying to identify? ____________________

____

2. What characteristics should you examine if you were asked to determine whether the organism was terrestrial or aquatic? ____________________

____

3. What is one advantage of using a preserved specimen instead of a live specimen in this investigation? ____________________

____

____

4. What is one disadvantage of using a preserved specimen instead of a live specimen in this investigation? ____________________

____

____

5. If you were unsure of one of your classifications, what could you further examine to make sure of your identification? ____________________

____

____

Critical Thinking and Application

1. What adaptations in body form are found in motile organisms that are not found in sessile organisms? ____________________

____

____

2. How might radial symmetry be advantageous? ____________________

____

Name ______________________ Class __________ Date __________

3. How might bilateral symmetry be advantageous? ______________________

4. How is being a hermaphrodite an advantage to certain invertebrates?

5. List two ways in which invertebrates are of benefit to humans. ______________________

6. List two ways in which invertebrates are harmful to humans. ______________________

Going Further

1. Collect living specimens of some of the different invertebrate phyla. Observe differences in their means of locomotion, food preferences, feeding behaviors, and other behavioral characteristics. Construct a data table to record your observations.
2. Invertebrates are incredibly diverse organisms. Prepare a report on one type of invertebrate that was not discussed in Chapters 26 through 30 of your textbook. What characteristics does this organism have that make it similar to other invertebrates? What characteristics does this invertebrate have that make it unusual or unique? Report your findings to the class.

Name ______________________ Class ____________ Date ____________

61

The Fish

Pre-Lab Discussion

Fishes are members of the phylum Chordata and subphylum Vertebrata. The largest class of fishes, class Osteichthyes, contains fishes with skeletons made of bone. Fishes exhibit many adaptations for life in an aquatic environment. The perch and the goldfish are representative members of the class Osteichthyes.

In this investigation, you will observe the movement and behavior of a live goldfish. You will also identify the external parts of a perch.

Problem

How are the structures of a fish adaptations for living in an aquatic environment?

Materials *(per group)*

Live goldfish
Preserved perch
Large glass jar or beaker
Water from an aquarium
Fish net
Stopwatch or clock with second hand
Dissecting tray
Probe
Forceps
Hand lens
Glass slide
Coverslip
Paper towels
Medicine dropper
Dissecting microscope

Safety

Put on a laboratory apron if one is available. Always handle the microscope with extreme care. You are responsible for its proper care and use. Use caution when handling glass slides as they can break easily and cut you. Handle all glassware carefully. Always use special caution when working with laboratory chemicals, as they may irritate the skin or cause staining of the skin or clothing. Never touch or taste any chemical unless instructed to do so. Be careful when handling sharp instruments. Follow your teacher's directions and all appropriate safety procedures when handling live animals. Note all safety alert symbols next to the steps in the Procedure and review the meanings of each symbol by referring to the symbol guide on page 10.

Procedure

Part A. Observing the Behavior of a Live Fish

1. Fill a large glass jar or beaker three-quarters full with water from an aquarium.
2. With a fish net, carefully remove one live goldfish from the aquarium. Immediately place the fish into the glass jar filled with aquarium water.
3. Observe the goldfish. Using a stopwatch or clock with a second hand, count how many times the fish opens and closes its mouth in one minute. Count how many times the gill covers move in one minute. Answer question 1 in Observations.
4. Observe how the goldfish swims. Carefully watch the movements of the body, fins, and tail. Answer question 2 in Observations.
5. Locate the goldfish's dorsal, caudal, pectoral, pelvic, and anal fins. Answer question 3 in Observations. Complete the Data Table in Observations after you have observed the function of each fin as the fish swims.
6. With a fish net, carefully return the goldfish to the aquarium.

Part B. Examining the External Anatomy of a Fish

1. Obtain a preserved perch. Rinse the fish under running water to remove excess preservative. **CAUTION:** *The preservative used on the fish can irritate your skin. Avoid rubbing your eyes while working with the fish. Leave a piece of wet paper towel at your work area to wipe your fingers after contact with the fish.* Dry the fish with paper towels and place it in a dissecting tray. Position the fish in the tray with the dorsal surface away from you and the head of the fish pointing left.
2. Observe the coloring of the fish. Compare the color of the dorsal and ventral surfaces. Answer question 4 in Observations.
3. Locate the three regions of the fish's body: the head, the trunk, and the tail.
4. Locate the eye and the nostrils on the head of the fish. Carefully insert a probe into one of the nostrils to determine where the nostrils lead. Answer question 5 in Observations.
5. Insert a probe into the mouth. Carefully pry open the mouth. Locate the upper jaw, or maxilla, and the lower jaw, or mandible. The perch possesses powerful jaws and strong teeth for catching and eating prey.
6. Carefully insert your index finger into the open mouth. Feel the location of the teeth and the texture and location of the tongue. **CAUTION:** *Touch the teeth very gently. They are very sharp.* Answer question 6 in Observations.
7. Locate the operculum, the protective bony covering over the gills. With a probe, carefully lift the operculum away from the gills lying underneath. Observe the flat, scalelike bones supporting the operculum.
8. While holding the operculum back, use a hand lens to examine the gills carefully. Locate the spaces between the gills. Also locate the gill arch—a cartilage support for each gill, the gill filaments—soft, fingerlike projections that make up the gills, and the gill rakers—hard projections located on the inner surface of the gill arch. In the space provided in Observations, draw and label a gill. Label the gill arch, gill filaments, and gill rakers.
9. Observe the scales that cover the fish's body. With forceps, carefully remove a single scale. Prepare a wet-mount slide of the scale and observe it under a dissecting microscope. Observe the growth rings on the scale. As a fish grows, the scales grow larger. Each growth ring on the scale represents one year's growth. Answer question 7 in Observations. In the space provided in Observations, draw a scale as seen under a dissecting microscope. Label the growth rings on your drawing. Record the magnification of the microscope.

Name ______________________ Class ____________ Date ____________

10. Locate the lateral line system—a series of sensory pits forming a line on each side of the body from the operculum to the tail.

11. Locate the anus, a small opening anterior to the anal fin. In a male perch, a small opening called the urogenital pore is located behind the anus. Sperm cells produced by the testes are released through the urogenital pore. In a female perch, the genital pore is located behind the anus. Eggs from the ovaries are released through the genital pore. The female fish also has a separate urinary pore located behind the genital pore.

12. On the diagram in Observations, label the following regions and parts of the external anatomy of a perch: head, trunk, tail, anterior dorsal fin, posterior dorsal fin, anal fin, caudal fin, pectoral fin, pelvic fin, eye, nostril, mouth, maxilla, mandible, operculum, gills, lateral line system, anus, urogenital pore, and scales.

13. When you have finished examining your specimen, follow your teacher's instructions for storing the perch for further use.

Observations

Data Table

Type of Fin	Function
Anterior dorsal	
Posterior dorsal	
Anal	
Caudal	
Pectoral	
Pelvic	

Perch Gill

Perch Scale

Magnification ____________

External Anatomy of the Perch

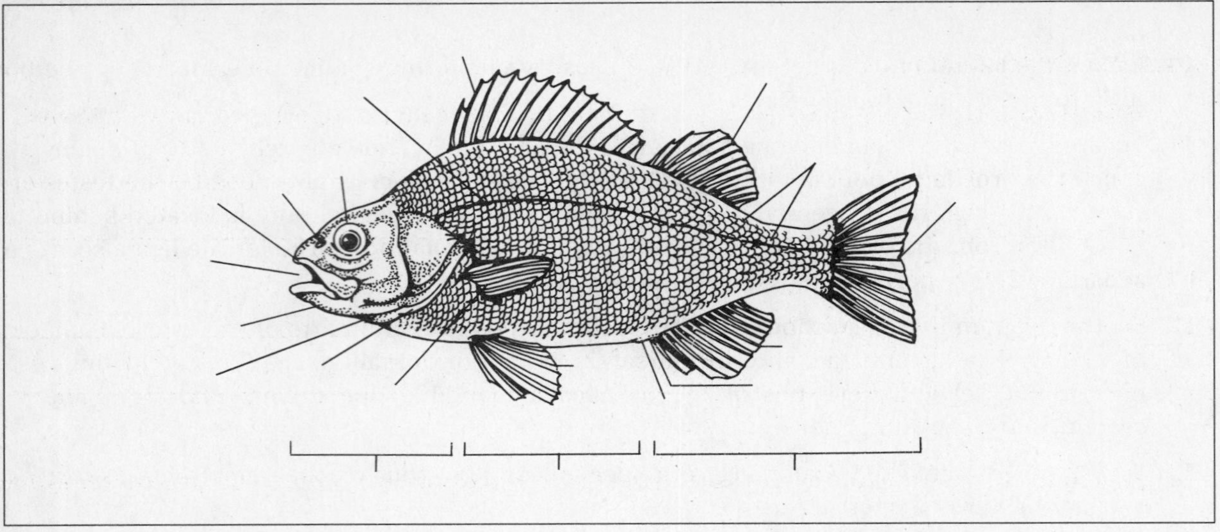

1. a. How many times does the fish open and close its mouth in one minute?

   ____

   b. How many times do the gills move in one minute? ______________________________

2. Describe the motion of the fish as it swims. ______________________________

   ____

3. a. How many individual fins are there on the goldfish? ______________________________

   b. Which fins occur in pairs? ______________________________

   ____

4. Compare the color of the dorsal and ventral surfaces of the fish. ______________________________

   ____

   ____

5. Where do the perch's nostrils lead? ______________________________

   ____

6. Describe the location and texture of the perch's teeth. ______________________________

   ____

   ____

7. How old is your fish? ______________________________

Name ______________________ Class ____________ Date ____________

Analysis and Conclusions

1. What is the relationship between the number of times the mouth opens and closes and the number of times the gill covers move? ______________________

2. How is the fish's body shape an adaptation to its environment? ______________________

3. How does the coloring of the dorsal and ventral surfaces of the fish provide protection against predators? ______________________

4. How are the perch's teeth adapted to their function? ______________________

5. What is the function of the gill rakers of a perch? ______________________

6. What is the function of the lateral line of a perch? ______________________

7. What structures on the perch make it adapted for living in an aquatic environment? ______________________

Critical Thinking and Application

1. While many invertebrates have an exoskeleton, or hard shell covering, vertebrates such as fishes have an endoskeleton. Of what advantage to the fish is the endosketeton?

2. The perch possesses a gas-filled internal structure called the swim bladder. What is the function of the swim bladder?

3. Certain species of fishes that live deep in ocean waters have chemicals within their skin that make them luminescent, or glow-in-the-dark. Of what advantage is this characteristic to these fishes?

4. The perch fertilizes its eggs externally and leaves the eggs exposed on rocks. The guppy fertilizes its eggs internally and gives birth to live young. Which fish probably produces fewer eggs? Compare the survival rate of these two species.

Going Further

1. Examine the skeleton of a large bony fish. To do so, boil the fish in water to make the flesh easier to remove. Then carefully remove the flesh and reassemble the skeleton. Sketch the fish's skeletal system and, using reference materials, label all of the bones.
2. Obtain a preserved shark or other cartilaginous fish. Compare the external structures with those of the perch.
3. Find out what you can about *Latimeria,* a lobe-finned fish that is called a coelacanth. Write a report on its discovery.

Name ______________________ Class __________ Date __________

The Frog

Pre-Lab Discussion

Frogs belong to the class Amphibia. Amphibians have adaptations for living in terrestrial as well as aquatic environments. Frogs are among the most commonly studied organisms in biology. Although many differences exist between humans and frogs, the basic body plans are similar. Humans and frogs both belong to the phylum Chordata. By studying the anatomy of the frog, you will be better able to understand your own body.

In this investigation, you will observe the behavior of a live frog. You will examine the external features of a frog and identify parts of its external anatomy. In addition, you will dissect a preserved frog to observe its internal anatomy.

Problem

How is a frog structured for survival?

Materials *(per group)*

Live frog
Large glass jar
Aquarium with water
Preserved frog
Dissecting tray
Scissors
Probe
Forceps
Hand lens
Plastic bag
Scalpel
Dissecting pins
Dissecting needle
Paper towels
Medicine dropper
Clock with second hand

Safety

Put on a laboratory apron if one is available. Handle all glassware carefully. Be careful when handling sharp instruments. Always use special caution when working with laboratory chemicals, as they may irritate the skin or cause staining of the skin or clothing. Never touch or taste any chemical unless instructed to do so. Follow your teacher's directions and all appropriate safety procedures when handling animals. Note all safety alert symbols next to the steps in the Procedure and review the meanings of each symbol by referring to the symbol guide on page 10.

Procedure
Part A. Observing a Live Frog

1. Line a large glass jar with moist paper towels. Place a live frog in the jar. Cover the jar so that the frog will not escape. **Note:** *Do not let the paper towels become dry. Sprinkle water on the towels if they begin to dry out.*
2. Observe the frog without disturbing it. Answer question 1 in Observations.
3. Closely observe the movement of the throat and nostrils as the frog breathes. Also observe the sides of the body. Answer questions 2 and 3 in Observations.
4. With your fingers, gently prod the frog and observe how it jumps. Observe the function of the front and back legs as the frog jumps and lands. Answer questions 4 and 5 in Observations.
5. Moisten your hands with water. Carefully pick up the frog and place it in an aquarium partially filled with room-temperature water. Observe how the frog swims and floats. Answer questions 6 through 10 in Observations.
6. Return the live frog to your teacher when you have completed your observations.

Part B. External Anatomy of the Frog

1. Obtain a preserved frog. Rinse the frog with water to remove excess preservative. **CAUTION:** *The preservative used on the frog can irritate your skin. Avoid touching your eyes while working with the frog.* Dry the frog with paper towels and place it in a dissecting tray.
2. Identify the dorsal and ventral surfaces and the anterior and posterior ends of the frog. Answer question 11 in Observations.
3. Locate the forelegs and the hindlegs. Each foreleg, or arm, is divided into four regions: upper arm, forearm, wrist, and hand. Each hindleg also has four regions: thigh, lower leg, ankle, and foot. Identify the parts of the forelegs and hindlegs. Examine the hands and feet of the frog. If the hands have enlarged thumbs, the frog is a male. Answer questions 12 through 14 in Observations.
4. Locate the two large, protruding eyes. Lift the outer eyelid using a probe. Beneath the outer lid is an inner lid called the nictitating membrane. Answer question 15 in Observations.
5. Posterior to each eye is a circular region of tightly stretched skin. This region is the tympanic membrane, or eardrum. Locate the tympanic membranes on both sides of the head.
6. Anterior to the eyes, locate two openings called the external nares (singular, naris), or nostrils.
7. In the appropriate place in Observations, label the following external areas and structures of a frog: anterior, posterior, dorsal, ventral, forelimb, hand, hindlimb, foot, tympanic membrane, external nares, eye, nictitating membrane, and mouth.
8. Hold the frog firmly in the dissecting tray. Using scissors, make a small cut at each of the hinged points of the jaw, as shown in Figure 1. **CAUTION:** *To avoid injury, cut in a direction away from your hands and body.* Open the mouth as much as possible. Under running water rinse away any excess preservative.

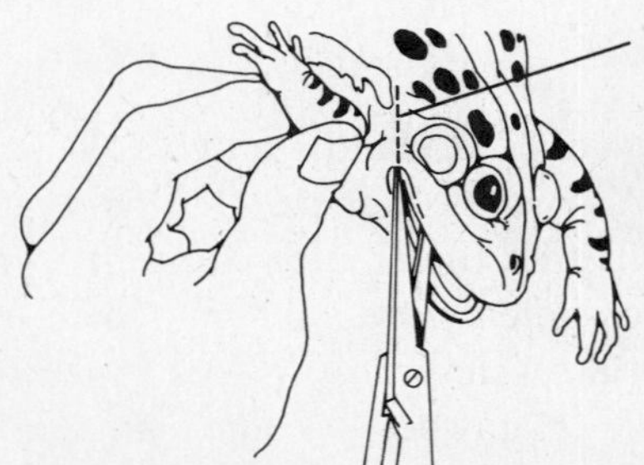

Figure 1

9. The tongue is the most noticeable structure in the mouth. Observe where the tongue is attached and note the two projections at the free end. Answer question 16 in Observations.

10. At the back of the mouth, locate the large horizontal opening, the gullet opening. In front of the gullet opening, find a vertical slit, the glottis.

11. Look for two openings on the back sides of the floor of the mouth. These are the openings to the vocal sacs. They are present in male frogs but not in female frogs.

12. Examine the roof of the mouth. Near the front center of the roof of the mouth are two small bumps. These bumps are the vomerine teeth. On either side of the vomerine teeth are the openings of the internal nares. Behind the vomerine teeth, observe two large bulges. These bulges are the eye sockets. Run your fingers along the top jaw. The teeth you feel are the maxillary teeth. The openings of the Eustachian tubes are on either side near the back of the mouth. Insert a probe into an opening of one Eustachian tube. Note where the probe stops.

13. In the appropriate place in Observations, label the following parts of a frog's mouth: vomerine teeth, internal nares, maxillary teeth, eye sockets, openings to Eustachian tubes, tongue, gullet opening, glottis, and openings to vocal sacs.

Part C. Internal Anatomy of the Frog

1. Place your preserved frog in a dissecting tray with the ventral surface up. With dissecting pins, securely pin the frog's feet and hands to the bottom of the dissecting tray as shown in Figure 2. Angle the pins away from the body of the frog so that they will not interfere with your dissection.

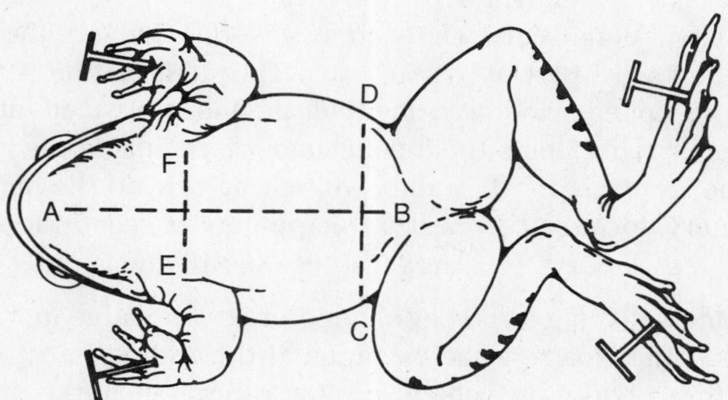

Figure 2

2. With forceps, lift the loose skin of the abdomen. Carefully insert the tip of a pair of scissors beneath the skin. **CAUTION:** *To avoid cutting yourself, cut in a direction away from your hands and body.* Cut the skin along line *AB as* shown in Figure 2. Using forceps and scissors, continue cutting the skin along lines *CD* and *EF*.

3. With your fingers, carefully separate the skin from the underlying muscles. Open the flaps of skin as far back as possible and pin them to the bottom of the dissecting tray. Angle the pins away from the body of the frog so that they will not interfere with your dissection. Notice the blood vessels branching throughout the inner lining of the skin. Observe the abdominal and pectoral muscles. Note the direction of the muscle fibers.

4. Carefully lift the abdominal muscles with the forceps. Cut a second *AB* incision. **Note:** *Keep the cut through the muscles shallow so as not to damage underlying organs.* As the incision is made in the chest, or pectoral area, you will need to cut through bone. This bone is part of the pectoral girdle. **Note:** *Use extra force with the scissors when cutting through the bone. Be careful not to damage any of the internal organs below the bone.* Make cuts *CD* and *EF* through the abdominal muscle.

5. Remove the pins holding the skin in place. Stretch the abdominal opening as much as possible. At this time the hands and feet of the frog may need to be repinned.

6. Study the positions of the exposed organs. Notice that most of the organs are held in place by thin, transparent tissues called mesenteries.

7. If the frog is a mature female, the most obvious organs will be the ovaries. The ovaries are white sacs swollen with tiny black-and-white eggs. Carefully lift the ovaries from the body cavity, cut the attachments with scissors, and remove the ovaries from the frog. **Note:** *Be careful not to rupture the ovaries with the scissors. If the ovaries are ruptured, they can spill out a mess of eggs.*

8. The large reddish-brown organ in the upper part of the abdominal cavity is the liver. Answer question 17 in Observations.

9. With your fingers or a probe, lift and separate the lobes of the liver upward. Behind the middle lobe, look for a greenish, finger-shaped gland. This gland is the gallbladder. You may be able to locate the bile duct leading from the liver to the gallbladder.

10. With scissors, carefully remove the liver and gallbladder from the body. The remaining organs of the digestive system are easier to see with the liver removed.

DIGESTIVE SYSTEM

11. Locate the esophagus, which is a white tube leading from the mouth and connecting to the upper part of the white, muscular stomach. Notice the shape of the stomach. Look for a constriction at the lowest part of the stomach. This constriction is the pylorus. The pylorus leads into the long, coiled small intestine. Pull the loops of small intestine away from the body. Notice the mesentery that holds the intestines in place. Inside the first loop of the small intestine near the stomach, locate a thin, white organ called the pancreas. Also in the intestinal mesentery, locate a brown bean-shaped organ called the spleen. **Note:** *The spleen is an organ of the circulatory system.* Answer questions 18 and 19 in Observations.

12. The small intestine ends in a large bag-shaped organ, the large intestine. The last organ of the digestive system is the cloaca, a saclike organ at the end of the large intestine. Undigested food leaves the frog's body through an opening called the anus.

13. With scissors, cut the esophagus near the stomach. Cut through the large intestine just above the cloaca. With your fingers, carefully remove the digestive system from the body.

14. Stretch out the digestive system on the dissecting tray. With scissors, cut open the stomach along its outside curve. Open the stomach and examine its structure and contents. Answer questions 20 through 22 in Observations.

15. Dispose of the digestive system, liver, and ovaries according to your teacher's instructions.

16. In the appropriate place in Observations, label the following parts of the frog's digestive system and related organs: esophagus, stomach, pylorus, small intestine, large intestine, cloaca, liver, gallbladder, pancreas, mesentery, anus, and spleen.

Name ______________________________ Class ____________ Date ______________

UROGENITAL SYSTEM

17. The reproductive system and urinary system of the frog are closely connected and can be studied as the combined urogenital system. The two kidneys are reddish-brown organs located on the dorsal posterior wall of the abdominal cavity. The kidneys lie on either side of the backbone. **Note:** *The kidneys may be covered with a thin membrane. If so, carefully tear open the membrane with the point of a dissecting needle.* The yellow, fingerlike lobes attached to the kidneys are fat bodies. A small, twisted tube called the ureter leads from each kidney into the saclike urinary bladder. The bladder is connected to the cloaca.

18. Locate the reproductive organs of the frog. If your frog is a male, it possesses testes, tiny white or yellow oval organs found on the ventral surface of the kidneys.

19. If your frog is female, it possessed egg-filled ovaries that were removed in step 7. If your frog is an immature female, the pale oval ovaries are located ventral to the kidneys. Leading from each ovary is a long, coiled tube called the oviduct. The oviduct extends along the side of the body cavity. The oviduct eventually joins the cloaca.

20. In the appropriate place in Observations, label the following parts of the male and female urogenital systems: kidney, fat body, ureter, urinary bladder, cloaca, testes, ovary filled with eggs, and oviduct.

RESPIRATORY SYSTEM

21. Locate the two lungs. They are small, spongy brown sacs that lie to the right and left of the heart. Look for the bronchial tubes that extend from the anterior part of the lungs and join with the trachea, or windpipe.

22. Insert a dropper into the glottis of the frog. Pump air into the lungs and observe what happens. Answer question 23 in Observations.

23. With scissors and forceps, carefully remove the lungs from the frog's body. Dispose of the lungs according to your teacher's instructions.

CIRCULATORY SYSTEM

24. Locate the heart. The heart is encased in a membranous sac called the pericardium. With the tip of the scissors, carefully cut open the pericardium.

25. Note the vessels attached to the heart. The large artery on the ventral surface of the heart is the coronary artery. **Note:** *If the frog has been injected with red and blue latex paint, the veins and arteries will be obvious.*

26. Carefully cut the blood vessels leading to and from the heart. Remove the heart from the frog. Place the heart in the dissecting tray with the dorsal surface facing up. Identify the right and left atria and the ventricle. Touch and compare the walls of the two atria and the ventricle. Answer question 24 in Observations.

27. Observe the dorsal surface of the heart. Locate the thin-walled triangular sac called the sinus venosus. Locate the two veins leading from the top and the one vein leading from the bottom of the sinus venosus.

28. With a scalpel, cut the heart into anterior and posterior halves. Note the thickness of the walls and the types of heart chambers. **CAUTION:** *Be careful when using a scalpel. Always cut in a direction away from your hands and body.*

29. In the appropriate place in Observations, label the following structures of the frog's heart: right atrium, left atrium, ventricle, coronary artery, and sinus venosus.

MUSCULAR SYSTEM

30. Remove the pins from the frog's feet and hands.

31. Cut the skin completely around the upper thigh of one leg, as if cutting off the leg of a pair of pants. With forceps, carefully pull the skin downward to the foot. Expose the thigh muscles, the knee, and the calf muscles.

32. Move the lower leg up and down to simulate the leg movement during a jump. Observe the various leg muscles involved in the leg movement. Answer question 25 in Observations.

33. Follow your teacher's instructions for storing the frog for further use or properly disposing of the frog and its parts. Thoroughly wash, dry, and put away your dissecting tray and tools. Wash your hands with soap and water.

Observations

External Anatomy of the Frog

Name ______________________ Class __________ Date __________

Mouthparts of the Frog

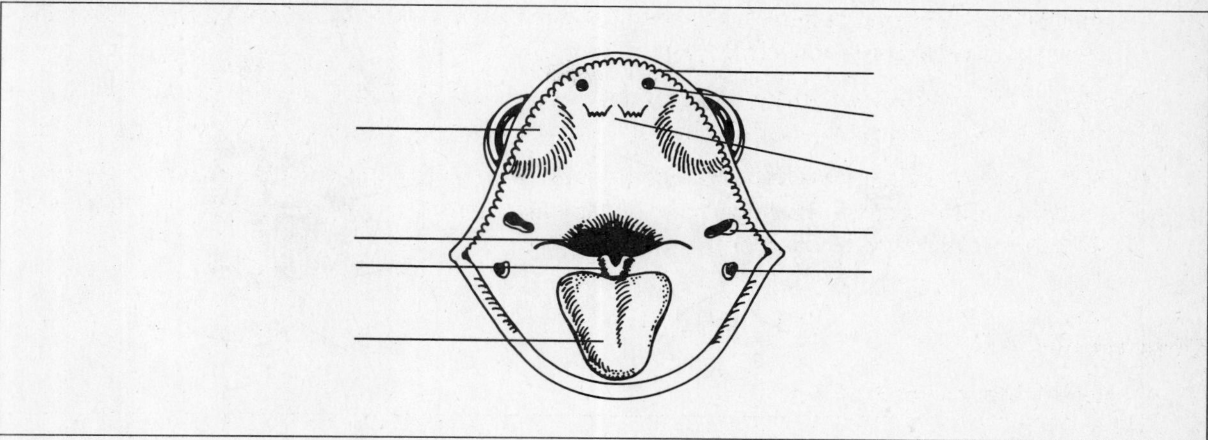

Digestive System and Other Parts of the Frog

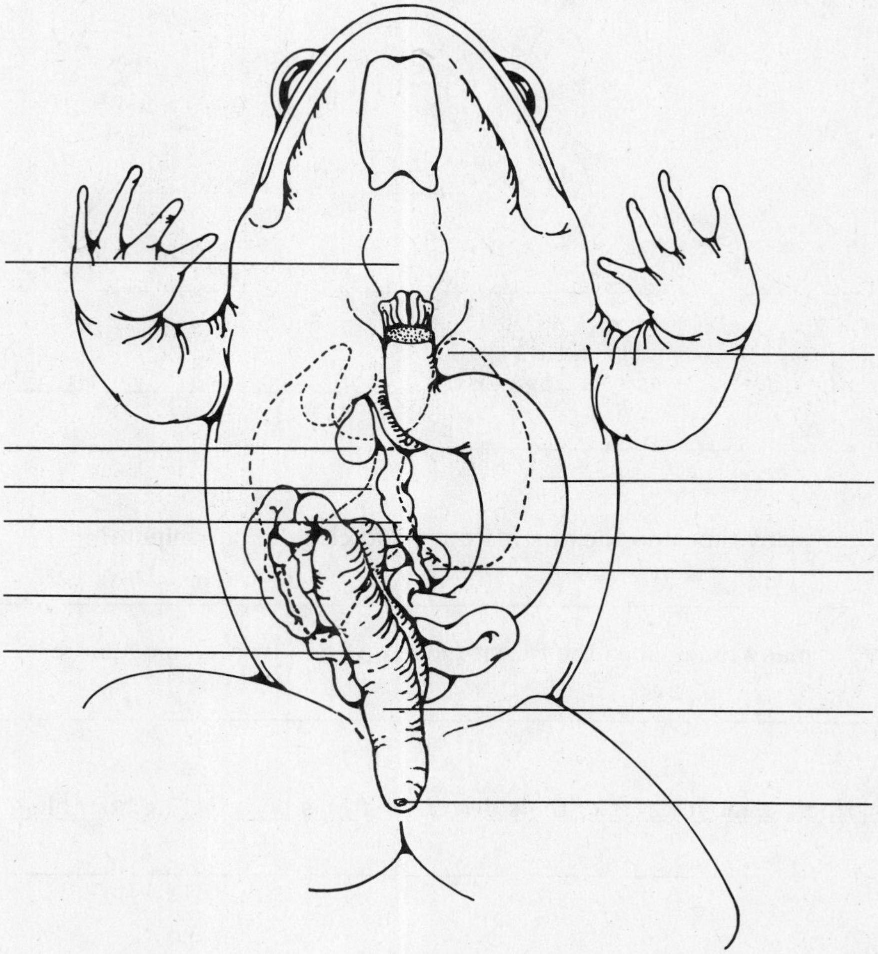

Urogenital System of the Frog

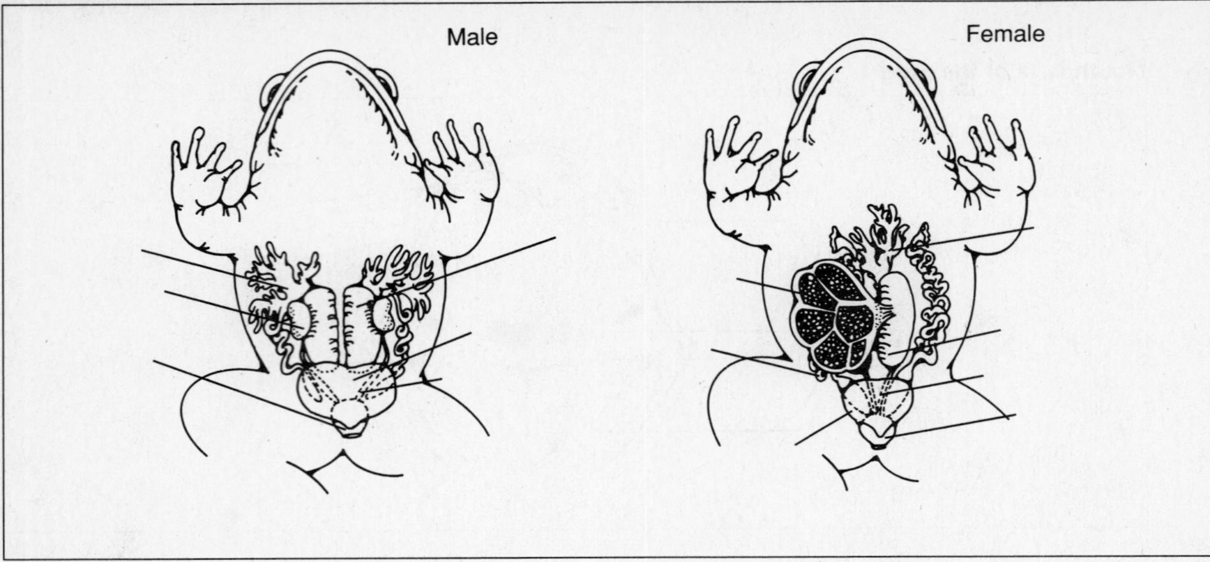

The Frog Heart

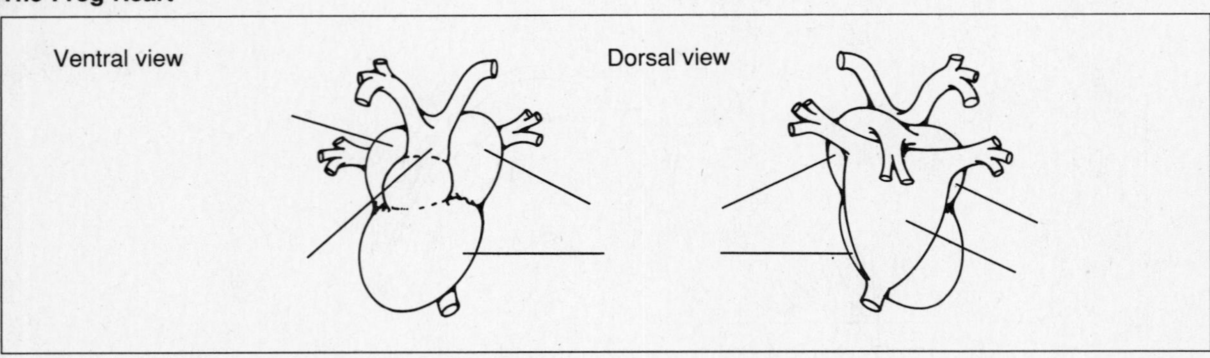

1. Describe how the eyes of a frog close. ______________________________

____

2. a. How many times do the nostrils open and close in one minute? ______________

____

b. How many times does the throat move up and down in one minute?

____

3. Describe the movements of the sides of the frog's body during breathing.

____

4. Describe how a frog jumps. ______________________________

____

Name ______________________ Class ____________ Date ____________

5. a. Are the hindlegs or forelegs more important in jumping? ______________________

____

b. Are the hindlegs or forelegs more important in landing? ______________________

____

6. Is the skin of the frog smooth or rough? Moist or dry? ______________________

____

7. Why is the frog difficult to hold? ______________________

____

8. Describe the position of a frog when it floats. ______________________

____

9. What parts of the body remain above the surface when the frog floats?

____

10. How does the frog use its legs while swimming? ______________________

____

11. Describe the color of the dorsal and ventral surfaces of the frog. ______________________

____

12. How many digits are on each of the frog's hands? ______________________

____

13. How many digits are on each of the frog's feet? ______________________

____

14. Is your frog male or female? How can you tell? ______________________

____

15. Where is the nictitating membrane attached? ______________________________

____

16. Where is the tongue attached to the mouth? ______________________________

____

17. How many lobes does the liver contain? ______________________________

____

18. What is the shape of the stomach? ______________________________

____

19. Describe the mesentery that holds the intestines. ______________________________

____

20. Describe the general shape, or plan, of the frog's digestive system.

____

21. Describe the inside wall of the stomach. ______________________________

____

22. Describe the contents of the frog's stomach. ______________________________

____

23. What happens when air is pumped into the lungs? ______________________________

____

24. Compare the walls of the two atria and the ventricle. ______________________________

____

25. Describe the movement of the leg muscles as the leg is bent and straightened.

____

Name ______________________ Class ____________ Date ____________

Analysis and Conclusions

1. How does the number of times the nostrils open and close in one minute compare with the number of times the throat moves up and down in one minute? ____________________

2. How are the feet of a frog adapted for swimming? ____________________

3. How is the coloration of the frog an adaptation to its habitat? ____________________

4. How is the location of the nares an adaptation to living in water? ____________________

5. The tip of the tongue in a live frog is sticky. What would be the advantage of this?

6. How does the length of the small intestine relate to its function in absorbing digested food?

7. Explain why in a frog's heart the ventricle has a thicker wall than the atrium.

Critical Thinking and Application

1. Frogs are insect eaters. How is the frog's tongue designed for the type of food it eats?

2. List three adaptations that permit the frog to live on land successfully.

____

____

3. List three adaptations that permit the frog to live in water successfully.

____

____

____

4. The frog's sense organs are located on top of the head. How does this help the frog when it is in the water? ____

____

Going Further

1. Construct a terrarium with a habitat that is favorable to a frog. Obtain a live frog and place it in the terrarium. Feed the frog several live insects and observe its feeding behavior.
2. Examine the skeleton of a preserved frog. To do this, carefully remove the skin, muscles, and organs of the frog. Work in one area of the body at a time until the bones are exposed. Using reference materials, identify as many bones as possible.
3. Carefully remove the brain and spinal cord of a preserved frog. To do this, remove the skin from the dorsal surface of the frog. Using a probe and a dissecting needle, carefully chip away the bone of the cranial cavity and the vertebral column. **Note:** *To remove the brain and spinal cord in one piece, work slowly and carefully.* Using reference materials, identify the parts of the brain. Sketch and label the parts of the brain.

Name ______________________ Class ____________ Date ____________

Adaptations in Lizards

Pre-Lab Discussion

Members of class Reptilia, which are called reptiles, are ectothermic, or "cold-blooded," vertebrates that are covered with dry scaly skin. They are adapted for reproduction on land. Most species are land-dwelling, but some species spend much time in water. Turtles, snakes, lizards, crocodiles, and alligators are reptiles. Reptilian skin is dry, thick, and waterproof, protecting the body from drying out even in very arid climates. The skin is covered by tough scales that protect the animal from injury.

Anolis carolinensis, or the anole or American chameleon, is a common lizard found throughout the southern United States. It can be found on shrubs, trees, and fences and on the ground. Often it is found around homes, and it seems to thrive in areas inhabited by humans. The anole is best known for the variability of its skin color, which can change from green to brown or gray. This change in skin color is thought to be stimulated by changes in light intensity, temperature, and emotional state.

In this investigation, you will observe the external structures of an anole. You will also observe its response to environmental change.

Problem

What are the external structures of a lizard? What changes occur in a lizard's coloration in different environments?

Materials *(per group)*

Preserved anole
Live anole
Aquarium or large glass jar
6 sheets of construction paper (green, yellow, brown, red, black, and white)
Metric ruler
Dissecting tray
Paper towels

Safety

Put on a laboratory apron if one is available. Handle all glassware carefully. Always use special caution when working with laboratory chemicals, as they may irritate the skin or cause staining of the skin or clothing. Never touch or taste any chemical unless instructed to do so. Follow your teacher's directions and all appropriate safety procedures when handling live animals. Note all safety alert symbols next to the steps in the Procedure and review the meanings of each symbol by referring to the symbol guide on page 10.

Procedure

Part A. External Anatomy of the Anole

1. Obtain a preserved anole. Rinse the anole with water to remove excess preservative. **CAUTION:** *The preservative used on the anole can irritate your skin. Avoid touching your eyes while working with the anole.* Dry the anole with paper towels and place it in a dissecting tray.

2. Measure the entire body length of the anole in centimeters. Record this measurement in Data Table 1. Also measure the length of the tail alone in centimeters. Record this measurement in Data Table 1. Obtain the measurements of four other groups of students and record this information in Data Table 1. Determine the average length of the five anoles and their tails and record this information in Data Table 1.

3. Observe the texture and color of the anole's skin. Answer questions 1 and 2 in Observations.

4. Identify the sex of your anole. Males are generally larger than females and have a dewlap, or fold of skin under the neck. Females have only a small, primitive dewlap. Answer question 3 in Observations.

5. Locate the head, trunk, and tail of the anole.

6. Examine the head of the anole. Look for the presence of eyelids and external ear openings. Look for the presence of nostrils.

7. Open the mouth of the anole. Look for the presence of teeth.

8. Examine the feet of the anole.

9. In the appropriate place in Observations, label the following external structures of the anole: head, trunk, tail, dewlap, foot, nostrils, and external ear opening.

10. Follow your teacher's instructions for storing the anole for further use.

Part B. Protective Behavior of an Anole

1. Place a sheet of white construction paper on the floor of an aquarium or a large glass jar.

2. Obtain a live anole. Place the anole in the aquarium or jar on the sheet of paper. Record the original color of the anole in Data Table 2. Observe any color changes in the skin of the anole. Record your observations in Data Table 2. **Note:** *Allow the anole to calm down for several minutes before observing any changes in color.*

3. Repeat steps 1 and 2 for each of the colored sheets of paper.

4. Return the anole to your teacher.

Name ______________________ Class ____________ Date ____________

Observations

External Anatomy of the Anole

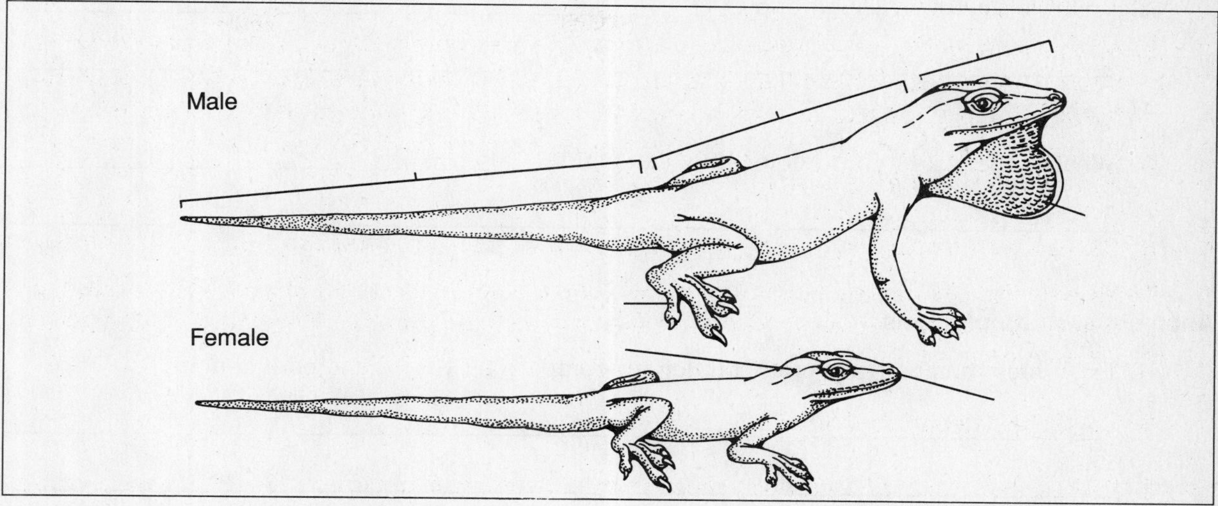

Data Table 1

Anole	Entire Body Length (cm)	Tail Length Only (cm)
1		
2		
3		
4		
5		
Average		

Data Table 2

Background Color	Original Color of Anole	Anole Color After Several Minutes

1. Describe the texture and appearance of the anole's skin. ______________________

____

2. What is the color of your anole? ______________________

____

3. What is the sex of your anole? How can you tell? ______________________

____

Analysis and Conclusions

1. How does the anole's average tail length compare to its average body length?

____

____

2. What is a possible function of the dewlap? ______________________

____

____

3. What is the function of the eyelids on the anole? ______________________

____

4. What is the function of the teeth of the anole? ______________________

____

5. How are the anole's feet adapted for life on land? ______________________

____

6. In what way do bright and dark background colors affect the intensity of reflected light?

____

____

Name ______________________ Class __________ Date __________

7. In general, what color does the anole become when placed against dark backgrounds? Against light backgrounds? ______________________

____

____

Critical Thinking and Application

1. The anole has the ability to lose its tail and then regenerate a new one. How is this a useful adaptation for the anole? ______________________

____

____

2. Only two species of lizard are poisonous. Why do most lizards not need to produce poisonous venom for survival? ______________________

____

____

3. Name three adaptations that anoles have for living on land. ______________________

____

____

4. Why are reptiles not found in arctic regions? ______________________

____

____

Going Further

1. Observe living anoles for several weeks. Record and compare the behaviors of males and females. Observe movements, social structure, territoriality, and feeding habits.
2. Obtain a liquid-crystal temperature strip. Under the supervision of your teacher, hold the strip against the skin of a live anole. Record the surface temperature of the anole. Place the anole in the sun for 5 minutes and check the temperature again. Carry out the same procedure with a small mammal such as a gerbil or hamster and compare the results.

Name ______________________ Class ____________ Date ____________

64

Examining Bird Adaptations

Pre-Lab Discussion

All birds are grouped in the class Aves. Birds are endothermic, or warm-blooded, vertebrates. The streamlined shape of their feather-covered bodies enables them to fly efficiently. Although a number of modern birds cannot fly, all birds are descended from ancestors that were capable of flight. The forelimbs of birds are wings, which are used for flight in most birds. Birds also have a four-chambered heart and a closed circulatory system.

There are many different types of birds. Each type of bird has special adaptations that enable it to live successfully in its environment. The mouth of a bird is formed by a toothless projecting beak. The various shapes and sizes of beaks are adaptations for eating different kinds of foods. The structure of a bird's legs and feet also exhibits adaptations based on the bird's native environment. Birds can also have different types of feathers. These feathers, which are actually modified scales, serve many different functions.

In this investigation, you will examine some of the general characteristics of birds. You will also observe how the feathers, legs, feet, and beaks of various birds enable them to survive in their environments.

Problem

In what ways are various birds adapted to different environments?

Materials *(per group)*

Contour feather
Down feather
Pinfeather
Hand lens
Chicken leg bone (femur)
Scalpel or single-edged razor blade

Safety

Put on a laboratory apron if one is available. Handle all glassware carefully. Be careful when handling sharp instruments. Note all safety alert symbols next to the steps in the Procedure and review the meanings of each symbol by referring to the symbol guide on page 10.

Procedure

Part A. Examining Feathers

1. Obtain a contour feather. Use Figure 1 to locate the different parts of the feather. The shaft of the feather has two parts: the quill and the rachis. Observe that the rachis supports the hairlike barbs that make up the vane. Notice that the portion of the vane on one side of the shaft is narrower than the vane portion on the other side of the shaft. In a bird's wing, the narrow portion of the vane on one feather overlaps the wide portion of the vane on the feather next to it. This slight overlap allows for a smooth, continuous wing surface.

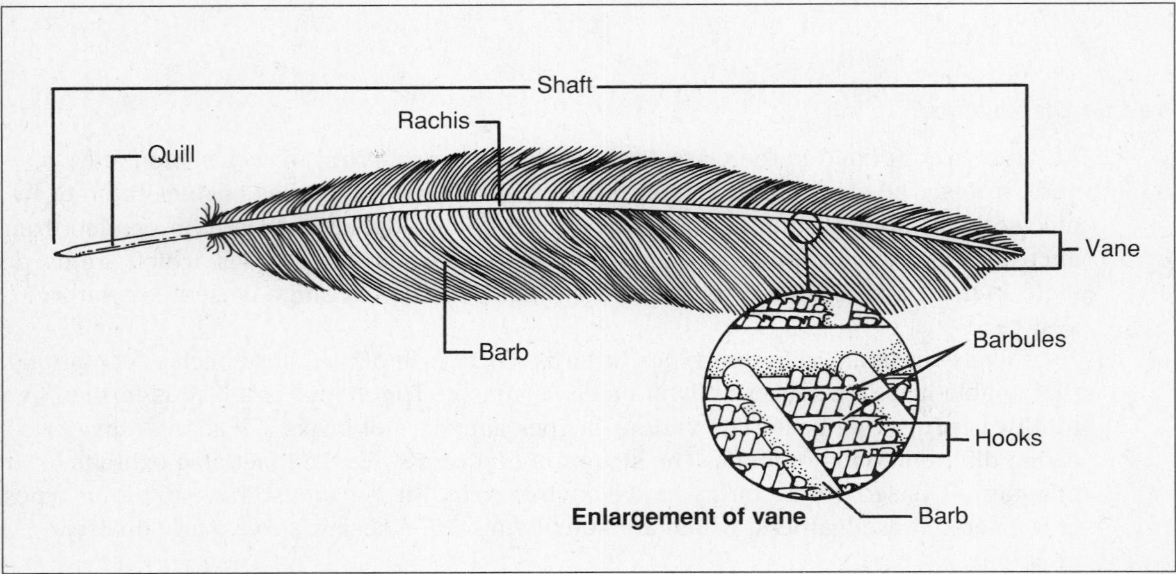

Figure 1

2. Use the hand lens to closely examine the quill. Answer question 1 in Observations.
3. Use the hand lens to examine the feather's vane. Gently ruffle the edge of the feather, and find the barbules on the barbs. Smooth the feather with your finger. Notice how easily the barbs can be smoothed back into place.
4. Obtain a down feather. Identify the quill, rachis, and barbs of this feather. In the space provided in Observations, draw the down feather. Label the quill, rachis, and barbs.
5. Notice the length, width, and flexibility of the shaft of the down feather. Answer question 2 in Observations.
6. Examine the down feather with a hand lens. Try to smooth the down feather as you did the contour feather. Answer question 3 in Observations.
7. Obtain a pinfeather, or filoplume. Examine the pinfeather with a hand lens. The pinfeather consists of a single, hairlike shaft that ends in a few barbs. In the space provided in Observations, draw a pinfeather. Label the shaft and barbs.

Part B. Examining a Chicken Bone

1. Observe the relative weight of a chicken leg bone (femur). Note its flexibility.
2. Using a scalpel or single-edged razor blade, carefully cut the chicken femur in half. **CAUTION:** *Be very careful when using a scalpel or razor blade. To avoid injury, cut in a direction away from your hands and body.* Observe the internal structure of the bone. Answer question 4 in Observations.
3. Follow your teacher's instructions for the proper disposal or storage of the chicken bone.

Name ______________________ Class __________ Date __________

Part C. Examining Adaptations of Bird Feet

1. Observe the drawings of the birds in Figure 2 on this page and on the following page. Examine the toes of each bird. Count the number of toes on the foot of each bird and record this information in Data Table 1. Answer question 5 in Observations.

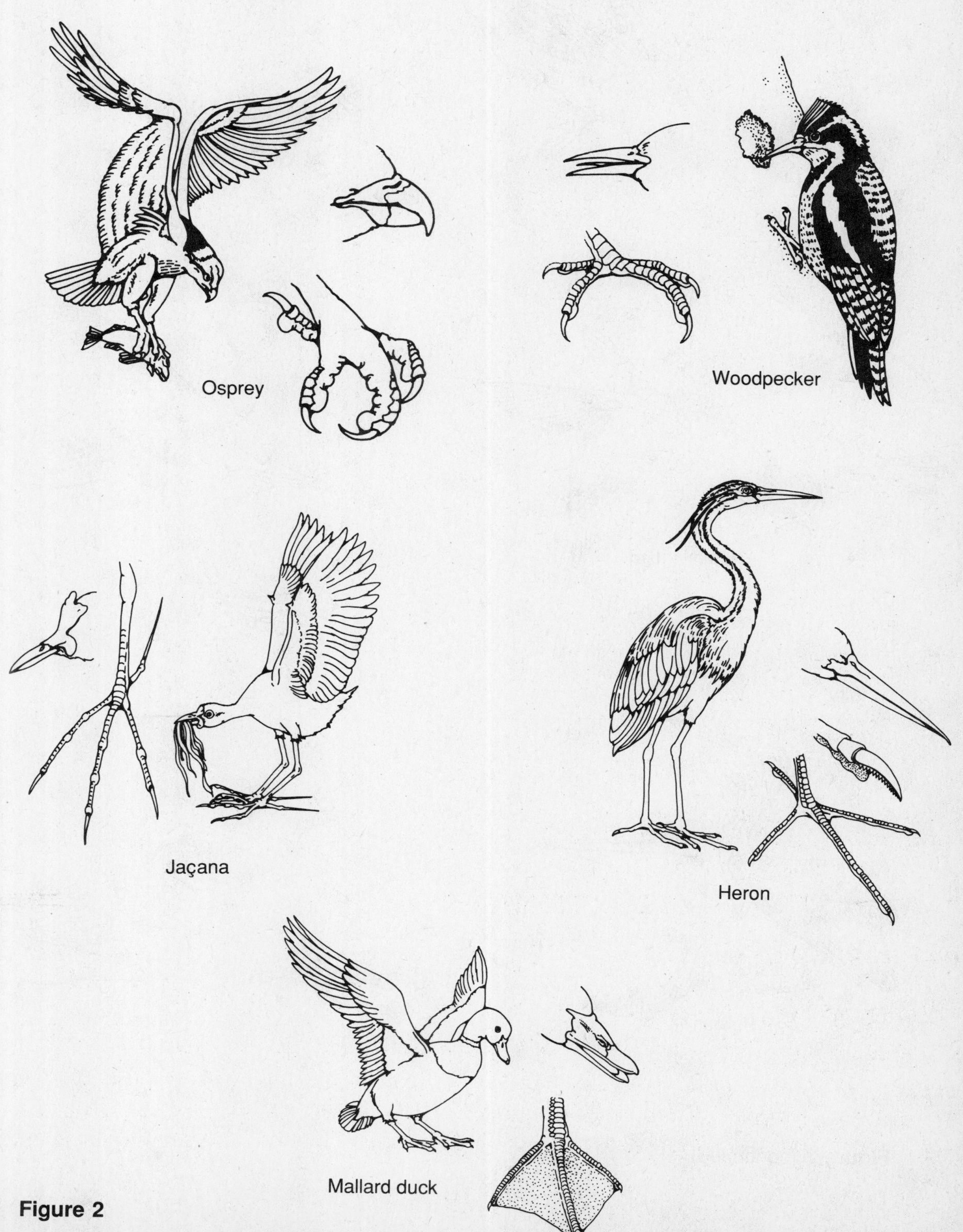

Figure 2

Figure 2 (continued)

2. Examine the foot of each bird in Figure 2 and indicate the position (front or back of foot) of the toes. Record this information in Data Table 1. Also examine the relative size of the talons, or nails. Describe them as large, medium, small, long, thin, etc. Record this information in Data Table 1.

3. Determine the structure and function of each foot from the following list. Record this information in Data Table 1.

   - Scratching foot: rakelike toes for finding food in soil
   - Perching foot: long back toe that can hold on to a perch tightly
   - Swimming foot: webbed, paddlelike
   - Running foot: three toes rather than four
   - Wading foot: large foot and long leg for wading in shallow water
   - Specialized foot: long talons and toes for running over leaves of large water plants
   - Climbing foot: two hind toes for support when climbing upward to prevent falling backward
   - Grasping foot: large curved claws to grab and hold such prey as fish, mice, and other small animals

Part D. Examining Adaptations of Bird Beaks

1. Examine the relative shape and size of the beak of each bird in Figure 2. Determine the structure and function of each beak from the following list. Record this information in Data Table 2.

   - Chisel: used for drilling into trees
   - Short and stout: used to eat insects, seeds, small crustaceans; multipurpose
   - Tubular: used to obtain nectar from flowers
   - Hooked: used to tear flesh
   - Flat, broad, and slightly hooked: used to strain algae and small organisms from water
   - Cracker: used to crack seeds; short and stout; sometimes curved upper portion
   - Scoop: used to scoop fish from water; long and stout
   - Spear-shaped and stout: used to spear fish
   - Trap: used to trap insects in midair

Observations

Down Feather	Pinfeather

Data Table 1

Bird	Number of Toes	Toe Positions	Size of Talons	Type or Function
Heron				
Osprey				
Woodpecker				
Duck				
Jaçana				
Quail				
Pelican				
Hummingbird				
Rhea				
Whippoorwill				

Data Table 2

Bird	Structure of Beak	Function of Beak
Heron		
Osprey		
Woodpecker		
Duck		
Jaçana		
Quail		
Pelican		
Hummingbird		
Rhea		
Whippoorwill		

1. Is the quill of the contour feather solid or hollow? ______________________________

Name ______________________ Class ____________ Date ____________

2. How does the down feather shaft compare to the contour feather shaft?

3. Do the barbs of the down feather stick together when you smooth it?

4. Describe the external and internal features of the chicken femur.

5. How many toes do most birds have?

Analysis and Conclusions

1. How does the structure of a contour feather shaft make it well adapted for flight?

2. How do hooks increase the strength of a contour feather?

3. Down feathers are found underneath a bird's contour feathers. What is the function of this type of feather?

4. Why is the filoplume also called a pinfeather?

5. How are a bird's bones well adapted for flight?

6. In what way are the feet of the woodpecker adapted to its feeding position?

____

____

Critical Thinking and Application

1. What characteristics of the down feather make it useful as a stuffing for pillows and sleeping bags? ____

____

____

2. The skeleton and feathers of a bird are found. The bones are solid and heavy, and the forelimbs are short. Do you think that this bird was able to fly well when it was alive? Explain your answer. ______________________________________

____

____

3. Write a description of the beak and feet of each of the following:

   a. An aquatic bird that strains plankton from the water for food ______________

   ____

   ____

   b. A bird that eats insects out of the cracks in trees ______________________

   ____

   ____

Going Further

1. Complete a nature study of birds. Obtain a field guide to birds. Use the guide to help identify the birds in your area. A pair of binoculars will also be helpful. Observe the movements, feeding behavior, coloration, nest shape, and location of the birds.
2. Many bird species have become extinct in recent years. Find out about the extinction of the passenger pigeon, dusky seaside sparrow, Hawaiian o'o, Carolina parakeet, and moa.

Name ______________________ Class ____________ Date ____________

The Most Intelligent Mammal

Pre-Lab Discussion

Mammals have the most highly developed brain of all animals. This enables them to perform a wide range of complicated behaviors.

Some of a mammal's behaviors are unlearned. This means that the mammal is capable of performing the behavior without being taught. Unlearned behaviors include reflexes and instincts. Reflexes are simple, quick, automatic responses—jerking away from contact with a hot object, for example. Instincts are inborn behavioral patterns that can be modified very little, if at all. A spider's web-building, a crane's courtship dance, and a newborn mammal's suckling are examples of instincts.

Many of a mammal's behaviors are learned, or acquired through experience. Because mammals have a large, highly sophisticated cerebrum—the part of the brain involved with learning, memory, and thinking—they are capable of performing a wide array of learned behaviors. Some examples of learned behaviors in humans include habits and solving visual or word problems. The most advanced type of learned behavior is known as insight learning, or reasoning. Reasoning is the ability to apply previous learning to a totally new situation. Reasoning is rare in animals other than primates and is found most often in humans, which are the most intelligent of all mammals.

In this investigation, you will study learned and unlearned human behavior.

Problem

Which types of human behaviors are unlearned and which are learned? What roles do reflexes, conditioned responses, habits, trial-and-error learning, and reasoning play in human behavior?

Materials *(per pair of students)*

Paper
Pencil
Clock or watch with second hand

Procedure

Part A. Reflexes

1. A reflex is a simple, automatic response to a stimulus. A reflex usually involves only part of the body. Working with a partner, you will alternate as subject and helper while you test two human reflexes. You will record the responses in Data Table 1.
2. Close your eyes and cover them with your hands. At the end of one minute, remove your hands and open your eyes while your partner watches your eyes closely. Record the response in Data Table 1.

3. Stand with your side to a wall. Hold your arm down at your side and slightly away from your body. Tightly press the back of your hand against the wall until your shoulder begins to ache. After one minute, step away from the wall while still holding your arm stiff. Record the response in Data Table 1.

Part B. Conditioned Responses

1. Read the following instructions to your partner: "Each time I say 'Write,' I want you to make a tally mark on a sheet of paper. Then place your pencil in position to make the next mark."
2. Give the command to write several times in succession. For most of the commands, hit your pencil on the desk at the same time that you say the word "Write."
3. Occasionally, hit your pencil on the desk but do not give the command to write. Try this several times. Answer question 1 in Observations.

Part C. Habits

1. Divide a sheet of lined paper in half lengthwise. Label one column "Normal Hand." Label the other column "Other Hand." Now use your normal writing hand to write your name as many times as you can in 30 seconds. Follow the same procedure in the second column using your other hand. Answer questions 2 and 3 in Observations.
2. Dictate the following passage to your partner while he or she writes it down. Dictate at a fairly rapid pace. "Habits are often useful in allowing routine activities to be carried out quickly. But most of us have some habits that we would like to break. Breaking a habit is not a simple thing to do."
3. Dictate the same passage again, but this time instruct your partner not to cross any *t*'s or dot any *i*'s. Answer question 4 in Observations.

Part D. Trial-and-Error Learning

1. Trial-and-error learning begins when an animal associates certain responses with favorable or unfavorable consequences. The animal then tries to repeat those behaviors that led to favorable results.
2. Find out how quickly you can successfully complete a path through Maze 1 in Figure 1 in Observations. Use a pencil to mark your path and have your partner time you. Record the time needed to complete Maze 1 in Data Table 2.
3. Complete the rest of the mazes in succession, timing each one as you go. Cover each completed maze as you finish so that you cannot look back at the completed mazes. Record your results in Data Table 2. Graph your results on the graph in Observations.

Part E. Reasoning

1. Reasoning is a type of learning that involves thinking, judgment, and memory. Reasoning enables humans to solve new problems without resorting to trial and error.
2. Use your ability to reason to solve the following problems.
   a. There are four separate, equal-sized boxes. Inside each box there are two separate small boxes. Inside each of the small boxes there are three even smaller boxes. Answer question 5 in Observations.

Name ______________________ Class __________ Date __________

b. Figure 2 in Observations represents nine bears in a square enclosure at the zoo. Build two more square enclosures within this one so that each bear is in a pen by itself. Draw the borders of the two new enclosures directly on Figure 2.

c. There are five girls: Maureen, Sue, Jill, Robin, and Pam. They are standing in a row, but not necessarily in the order named. Neither Maureen nor Sue is next to Robin. Neither Sue nor Maureen is next to Pam. Neither Robin nor Sue is next to Jill. Pam is just to the right of Jill. Name the girls from left to right. Answer question 6 in Observations.

Observations

Data Table 1

Stimulus	Response
Light	
Pressure on arm muscle	

1. What happened when you hit the pencil on the desk but did not give the command to write?

____

____

2. How many times did you write your name with your normal writing hand?

____

3. How many times did you write your name with your other hand?

____

4. How many times did you cross *t*'s or dot *i*'s in the second passage?

____

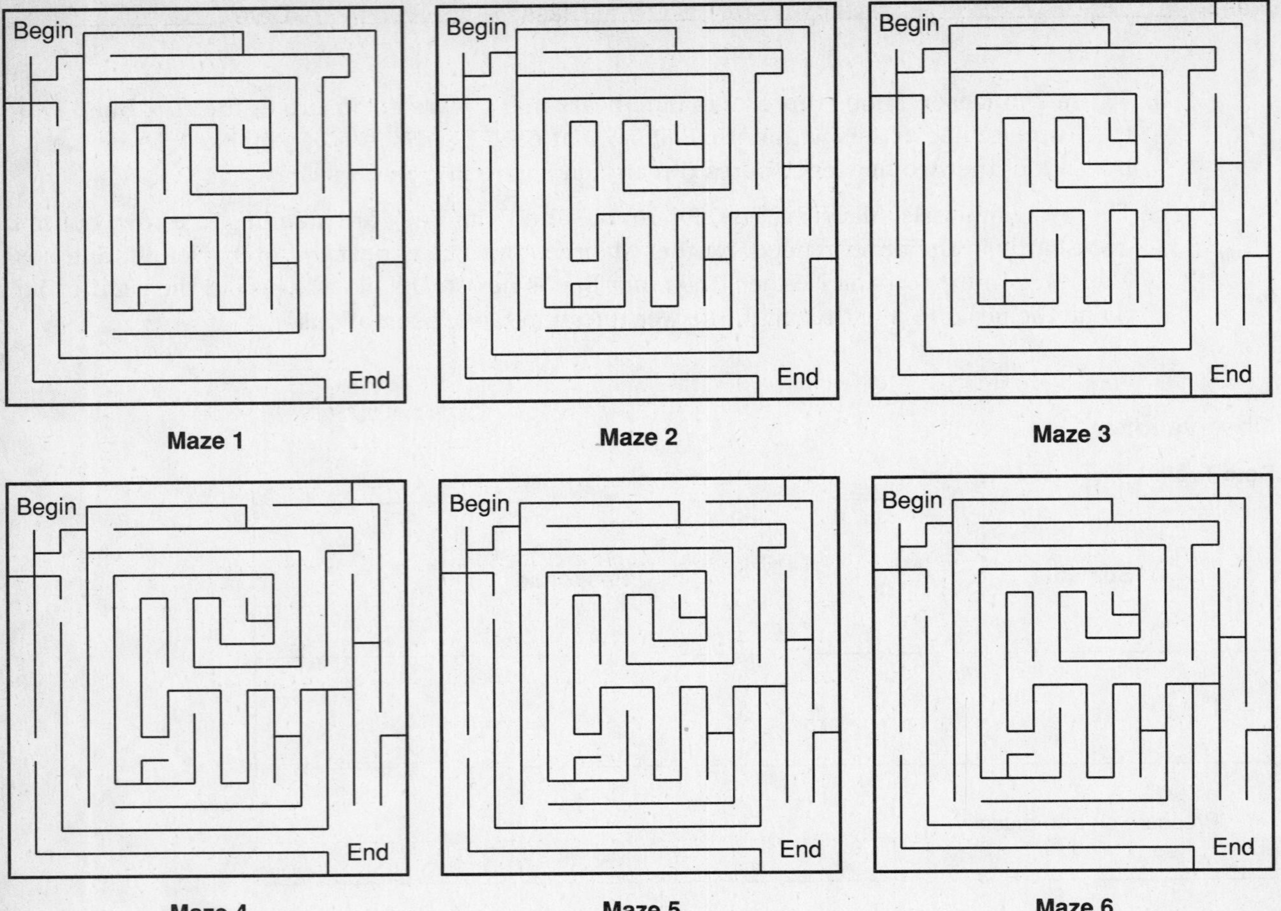

Figure 1

Data Table 2

Maze Number	**Time** (sec)
1	
2	
3	
4	
5	
6	

Name ______________________ Class ______________ Date ______________

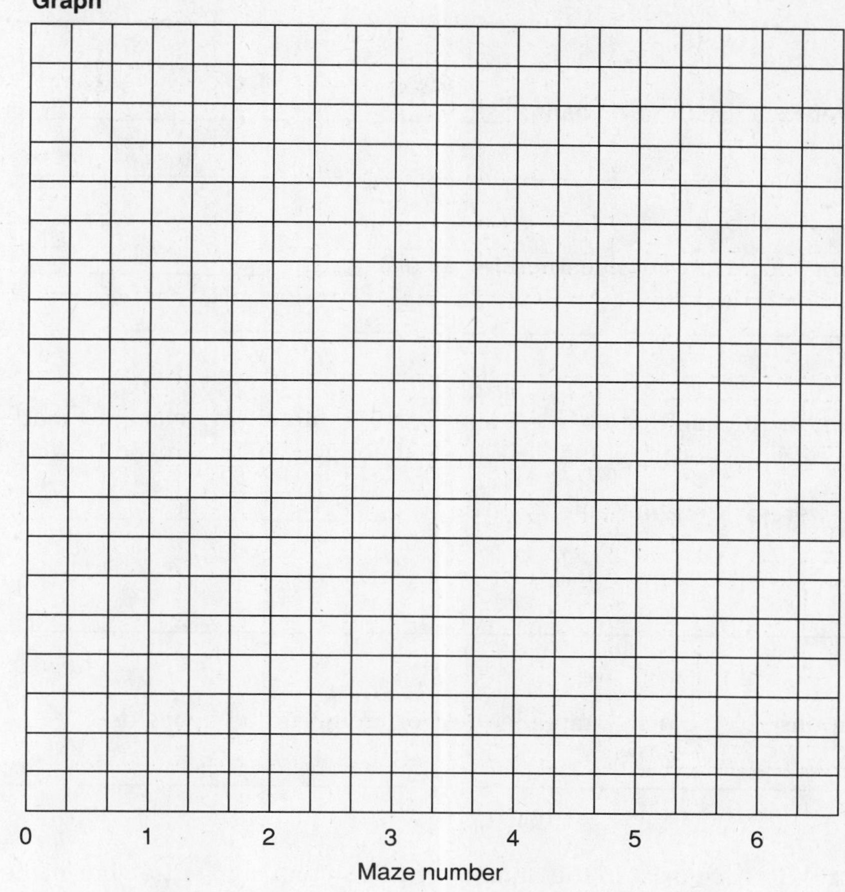

5. What is the total number of boxes? ______________________

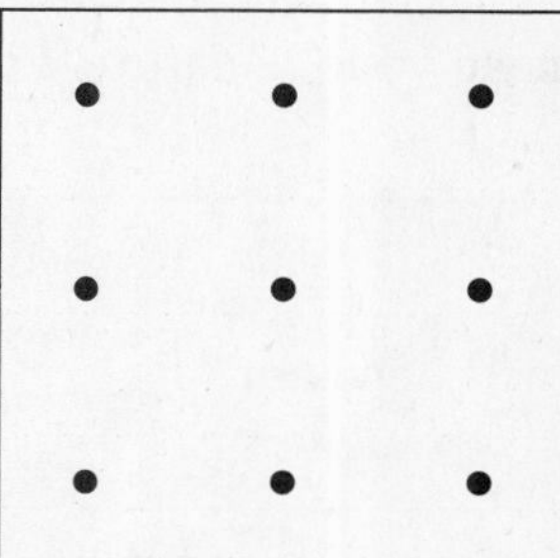

Figure 2

6. Name the girls from left to right: ______________________

Analysis and Conclusions

1. Name three other reflexes in humans. ______________________________

____

2. How are reflexes useful to humans? ______________________________

____

3. Is a reflex a learned or an unlearned behavior? ______________________________

____

4. Were you able to condition or "fool" your partner into always making a mark on the paper when you hit your pencil on the desk but did not give the command to write? Explain why you think you were or were not able to do so. ______________________________

____

____

5. Is the response to the pencil tap a learned or an unlearned behavior?

____

6. Why did a difference exist in the number of times you could write your name with your normal writing hand and with your other hand? ______________________________

____

7. What would you have to do to make your signature written by both hands the same?

____

8. Is a habit a learned or an unlearned behavior? ______________________________

____

9. Did any learning take place in Part D of this investigation? Give evidence to support your answer. ______________________________

____

Name ______________________ Class __________ Date __________

Critical Thinking and Application

1. How is the blinking response a protective reflex? ______________________

2. Describe three conditioned reflexes that you exhibit while at school.

3. How might you go about breaking yourself of a particular habit? ______________________

4. Describe three situations in which you learned through trial and error.

5. Describe two situations in which the ability to reason was useful to you.

Going Further

1. Design and conduct an experiment to show how distractions, such as noise, television, or music, affect the time it takes to learn a short poem or solve an arithmetic problem. Construct a data table to record your results.
2. Cover Columns B, C, and D of the table on page 430 with a piece of paper. Study the words in Column A for one minute. Then write down as many words as you can remember. Record your score. Follow the same procedure for Column B, then C, and then D.

A	B	C	D
Zop	House	Purple	Sally
Wab	Tree	Gold	and
Dod	Shoe	Red	Bob
Jav	Sock	Blue	went
Cug	Dog	Yellow	to
Sor	Floor	Green	the
Duz	Rock	Orange	football
Tig	Father	Black	game
Wek	Candy	White	last
Foy	Picture	Pink	night

Which column was easiest to remember? How do you explain the difference in your ability to learn the words in each column? How can this experiment help you in studying?

Name ______________________ Class ____________ Date ____________

66

Comparing Primates

Pre-Lab Discussion

In *The Descent of Man*, the English naturalist Charles Darwin formulated the hypothesis that human beings and other primates have a common ancestor. A hypothesis is a suggested explanation for observed facts. All scientific hypotheses, including that of Darwin, are based on observations.

Darwin observed that human beings and other primates differ in many important ways. Although all primates have opposable thumbs, the human hand is capable of more refined and exact movements than those of other primates. The human braincase, or cranium, has more volume and more mass than those of other primates. In addition, human beings are bipedal, or able to walk on two limbs. Other primates use all four limbs for locomotion. Being bipedal frees the arms and hands for other tasks, such as toolmaking. Darwin regarded these human traits as adaptations, resulting from natural selection. The adaptations of other primates, he suggested, evolved differently.

More recently, biochemists have determined that certain proteins found in different primates contain many of the same amino acid sequences. Scientists reason that because proteins are produced by DNA, human beings and other primates must have a similar genetic makeup. Paleontologists have also found fossil remains that provide evidence that all primates came from a common ancestor. These and other observations lend support to Darwin's hypothesis of human origins.

In this investigation, you will develop skill in observing and interpreting data provided by biochemists, paleontologists, and other scientists.

Problem

How are human beings similar to and different from other primates?

Materials *(per student)*

Metric ruler
Pencil
Protractor

Procedure

Part A. Comparing Amino Acid Sequences in Vertebrate Proteins

1. Figure 1 shows the amino acids found in selected sites in the hemoglobin of different vertebrates.

Amino Acid Positions in the Hemoglobin of Some Vertebrates

	Human being	SER	THR	ALA	GLY	ASP	GLU	VAL	GLU	ASP	THR
	Chimpanzee	SER	THR	ALA	GLY	ASP	GLU	VAL	GLU	ASP	THR
Primate	Gorilla	SER	THR	ALA	GLY	ASP	GLU	VAL	GLU	ASP	THR
	Baboon	ASN	THR	THR	GLY	ASP	GLU	VAL	ASP	ASP	SER
	Lemur	ALA	THR	SER	GLY	GLU	LYS	VAL	GLU	ASP	SER
	Dog	SER	SER	GLY	GLY	ASP	GLU	ILU	ASP	ASP	THR
Nonprimate	Chicken	GLN	THR	GLY	GLY	ALA	GLU	ILU	ALA	ASN	SER
	Frog	ASP	SER	GLY	GLY	LYS	HIS	VAL	THR	ASN	SER
	Human being	PRO	GLY	GLY	ALA	ASN	ALA	THR	ARG	HIS	
	Chimpanzee	PRO	GLY	GLY	ALA	ASN	ALA	THR	ARG	HIS	
Primate	Gorilla	PRO	GLY	GLY	ALA	ASN	ALA	THR	LYS	HIS	
	Baboon	PRO	GLY	GLY	ASN	ASN	ALA	GLN	LYS	HIS	
	Lemur	PRO	GLY	SER	HIS	ASN	ALA	GLN	LYS	HIS	
	Dog	PRO	SER	ASN	LYS	ASN	ALA	ALA	LYS	LYS	
Nonprimate	Chicken	PRO	GLU	THR	LYS	ASN	SER	GLN	ARG	ALA	
	Frog	ALA	HIS	ALA	LYS	ASN	ALA	LYS	ARG	ARG	

Figure 1

2. Count the number of molecules of each amino acid in human hemoglobin. **Note:** *Be sure to count the molecules of amino acid in human hemoglobin in the second section titled Primate as well as in the first section.* Record these totals in the appropriate column in Data Table 1.

3. Count the number of molecules of each amino acid in the hemoglobin of other vertebrates in Figure 1. (Remember to count both sections.) Record these totals in the appropriate columns in Data Table 1.

4. Going from left to right, note the position of each amino acid. Count the number of similarities in the amino acid positions in human hemoglobin as compared with the hemoglobin of each of the other vertebrates in Figure 1. Record your observations in Data Table 2.

5. Reexamine Figure 1 and count the number of differences in the amino acid positions in human hemoglobin as compared with the hemoglobin of each of the other vertebrates in Figure 1. Record your observations in Data Table 2.

Part B. Comparing Primate Features

1. Determine the relative size of the lower jaw of each primate by measuring the length in millimeters of lines *ab* and *bc* in Figure 2. Record these lengths in Data Table 3. Record the product of these lengths in Data Table 3.

2. Determine the angle of the jaw by using a protractor to measure the angle *xy* in each primate skull in Figure 2. Record your observations in Data Table 3.

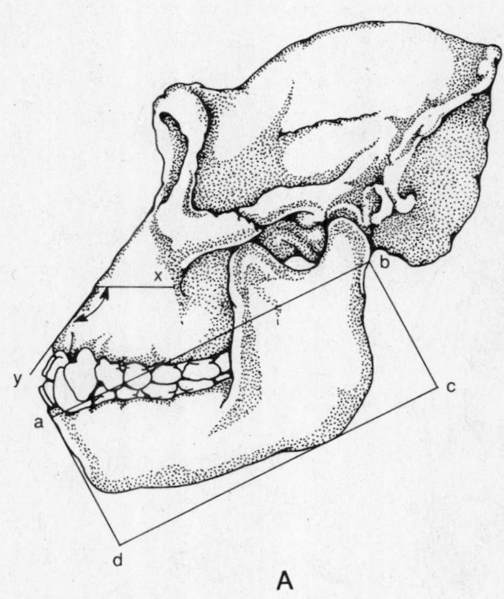

A

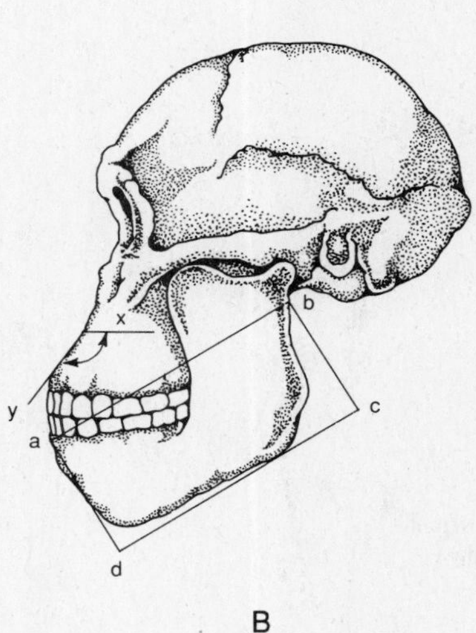

B

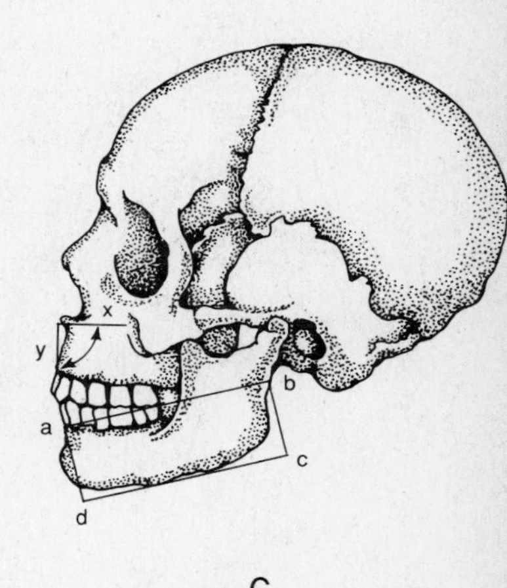

C

Primate Skulls

Figure 2

3. Examine the teeth of each of the three primates in Figure 3.

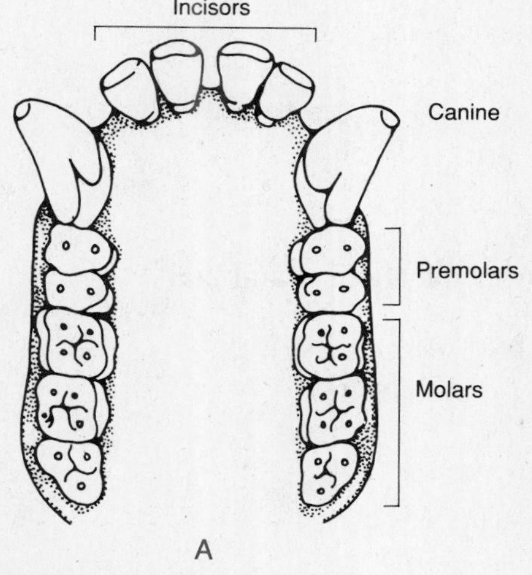

A

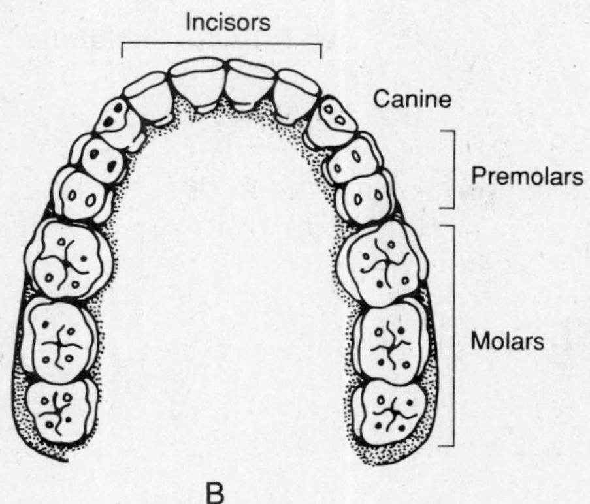

B

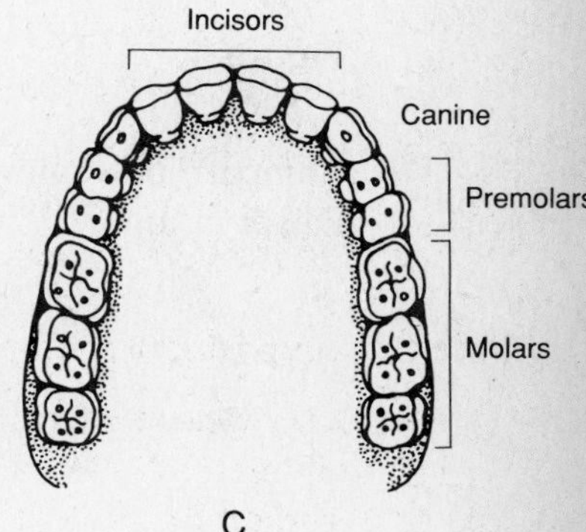

C

Teeth of Three Primates

Figure 3

4. Count the number of incisors, canines, premolars, and molars of each primate in Figure 3. Record your observations in the appropriate columns in Data Table 4.

5. Examine the two skeletons in Figure 4.

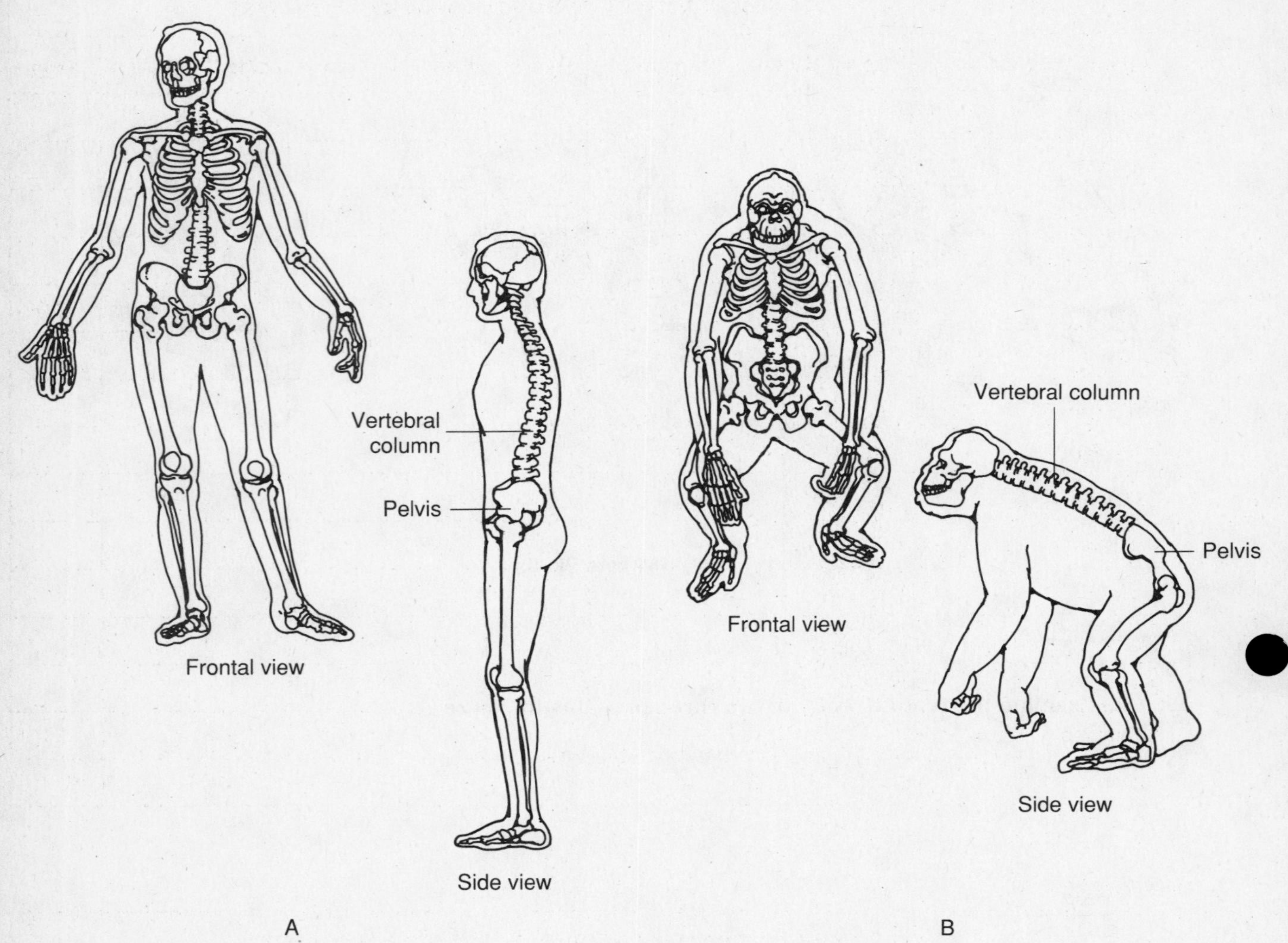

Two Primate Skeletons

Figure 4

6. Compare both views of skeleton A with those of skeleton B. Answer questions 1 and 2 in Observations.

Name ______________________ Class ______________ Date ______________

Observations

Data Table 1

Number of Molecules of Different Amino Acids in Some Vertebrates									
Amino Acid	**Abbreviation**	**Human**	**Chimpanzee**	**Gorilla**	**Baboon**	**Lemur**	**Dog**	**Chicken**	**Frog**
Alanine	ALA								
Arginine	ARG								
Asparagine	ASN								
Aspartic acid	ASP								
Glutamine	GLN								
Glutamic acid	GLU								
Glycine	GLY								
Histidine	HIS								
Isoleucine	ILU								
Leucine	LEU								
Lysine	LYS								
Proline	PRO								
Serine	SER								
Threonine	THR								
Valine	VAL								

Data Table 2

Similarities and Differences in Amino Acid Positions in Hemoglobin		
Organisms	**Number of Amino Acid Position Similarities**	**Number of Amino Acid Position Differences**
Human and chimpanzee		
Human and gorilla		
Human and baboon		
Human and lemur		
Human and dog		
Human and chicken		
Human and frog		

Data Table 3

Comparison of Three Primate Skulls				
Skull	**Length of Lower Jaw** (mm) (*ab*)	**Depth of Lower Jaw** (mm) (*bc*)	**Area of Lower Jaw** (mm^2) (*ab* x *bc*)	**Angle of Jaw**
A				
B				
C				

Data Table 4

Comparison of Primate Teeth			
Type of Teeth	**Number of Teeth**		
	A	**B**	**C**
Incisors			
Canines			
Premolars			
Molars			

1. Describe three differences between skeleton A and skeleton B. ______________________

____

____

____

2. Which type of primate skeleton is considered bipedal? ______________________

____

Analysis and Conclusions

1. a. From your observations in Data Table 2, which primate is most closely related to the human being? ______________________

____

b. Which primate is least closely related? ______________________

____

Name ______________________ Class ____________ Date ____________

2. a. Which nonprimate vertebrate listed in Data Table 2 is most closely related to the human being? ______________________

____

b. Which nonprimate is least closely related? ______________________

____

3. a. Which of the three primates shown in Figure 2 has the largest brain? What do you think is the name of this primate? ______________________

____

b. Which of the three primates shown in Figure 2 has the smallest brain? What do you think is the name of this primate? ______________________

____

4. What is the relationship between jaw size and brain size in these three primates?

____

____

5. From your observations in Data Table 4, what dental characteristics do the primates have in common? ______________________

____

6. Reexamine Figure 3. How would the diet of primate A differ from the diet of primate C?

____

____

7. What is an advantage of being bipedal? ______________________

____

Critical Thinking and Application

1. Many primates use tools. For example, chimpanzees often use sticks to probe ant hills when searching for food. They also use leaves as sponges to collect drinking water. How is the use of tools by humans different from that of other primates? ______________________________

____

____

____

2. The brain of the human being is larger than that of other primates. How would this relate to the different methods of communication displayed by humans and other primates?

____

____

____

3. Certain fossil evidence indicates that the primate ancestors of humans lived in areas where trees were scattered instead of clustered together. How might this type of environment have affected the development of bipedalism in humans? ______________________________

____

____

4. Describe three physical characteristics that are unique to human beings.

____

____

____

Going Further

Visit a local zoo to observe the behavior of gorillas, chimpanzees, baboons, and other primates. Observe the ways in which the animals communicate and interact with one another. What similarities and differences do you observe between the behaviors of the primates you studied and those of human beings? Use a notebook to record your observations.

Name ______________________ Class __________ Date __________

The Hands of Primates

Pre-Lab Discussion

One of the differences between human beings and other primates is the structure of the hand. Being bipedal, or able to walk on two limbs, has freed the arms and hands of human beings for other tasks, such as making and using sophisticated tools. The human hand has a totally opposable thumb adapted for refined movements. In addition, the human hand can be rotated. The opposable thumb and the rotation of the hands enable humans to grasp and hold a variety of objects in many different ways.

In this investigation, you will compare the hands of several different primates. You will also examine the usefulness of a totally opposable thumb.

Problem

What are the similarities and differences between the hands of different primates? How is a totally opposable thumb useful to a human being?

Materials *(per pair of students)*

Masking tape
Comb
Bottle with screw-on cap
Jacket or sweater
Sewing needle
15-cm long piece of sewing thread
Clock or watch with second hand

Safety

Handle all glassware carefully. Be careful when handling sharp instruments. Note all safety alert symbols next to the steps in the Procedure and review the meanings of each symbol by referring to the symbol guide on page 10.

Procedure

Part A. Comparing the Hands of Different Primates

1. Study the six different primate hands shown in Figure 1.

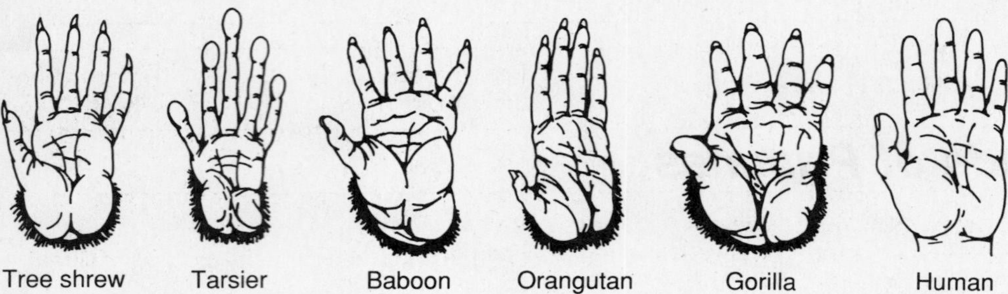

Figure 1

2. After you have carefully examined the diagrams in Figure 1, answer questions 1 and 2 in Observations.

Part B. Examining the Usefulness of the Totally Opposable Thumb

1. Working with a partner, you will alternate as subject and helper throughout this part of the investigation.
2. Do each of the following activities and have your partner time how long it takes you to do each of them. Record this information in the appropriate column of the Data Table.
   - **a.** Remove one shoe
   - **b.** Put your shoe on again, and tie the shoelace
   - **c.** Put on a jacket or sweater
   - **d.** Unbutton a button and button it again
   - **e.** Unscrew a bottle cap
   - **f.** Open a door
   - **g.** Write your name and address
   - **h.** Comb your hair
   - **i.** Take off your wristwatch and put it back on again
   - **j.** Thread a sewing needle
3. Using masking tape, have your partner tape each of your thumbs to its adjacent index finger.
4. After your thumbs are securely taped, try each of the activities listed in step 2 again. Have your partner time how long it takes you to do each without the use of your thumbs. Record this information in the appropriate column of the Data Table. If an activity takes longer than 5 minutes, record the word "Unsuccessful."

Name ______________________ Class __________ Date __________

Observations

Data Table

Activity	Time Needed with Thumbs	Time Needed without Thumbs
Remove one shoe		
Put shoe on again and tie shoelace		
Put on a jacket or sweater		
Unbutton a button and button it again		
Unscrew a bottle cap		
Open a door		
Write your name and address		
Comb your hair		
Take off wristwatch and put it back on		
Thread a sewing needle		

1. List two features that all of the hands in Figure 1 have in common.

____

____

2. List one feature that is unique to each of the hands. ______________

____

____

____

____

Analysis and Conclusions

1. List two features of the tree shrew's hand that make it well adapted for the environment in which it lives. ______________________________

____

2. Based on the structure of its hand, in what type of environment would you expect the tarsier to live? Explain your answer. ______________________________

____

3. The amino acid sequence of the hemoglobin molecules in humans and gorillas is very similar, indicating an evolutionary closeness. Do the structures of the gorilla and human hands support this idea? Explain your answer. ______________________________

____

____

4. Based on the activities that you tried and timed with your thumbs free and taped down, summarize your conclusions about the usefulness of your opposable thumb.

____

____

Critical Thinking and Application

1. Many primates also have an opposable toe on each foot. How is this a useful adaptation?

____

2. The hands of many primates are connected to very long arms. For what type of environment are these primates well adapted? ______________________________

____

3. Besides the activities conducted in this investigation, list three other activities that would be difficult for you to accomplish without opposable thumbs. ______________________________

____

Going Further

Using wildlife magazines and other reference materials, observe the structure of the hands of other species of primates not shown in this investigation. How are the hands of each type of primate adapted for life in its particular environment?

Animal Behavior

Pre-Lab Discussion

The way in which an organism responds in its environment is called *behavior.* Behaviors can be classified as either *inborn* or *learned.* Inborn behavior is also referred to as involuntary, unlearned, or instinctive behavior. Inborn behaviors are predictable, automatic responses to physical or psychological stimuli. The simplest inborn behaviors are known as reflexes. Some reflexes coordinate internal body processes, such as the slowing or quickening of the heartbeat. Other reflexes protect the organism. For example, when a cat is frightened by another animal, it will arch its back and fluff up its fur. This response makes the cat appear larger in an attempt to scare away the other animal. More complex inborn behaviors are known as instincts. Instincts control nest-building in birds, migrations in animals such as wildebeest and caribou, courtship "dances" in stickleback fish, and many other complicated animal behaviors.

Learned behavior is also referred to as acquired, voluntary, or conditioned behavior. This type of behavior depends on memory and repetition. When a conditioned behavior pattern is repeated so often that it becomes almost automatic, it becomes a habit. Habits permit organisms to perform familiar activities without consciously thinking about them. For this reason, habits are often confused with reflexes.

In this investigation, you will be observing both inborn and learned behaviors of a small mammal—a rodent.

Problem

What is a behavior? How do inborn and learned behaviors differ?

Materials *(per group)*

Rodent
Thick gloves
Small animal cage or cardboard box
Clay flower pot or glass container, 10 cm high
Animal maze
Clock or watch with second hand

Safety

Put on a laboratory apron if one is available. Handle all glassware carefully. Follow your teacher's directions and all appropriate safety procedures when handling live animals. Note all safety alert symbols next to the steps in the Procedure and review the meanings of each symbol by referring to the symbol guide on page 10.

Procedure

Part A. Observing Inborn Behaviors

1. Obtain a rodent from your teacher. Observe the physical characteristics of your rodent and answer questions 1 and 2 in Observations.
2. Observe any behavioral responses of the rodent as it moves around the cage or box. Record your observations in Data Table 1.
3. Working with another group of students, carefully pick up your rodent and place it in the box of another rodent of the same species. Observe the two rodents' behaviors as they interact with each other. Record your observations in Data Table 1. **CAUTION:** *Handle the animals carefully and gently, without frightening them. Mice and rats may by picked up gently by their tails. Do not pick up gerbils or hamsters by their tails because the tips of their tails will break off. Wear leather gloves when handling rodents. Separate the animals immediately if one attacks the other.*
4. Return your rodent to its cage or box and allow it to readjust to its surroundings for several minutes.
5. Gently pick up your rodent and place it in an empty clay flower pot or glass container with slanted sides, as shown in Figure 1. Observe the rodent as it explores the flower pot and record your observations in Data Table 1. Record the time it takes the rodent to explore the flower pot, seek a way out and escape. Record the number of times it attempts to escape before succeeding.

Figure 1

6. Return the rodent to its original cage or box and allow it to readjust to its surroundings for several minutes before beginning Part B.

Part B. Observing Acquired Behaviors

1. Obtain a maze from your teacher. Place a piece of rodent food at the end of the maze. Carefully place the rodent at the beginning of the maze, as shown in Figure 2.

Top View of Animal Maze

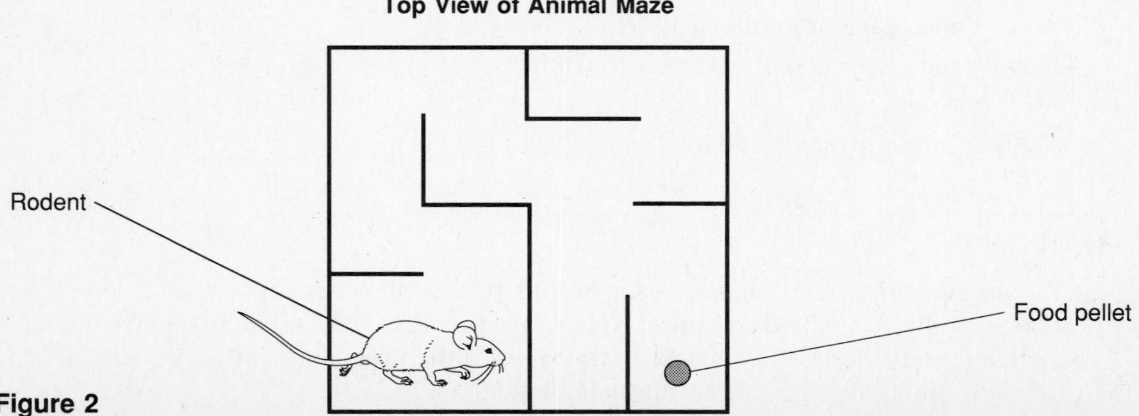

Figure 2

Name ______________________ Class ____________ Date ____________

2. In Data Table 2, record the time it takes the rodent to successfully complete the maze the first time.
3. Repeat steps 1 and 2 four more times and record the times in Data Table 2.
4. Return the rodent to its original cage or box. Return the rodent to your teacher.

Observations

Data Table 1

Type of Environment	Animal Behaviors
Alone in large area	
In an area with another member of the same species	
Alone in small, confining area	

Data Table 2

Trial	Time Needed to Complete Maze
1	
2	
3	
4	
5	

1. What type of rodent are you using to conduct this investigation? ______________________

____

2. List the physical characteristics of the rodent you are using in this investigation.

____

____

____

Analysis and Conclusions

1. In what ways did your rodent's behavior change when it came in contact with another rodent after having been alone? ______________________

____

____

2. Why might an animal's behavior change when it comes in contact with another animal of its own species? ______________________

____

____

3. In what ways did your rodent's behavior change when it was placed in a small, confining place? ______________________

____

____

4. Why might an animal respond differently to a small confining area than to a large nonconfining area? ______________________

____

____

5. Why is the information you collected in Data Table 1 considered to be inborn behavior?

____

____

Name ______________________ Class __________ Date __________

6. How did the time it took the rodent to run the maze the first time compare with the time it took the rodent to run it the fifth time? Why might a change in time have occurred?

7. Why is the information you collected in Data Table 2 an example of acquired behavior?

Critical Thinking and Application

1. Why is the study of animal behavior an important part of biology?

2. How are behavioral responses to a stimulus of benefit to both the individual organism and the species?

3. How is inborn behavior important to an organism?

4. You enter the house of a new friend for the first time. You notice that your friend has a dog. What behaviors would you look for to determine if the dog was acting friendly or aggressively toward you?

Going Further

Due to the incredible variety of physical and behavioral adaptations they possess, insects have been successful in inhabiting even the harshest and remotest places on Earth. To observe the complex behaviors of ants, cut off the top of a 2-liter clear plastic soft drink bottle. Fill an individual-sized soft drink bottle with water and place it inside the 2-liter bottle. Fill the space between the two bottles with moistened sandy soil. Place a 10-centimeter-long piece of cotton in the small bottle to serve as a wick to provide water for the ants. Obtain about a dozen ants from your teacher and place them on the sand. To cover the outside of the 2-liter soft drink bottle, wrap a piece of plastic wrap over the opening and tape it to the sides of the bottle. Use a straight pin to make about 10 small holes in the plastic wrap. Place the entire setup in a pan of shallow water to keep the ants from escaping. Place food on the surface of the soil three times a week. Small pieces of lettuce and hard-boiled egg yolks make good ant food. Using a hand lens, observe the ants twice a week for one month. Use a notebook to record any social behaviors you observe among members of the ant colony.

Name ____________________ Class __________ Date __________

Competition or Cooperation?

Pre-Lab Discussion

A task can be completed or a problem solved using several different organizational strategies. A task can be completed by working alone or as part of a team. An animal might live, hunt, and raise and care for its young alone, or it might live with others of its kind to carry out these same tasks using cooperation.

Humans depend to a large extent on learned behaviors in order to process information, complete tasks, and solve problems. A person can complete a task or solve a problem by working independently or a group of people can cooperate to complete the assigned task.

In this investigation, you will examine different ways to complete an assigned task.

Problem

Is competition between individuals or cooperation among team members a more efficient means of completing a task?

Materials *(per group)*

Clock or watch with second hand.

Procedure

1. Decide which members of your group should play the following roles:
   a. Timekeeper
   b. Independent worker
   c. Cooperative team (two or more students, depending on group size, who will work together to complete a task)
2. The student who is working independently will complete each task with no outside assistance, whereas the cooperative team will complete each task by working together. The timekeeper is responsible for allowing only the designated amount of time to be used for a task and for recording the groups' observations in the Data Table. Each student will have the opportunity to be the timekeeper, the independent worker, and part of the cooperative team.
3. The independent worker should perform the task first. The cooperative team should neither observe the task being performed nor see the final product until they have completed the same task. Remember to change assignments following the completion of each task.

4. **Task A:** In 2 minutes count the number of triangles in the diagram provided by your teacher. The timekeeper should check the correctness of each count using the key provided by your teacher. Record your observations in the Data Table.
5. **Task B:** In 2 minutes count the number of boxes in the diagram provided by your teacher. Record the number of responses in the Data Table.
6. **Task C:** In 1 minute name and record as many words as you can that begin with the letter _____. Record the number of responses in the Data Table.
7. **Task D:** In 1 minute name and record as many _______________ as you can. Record the number of responses in the Data Table.

Observations

Data Table

Task	Number of Responses	
	Independent Worker	**Cooperative Team**
A		
B		
C		
D		

1. Which organizational strategy resulted in the greater number of correct responses?

____

____

2. What kinds of behaviors were observed as the cooperative team completed each task?

____

____

3. What cooperative efforts have you observed or participated in at school?

____

____

Analysis and Conclusions

1. Why is communication important during cooperative work? _______________

____

____

Name ______________________ Class __________ Date __________

2. How are most of your at-school tasks structured—cooperatively or competitively? Explain your answer. ______________________

3. Is cooperation an appropriate strategy for completing all the tasks expected of you in a normal school day? Explain your answer. ______________________

4. Does cooperation provide an advantage to animals in their natural habitats? Be sure to include an example as you explain your answer. ______________________

Critical Thinking and Application

1. Would there be an advantage to an animal of one species cooperating with an animal of another species? Explain your answer. ______________________

2. In what situations do you think cooperation would be more productive than competition?

Going Further

Design an experiment to illustrate the role of communication in completing a task where two or more people are cooperating. You might want to consider the role of both verbal and nonverbal communication as you design your experiment.

Name ______________________ Class __________ Date __________

Vertebrate Skeletons

Pre-Lab Discussion

The body plans of all vertebrates are similar in some ways. One characteristic common to all vertebrates is the presence of a skeleton. The skeleton of a vertebrate is an *endoskeleton,* or internal skeleton. It is made up, in part, of living cells and thus is able to grow. It is not shed as are many exoskeletons. The endoskeleton provides support, protects the internal organs, and is a site for the attachment of muscles. In jawless fishes—lampreys and hagfishes—and in sharks and rays, the endoskeleton is made of cartilage. Bony fishes, amphibians, reptiles, birds, and mammals have endoskeletons of bone with small amounts of cartilage present.

The examination of similar skeletal features in different types of vertebrates reveals important evolutionary links among members of this diverse group. Many biologists consider skeletal similarities to be evidence that these different animals have evolved from a common ancestor. Structures such as bones that have a common origin but different function are called *homologous structures.*

In this investigation, you will compare the skeletons of several different vertebrates and look for evidence of homologous structures. You will also classify unknown bone specimens based on their similarities to and differences from known vertebrates.

Problem

What are the similarities and differences among certain vertebrate skeletons? What homologous structures can be identified on these skeletons?

Materials *(per group)*

Set of "mystery" bones.

Procedure

1. Carefully examine the labeled human skeleton in Figure 1. The human skeleton contains more than 200 bones. Become familiar with the names and structures of the bones in Figure 1.

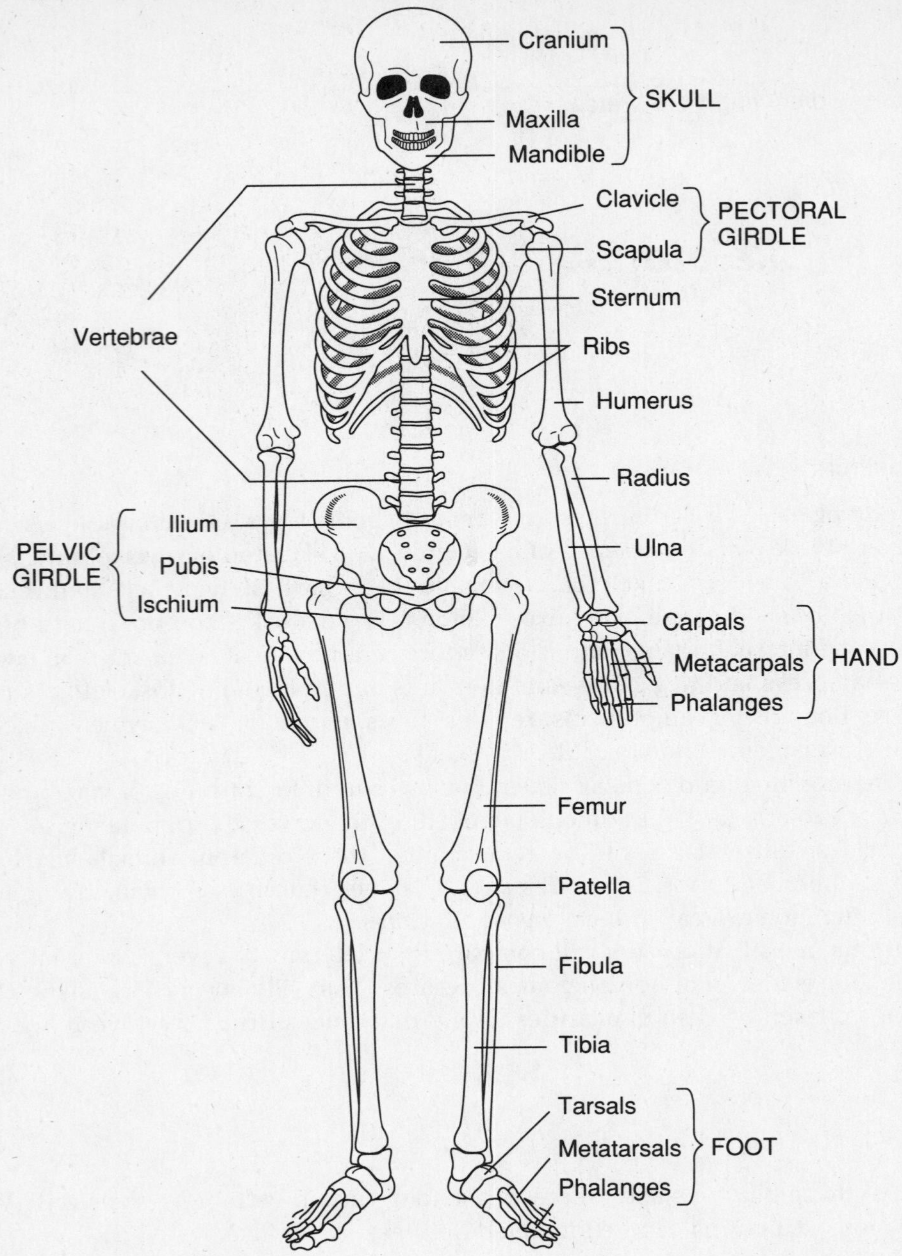

Human Skeleton

Figure 1

2. Look at the frog skeleton in Figure 2. As you examine the skeleton, compare it to the human skeleton in Figure 1. Label the bones of the frog skeleton using the names from Figure 1.
3. Repeat step 2 with the skeletons of the crocodile, pigeon, and cat in Figures 3, 4, and 5.
4. Obtain a set of "mystery" bones from your teacher. Identify the bones by comparing them to the bones of each skeleton observed in this investigation. Answer questions 9 and 10 in Observations.

Name ______________________ Class __________ Date __________

Observations

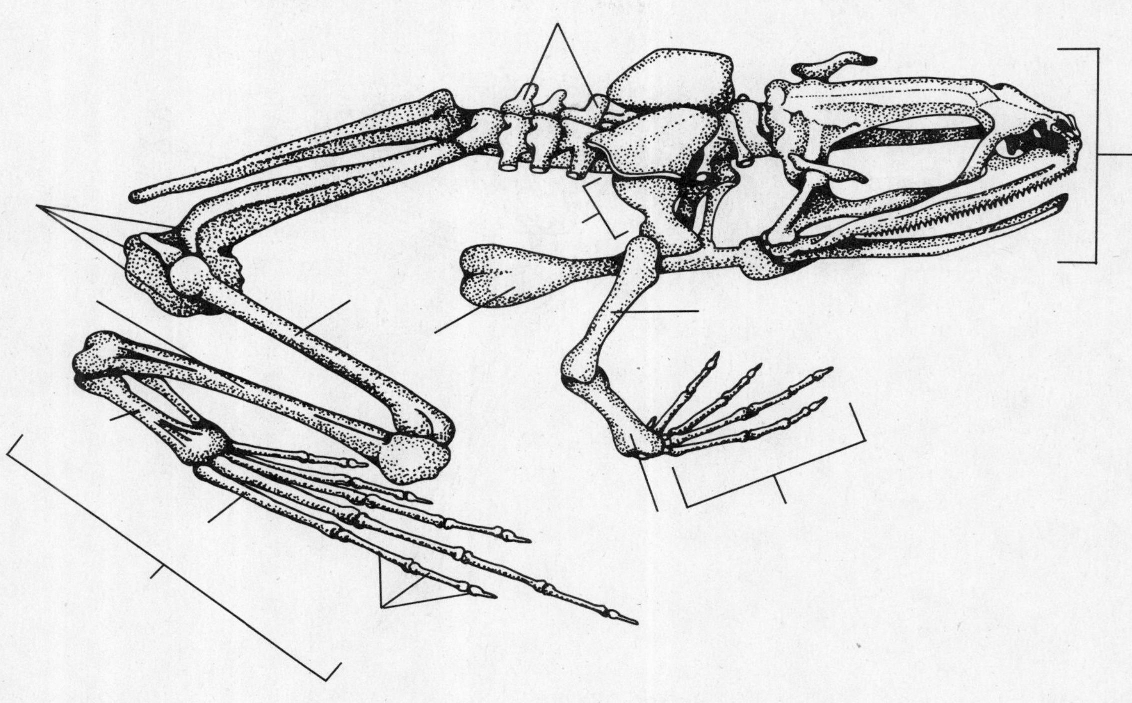

Figure 2 Frog Skeleton

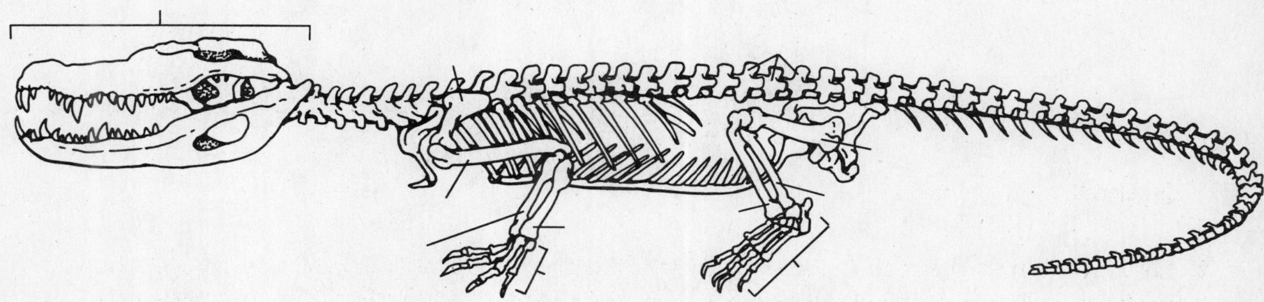

Figure 3 Crocodile Skeleton

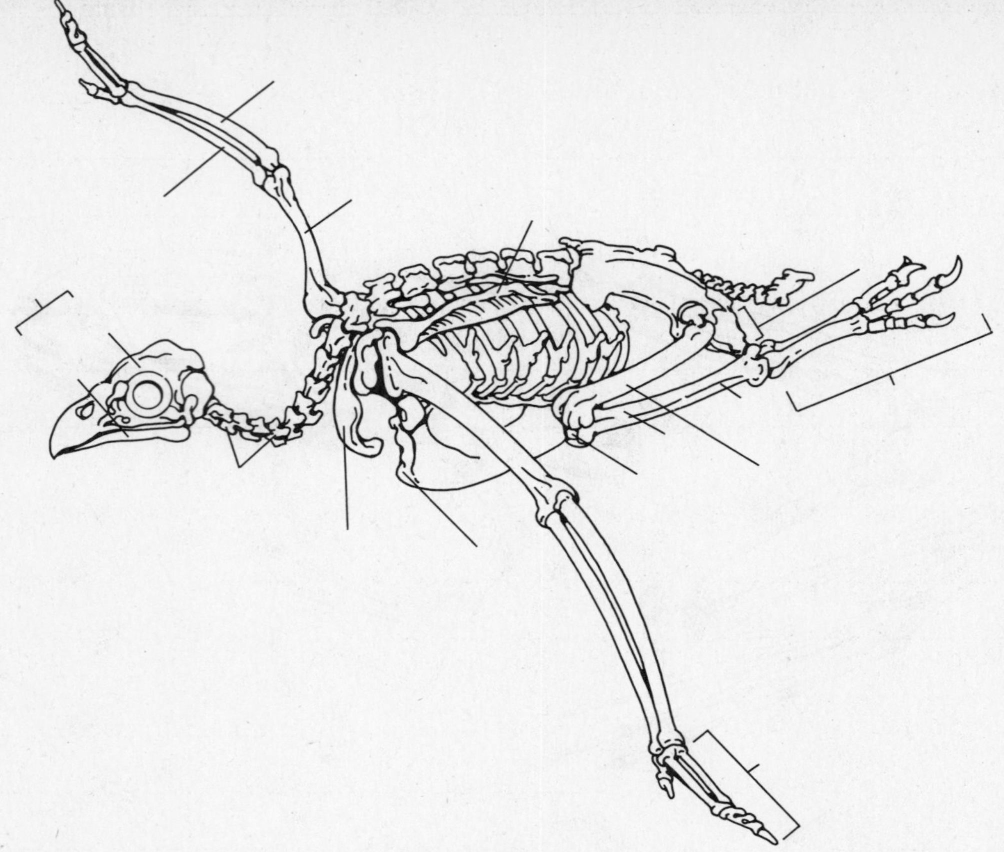

Figure 4 **Pigeon Skeleton**

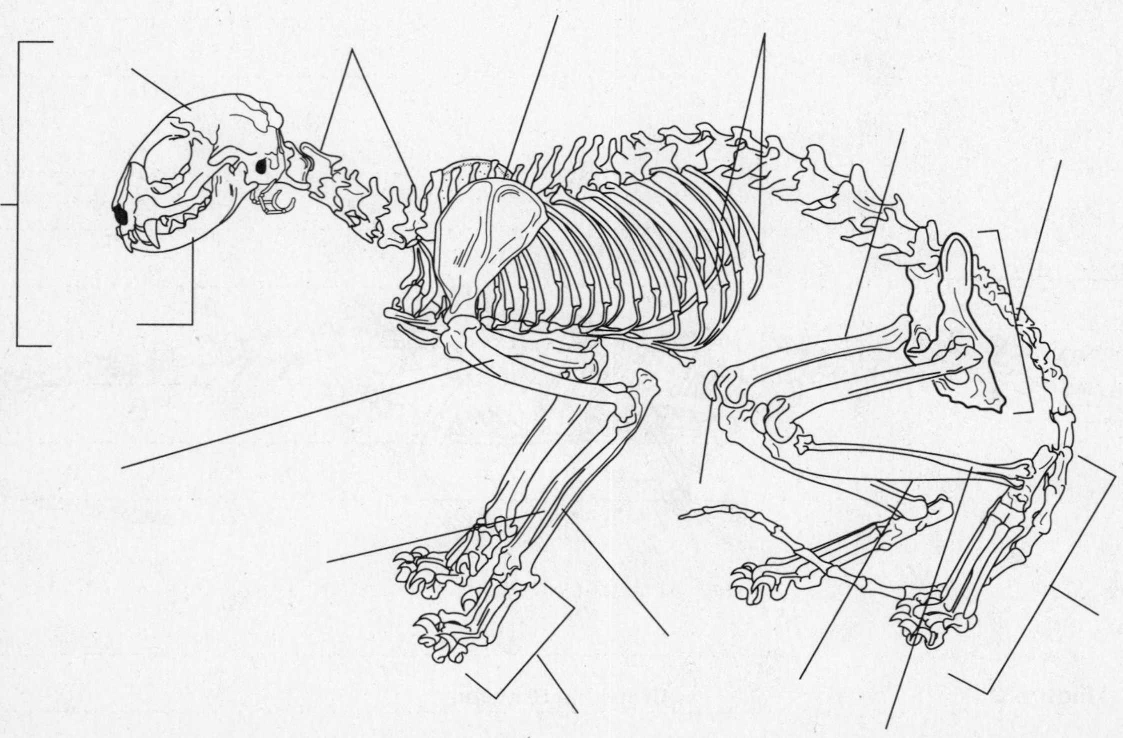

Figure 5 **Cat Skeleton**

Name ______________________ Class ____________ Date ____________

1. What are three characteristics that all of the skeletons share? ______________________

____

____

2. What are three differences that exist among the skeletons? ______________________

____

____

3. How do the forelimbs differ from the hindlimbs in terms of the way they bend?

____

____

4. How are the vertebral columns of the skeletons similar? ______________________

____

____

5. How are the vertebral columns of the skeletons different? ______________________

____

____

6. How are the hindlimbs similar? ______________________

____

____

7. How are the hindlimbs different? ______________________

____

____

8. How do the foot bones of these vertebrates differ from one another?

____

____

____

9. How do the bones in the limbs of the frog differ from those in the other four skeletons?

10. Describe the "mystery" bones in your collection.

11. To what parts of the skeleton do your "mystery" bones belong?

Analysis and Conclusions

1. How do the functions of the forelimbs differ among the five vertebrates you have examined?

2. What type of evidence would indicate that the human hand, pigeon wing, and cat paw are homologous structures?

3. Which two animals have backbones that are most alike? Explain your answer.

4. Which two animals have backbones that are least alike? Explain your answer.

5. Which two animals have forelimbs that are most alike? Explain your answer.

6. Which two animals have forelimbs that are least alike? Explain your answer.

7. Which two animals have hindlimbs that are most alike? Explain your answer.

8. Which two animals have hindlimbs that are least alike? Explain your answer.

9. To what type of animal do you think your "mystery" bones belong? On what evidence do you base your conclusion?

Critical Thinking and Application

1. Are bones that are similar in structure always similar in function? Give an example to defend your answer.

2. What evidence have you obtained in this investigation to support the theory that vertebrates evolved from a common ancestor?

3. Do you think bones should be named according to their structure or their function? Defend your answer. ______________________________

____

____

4. In terms of classification, why should the human and cat skeletons be most closely related? Do your observations in this investigation support this idea? ______________________________

____

____

____

Going Further

Using reference materials, find examples of other vertebrate skeletons. How are these skeletons similar to and different from those you have studied in this investigation? How are the skeletons of these other vertebrates adapted to the environments in which they live?

Name ______________________ Class __________ Date __________

Constructing a Model of a Nerve Cell

Pre-Lab Discussion

Nerve cells communicate with each other by sending electrical signals. The electrical activity of nerve cells was discovered more than 150 years ago by the Italian scientist Luigi Galvani. The electricity in nerve cells comes from the movements of charged particles into and out of the cells. These charged particles are supplied by minerals in the diet such as table salt. Even nerve cells that are not sending signals have a voltage, or difference in positively charged and negatively charged ions on each side of the cell membrane. If they did not, the nerve cells—like dead batteries—would not function. This voltage is called a *resting potential.*

In this investigation, you will construct a model of a nerve cell. You will also observe how the distribution of ions around the membrane generates a resting potential.

Problem

How do nerve cells develop a resting potential?

Materials *(per group)*

2 alligator clips
2 150-mL beakers
2 40-cm pieces of copper wire
DC millivolt meter
Dialysis membrane
2 glass stirring rods
2 15-cm pieces of nickel-chromium alloy wire
3 M potassium chloride solution
String
Rubber band
Scissors
Screwdriver
3 M sodium chloride solution

Safety

Put on a laboratory apron if one is available. Put on safety goggles. Handle all glassware carefully. Always use special caution when working with laboratory chemicals, as they may irritate the skin or cause staining of the skin or clothing. Never touch or taste any chemical unless instructed to do so. Be careful when handling sharp instruments. Observe proper laboratory procedures when using electrical appliances. Note all safety alert symbols next to the steps in the Procedure and review the meanings of each symbol by referring to the symbol guide on page 10.

Procedure

1. Using scissors, cut a 20-cm strip of dialysis membrane. **CAUTION:** *Be careful when using sharp instruments.* Place the dialysis membrane in a beaker of water. Put the beaker aside for now.
2. Tightly coil a 15-cm strip of nickel-chromium alloy wire around one end of a glass stirring rod. Allow 3 cm of the nickel-chromium alloy wire to remain uncoiled and parallel to the length of the stirring rod. See Figure 1. Repeat this procedure using the other piece of nickel-chromium alloy wire and stirring rod.

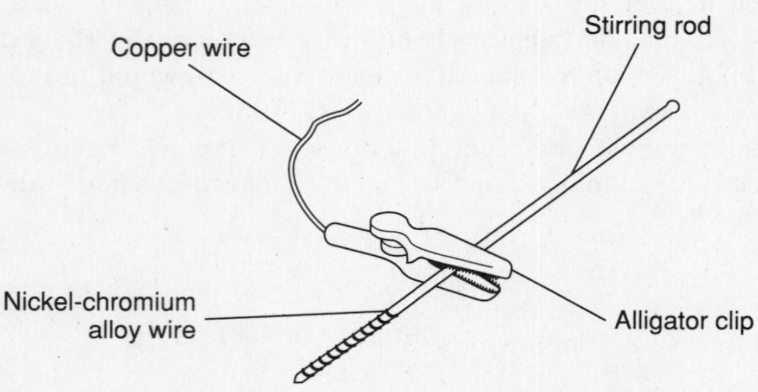

Figure 1

3. Attach an alligator clip to one end of a 40-cm length of copper wire by pulling the wire through the hollow end of the clip and looping it around the screw. Tighten the screw with a screwdriver. Repeat this procedure using the other alligator clip and piece of copper wire.
4. To construct an electrode, clamp each alligator clip to a stirring rod so that the jaws of the alligator clip make contact with the uncoiled portion of the nickel-chromium alloy wire. See Figure 1.
5. Attach one electrode to the positive terminal of the millivolt meter by looping the free end of the copper wire around the post and tightening the nut. Attach the other electrode to the negative terminal of the millivolt meter.
6. Half fill a clean 150-mL beaker with 3 M sodium chloride solution. **CAUTION:** *Be careful when using chemicals.* Place the electrode attached to the positive terminal of the millivolt meter in the beaker.

Name ______________________ Class ____________ Date ____________

7. Remove the dialysis membrane from the beaker. To open it, rub both ends between your fingers. Tie off one end of the dialysis membrane with the string to make a sac.
8. Half fill the dialysis membrane with 3 M potassium chloride solution. **CAUTION:** *Be careful when using chemicals.* The dialysis membrane sac represents a model of a nerve cell.
9. Insert the electrode attached to the negative terminal of the millivolt meter into the potassium chloride solution in the sac.
10. Use a rubber band to close the open end of the dialysis membrane and to hold the dialysis membrane around the stirring rod.
11. Place the dialysis membrane sac into the beaker containing sodium chloride. The beaker of sodium chloride represents the fluid surrounding the nerve cell. Make sure the sac does not touch the other electrode and the wires do not touch each other. See Figure 2.

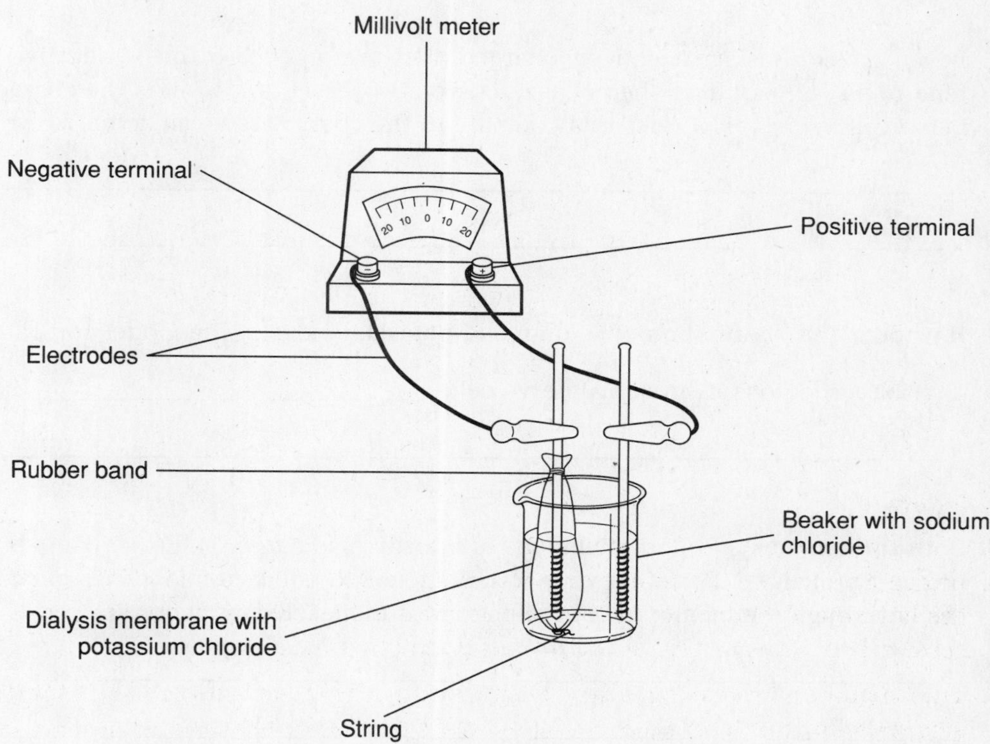

Figure 2

12. Observe the direction in which the indicator on the millivolt meter moves. Note the number of millivolts registered on the meter. Record this information in the appropriate place in Observations.

Observations

1. In which direction does the indicator on the millivolt meter move?

2. How many millivolts are registered on the millivolt meter?

Analysis and Conclusions

1. Assuming that the indicator on the millivolt meter moves in the same direction the electrons are flowing, do electrons flow toward or away from the potassium chloride solution?

2. Because electrons are negatively charged, they are repelled by other negative charges and tend to move away from them. Based on your observations, what is the charge inside the nerve cell model? How does this compare to the charge inside an actual nerve cell?

3. How does the distribution of sodium and potassium ions in the model compare to the distribution of ions in an actual nerve cell?

4. Both sodium ions and potassium ions are positively charged. Chloride ions, however, are negatively charged. Based on your observations, do potassium ions or sodium ions move more easily through the membrane of the nerve cell model? Explain your answer.

5. How do nerve cells develop a resting potential?

Name ______________________ Class __________ Date __________

Critical Thinking and Application

1. Unlike an actual nerve cell, the model does not have a sodium-potassium pump to maintain the distribution of ions. How would this affect the functioning of the model over time? Explain your answer. ______________________

2. The data below show how the speed of an action potential is affected by the diameter of the axon.

Diameter (μm)	**Speed** (m/sec)
2	10
4	20
7	30
9	40
20	90

Using the grid below, construct a line graph of the data. Plot the diameter of the axon on the horizontal axis and the speed of the action potential on the vertical axis. How does the diameter of the axon affect the speed of conduction of the action potential? Based on your graph, how fast would an action potential be propagated through an axon with a diameter of 14 μm? ______________________

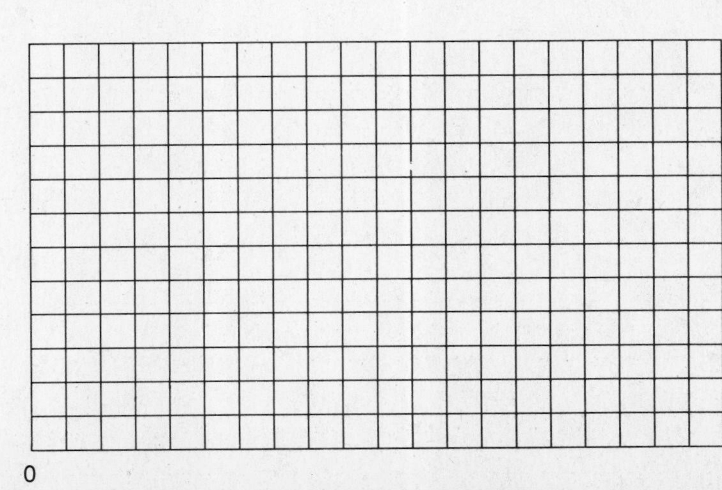

3. Too much sodium in the diet can cause high blood pressure in some individuals. What would happen to the functioning of the nervous system if there were no sodium in the diet? Explain your answer. ______

______

______

4. In the model nerve cell you constructed, how can you tell that the movements of potassium and sodium ions caused the voltage rather than the movement of chloride ions?

______

______

______

5. How is a resting nerve cell similar to a battery that is in use? ______

______

______

______

______

Going Further

Repeat this investigation using potassium chloride and sodium chloride solutions of varying concentrations. How does the concentration of the solutions affect the amount of voltage produced by the model nerve cell? How might this information be important to the study of an actual nerve cell?

Name ____________________ Class __________ Date __________

Skin Sensitivity

Pre-Lab Discussion

The senses receive information about the environment and relay it to the central nervous system, where the information is interpreted. The sense organs contain specialized neurons called *receptors,* each of which is adapted to receive a particular kind of stimulus. Like all neurons, receptors follow the *all-or-none principle.* That is, any stimulus that is weaker than the *threshold* will produce no impulse; any stimulus that is as strong as or stronger than the threshold will produce an impulse. The threshold is the minimum level of a stimulus that is required to activate a neuron.

The receptors for the sense of touch are scattered over the surface of the entire body. The receptors, however, are more closely grouped together in some areas of the body than in others. In addition, the receptors in one area may respond to a weaker threshold than receptors in another area.

In this investigation, you will study the distribution and sensitivity of the touch receptors of the human body.

Problem

How sensitive is the skin?

Materials *(per pair of students)*

3 bristles of various thickness
Blindfold
9 toothpicks
Masking tape
Metric ruler
Pencil
Blunt probe

Safety

Be careful when handling sharp instruments. Note all safety alert symbols next to the steps in the Procedure and review the meanings of each symbol by referring to the symbol guide on page 10.

Procedure

Part A. Determining Threshold

1. Blindfold your partner.
2. Gently touch your partner on the fingertip with one of the three bristles just until the bristle bends. Your partner should tell you whether he or she feels the touch of the bristle. **Note:** *Avoid parts of the fingertip that have calluses.*
3. Touch each of the three bristles to the fingertip five times. Try to alternate the bristles so that your partner will not know which bristle is being used. In Data Table 1, record the bristles that are felt by placing a check mark in the appropriate place. If the bristles are not felt, leave the space blank.
4. Repeat steps 2 and 3 for the palm, back of the hand, inside of the forearm, and back of the neck. Record the information in the appropriate places in Data Table 1.
5. Switch roles with your partner and repeat steps 2 through 4.

Part B. Distinguishing Between Two Stimuli

1. Obtain nine toothpicks. Tape two toothpicks together so that their tips are 5 millimeters apart. Use the metric ruler to ensure that the spacing between the toothpicks is accurate. Make sure that the tips of the toothpicks are even with each other. See Figure 1. Using the same procedure, tape three pairs of toothpicks together so that the tips of the first pair are 10 millimeters apart, the tips of the second pair are 15 millimeters apart, and the tips of the last pair are 20 millimeters apart. Allow the last toothpick to remain unpaired and untaped.

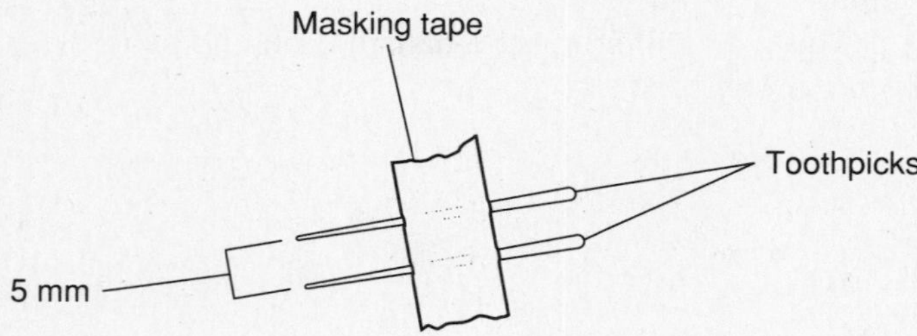

Figure 1

2. Blindfold your partner. Gently touch your partner's fingertip with any of the sets of toothpicks. **CAUTION:** *Be very careful when touching your partner's skin. Apply only a small amount of pressure.* Have your partner identify whether you are using one or two toothpicks. Record this information in the appropriate place in Data Table 2.
3. Repeat step 2 with the remaining sets of toothpicks. Use the sets of toothpicks in random order so that your partner will not know the pattern in the testing. **Note:** *Use the same pressure on the toothpicks in all trials.*
4. Repeat steps 2 and 3 using the palm, back of the hand, inside of the forearm, and back of the neck. Record the information in the appropriate places in Data Table 2.
5. Switch roles and repeat steps 2 through 4.

Part C. Locating a Stimulus

1. Blindfold your partner and give him or her a blunt probe to use as a pointer.
2. With a pencil, touch your partner's fingertip so as to make a visible mark on the skin. **CAUTION:** *Do not press so hard as to break the skin. Make sure that the pencil has a rounded, blunt tip.*

Name ______________________ Class ____________ Date ____________

3. Have your partner try to touch the same point on his or her skin with the blunt probe that you touched with the pencil.
4. With the metric ruler, measure the distance between your mark and the point of the probe. Record this measurement in the appropriate place in Data Table 3.
5. Repeat steps 2 through 4 two more times using points near, but not the same as, the first point. Record the information in the appropriate place in Data Table 3.
6. Repeat steps 2 through 5 on the palm, back of the hand, inside of the forearm, and back of the neck. Record the information in the appropriate places in Data Table 3.
7. Switch roles and repeat steps 1 through 6.
8. Calculate the average distance between the pencil mark and the probe for each of the five sets of data. Record the averages.

Observations

Data Table 1

Skin Surface	Small Bristle	Medium Bristle	Large Bristle	Trial
Fingertip				1
				2
				3
				4
				5
Palm of hand				1
				2
				3
				4
				5
Back of hand				1
				2
				3
				4
				5
Inside of forearm				1
				2
				3
				4
				5
Back of neck				1
				2
				3
				4
				5

Data Table 2

Body Part	Number of Stimuli Felt				
	One Toothpick	Toothpicks 5 mm Apart	Toothpicks 10 mm Apart	Toothpicks 15 mm Apart	Toothpicks 20 mm Apart
Fingertip					
Palm of hand					
Back of hand					
Inside of forearm					
Back of neck					

Data Table 3

	Distance Between Pencil Mark and Probe (mm)				
Trial	Fingertip	Palm of Hand	Back of Hand	Inside of Forearm	Back of Neck
1					
2					
3					
Average					

Analysis and Conclusions

1. Of the parts of the body you tested, which part had the lowest threshold? Which part had the highest threshold? ____

2. Of the parts of the body you tested, which part was best able to distinguish between the closest stimuli? Which part was least able to distinguish between the closest stimuli?

3. What does your answer to question 2 indicate about the distribution of touch receptors in the skin? ____

Name ______________________ Class __________ Date __________

4. On which part of the body you tested were you best able to locate a stimulus? On which part were you least able to locate a stimulus? ______________________

5. How are the results of Part C related to the results of Parts A and B?

6. Were there any differences between your results and those of your partner? If so, what might account for these differences? ______________________

Critical Thinking and Application

1. Braille is a special system of writing with raised dots. Blind people read Braille by touching these dots and recognizing patterns for each letter. The Braille pattern below is for the word *biology*.

How does the body's arrangement of nerve receptors for touch make the Braille system possible? ______________________

2. In addition to touch receptors, the skin contains receptors for pain. How are pain receptors helpful to us? ____

____

____

3. In what way is the concentration of touch receptors in the human hand related to its functions? ____

____

____

4. When you put on a wristwatch, you become aware of the pressure it applies to the touch receptors in your wrist. Why do you think that you will not be aware of the presence of the watch after a period of time? ____

____

____

Going Further

1. To determine the sensitivity of other areas of the body, such as the upper arm, the knee, the foot, or various parts of the face, repeat the procedures in Parts A, B, and C of this investigation. Record your observations and compare them with those in this investigation.
2. Use blunt probes warmed in warm water to determine the distribution of heat receptors in the skin. **CAUTION:** *Do not heat the probes to a temperature above 40°C.* Record your observations.
3. Use blunt probes cooled in ice water or in a refrigerator to determine the distribution of cold receptors in the skin. Record your observations.

Name ______________________ Class ____________ Date ______________

73

Bone Composition and Structure

Pre-Lab Discussion

Human bone contains living tissue and nonliving materials. The living tissue includes bone cells, blood vessels, fat cells, nerve cells, and cartilage. The nonliving materials include water, and minerals such as calcium and phosphorus. In fact, bones are hard and strong because they contain a great amount of calcium. Bone is composed of *compact bone* and *spongy bone.* Unlike compact bone, which is very hard and dense, spongy bone is soft and has many spaces in it.

In this investigation, you will examine the internal structure of compact bone. You will also determine the percentage of water in bone and observe how calcium gives strength to bone.

Problem

What is the internal composition and structure of bone?

Materials *(per group)*

Prepared slide of compact bone with Haversian system
Microscope
2 uncooked pork bones
Triple-beam balance
Felt-tip marking pen
Heat-resistant gloves
Heat-resistant pad
Large metal cookie sheet or cooking tray
100 mL of 20% hydrochloric acid solution
250-mL beaker
Oven
Tongs
Paper towels
100-mL graduated cylinder
Masking tape

Safety

Put on a laboratory apron if one is available. Put on safety goggles. Handle all glassware carefully. Use extreme care when working with heated equipment or materials to avoid burns. Always use special caution when working with laboratory chemicals, as they may irritate the skin or cause staining of the skin or clothing. Never touch or taste any chemical unless instructed to do so. Always handle the microscope with extreme care. You are responsible for its proper care and use. Use caution when handling glass slides as they can break easily and cut you. Note all safety alert symbols next to the steps in the Procedure and review the meanings of each symbol by referring to the symbol guide on page 10.

Procedure

Part A. Calculating the Percentage of Water in a Bone

1. Place a small piece of paper towel on the pan of a triple-beam balance. Move the rider on the front beam of the balance until the pointer of the balance points to zero. Find the mass of the paper towel to the nearest tenth of a gram.

2. Place an uncooked pork bone on the balance and find the mass of the bone and paper towel to the nearest tenth of a gram. To determine the mass of the pork bone, subtract the mass of the paper towel from the mass of the bone and paper towel. Record this information in Data Table 1.

3. Using the felt-tip marking pen, write the last name of one of your group members on the surface of the bone.

4. Place the bone, along with bones of the other groups, on a large metal cooking tray. Place the cooking tray in an oven that has been preheated to 149°C (300°F) for 25 minutes. **CAUTION:** *To avoid burns, place the tray in the hot oven carefully.*

5. Using heat-resistant gloves, remove the tray from the oven and place it on a heat-resistant pad to cool. **CAUTION:** *To avoid burns, use heat-resistant gloves to remove the hot tray from the oven slowly and carefully.*

6. After the bone has cooled for 10 minutes, place it on the piece of paper towel on the triple-beam balance and again determine its mass to the nearest tenth of a gram. Record this information in Data Table 1.

7. The loss in mass of the bone is due primarily to the evaporation of water and to some oxidation of the minerals in the bone. Calculate the percentage of water in the pork bone by using the following formula:

$$\frac{\text{mass before heating} - \text{mass after heating}}{\text{mass before heating}} \times 100 = \text{percentage of water}$$

8. Record the percentage of water in the bone in the appropriate place in Observations.

9. Follow your teacher's instructions for proper disposal of the bone.

Part B. Observing Bone Cells

1. Observe a prepared slide of compact bone under the low-power objective of a microscope. Notice the circular-patterned units in the cross section of the bone. Each of these circular units is a Haversian system.

2. Switch to high-power to observe the structures that make up each Haversian system. **CAUTION:** *When switching to the high-power objective, always look at the objective from the side of the microscope so that the objective does not hit or damage the slide.*

3. Focus on a group of concentric circles. The central, hollow core of these circles is called the Haversian canal. The Haversian canal contains nerves and blood vessels. The rings around the Haversian canal are called lamellae (singular, lamella). The small dark cavities between adjacent lamellae are called lacunae (singular, lacuna). Lacunae appear as long, dark areas between lamellae.

4. Within each Haversian system, the lacunae are interconnected by small, branching canals called canaliculi (singular, canaliculus). Canaliculi appear as thin, dark lines that resemble the spokes of a wheel. Fluids pass from one part of the bone to another through the canaliculi.

5. Look for darkly stained bodies within the lacunae. These are the osteocytes, or living bone cells. See Figure 1. Notice that osteocytes have fine branches that extend into the canaliculi. The osteocytes are responsible for controlling the life functions of the bone.

6. In the appropriate place in Observations, draw a section of bone tissue as seen through the high-power objective of the microscope. Label the following parts of the Haversian system: Haversian canal, lamella, lacuna, canaliculus, and osteocyte. Record the magnification of the microscope.

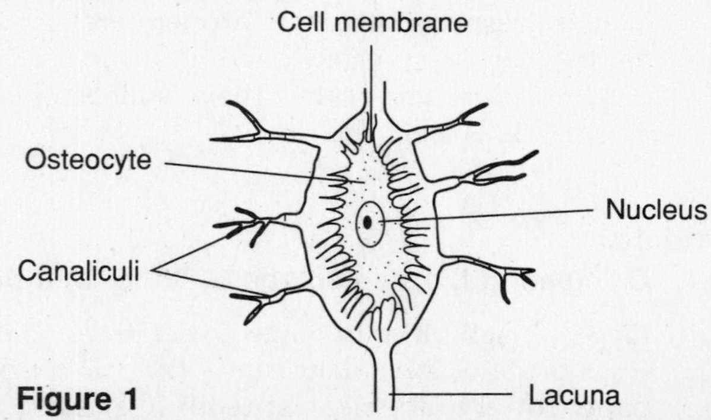

Figure 1

Name ______________________ Class __________ Date __________

Part C. Observing the Role of Calcium in a Bone

1. Obtain another uncooked pork bone. Observe the flexibility of the bone by trying to bend and twist it. Record your observations in Data Table 2.
2. Place the bone in a 250-mL beaker. Use masking tape to label the beaker with the name of one member of your group.
3. Carefully add 100 mL of a 20% hydrochloric acid solution to the beaker. **CAUTION:** *Wear safety goggles when using hydrochloric acid. Hydrochloric acid can burn the skin or clothing. If any acid spills on your skin or clothing, wash the affected area with water immediately. Notify your teacher.*
4. Put the beaker in a place where it will remain undisturbed for 3 days.
5. Using tongs, carefully remove the bone from the acid. Rinse the bone under running tap water to remove any acid solution.
6. Dry the bone with a paper towel. Again test the flexibility of the bone. Record your observations in Data Table 2.
7. Follow your teacher's instructions for the proper disposal of the bone and the acid solution.

Observations

Data Table 1

Object	Mass Before Heating (g)	Mass After Heating (g)
Pork bone		

Percentage of water in pork bone = ______________________

Magnification __________

Haversian System of a Human Bone

Data Table 2

Object	Description of Flexibility
Pork bone before being soaked in acid	
Pork bone after being soaked in acid	

Analysis and Conclusions

1. Is the percentage of water loss in your group's bone exactly the same as that of other groups? If not, why might the percentages vary? ______

2. Why is water a necessary substance in a bone? ______

3. Many people incorrectly think of bone as nonliving tissue. How is a bone similar to other living tissues? ______

4. What materials are removed from a bone when it is soaked in a hydrochloric acid solution for several days? What evidence do you have to support your answer?

Critical Thinking and Application

1. Suppose a person's diet lacks the mineral calcium. How would this deficiency affect the skeletal system? ______

2. Bone ash is a white, powdery material obtained from charred bones. Why do you think bone ash is used in fertilizers? ______

3. What is the function of the osteocytes when a bone is broken? ______

Going Further

1. Determine the percentage of water in other types of bones, such as cattle bones or chicken bones.
2. Use reference materials to research the degenerative bone disease known as osteoporosis. What is the cause of this disease? What are its symptoms? What age group does it normally affect? What preventive measures can be taken to combat osteoporosis?

Name ______________________ Class __________ Date __________

The Skeletal System

Pre-Lab Discussion

The skeleton is not just a framework of bones. It serves as an attachment for muscles, as support for the body, and as protection for vital organs. Bones also store certain minerals and contain special cells that form red and white blood cells. The human body contains 206 bones.

The bones of the human skeleton are grouped into two divisions: the *axial skeleton* and the *appendicular skeleton*. The axial skeleton includes the skull, vertebrae, and rib cage. The appendicular skeleton includes the rest of the skeleton: the arms, legs, shoulder girdle, and pelvic girdle.

In addition to bone, human skeletons also contain *cartilage*. In fact, the human skeleton is initially formed from cartilage, which is gradually replaced by bone. Cartilage is found in adult humans between the vertebrae of the spinal column, at the tips of ribs and other bones, and in the nose, ears, and larynx. Some lower vertebrates, such as the shark and the lamprey, have skeletons composed entirely of cartilage. Of the three types of cartilage found in the human body—hyaline, fibrous, and elastic—*hyaline cartilage* is the most common. It is pearly white and glassy or translucent in appearance. Hyaline cartilage is found at the ends of bones in movable joints.

In this investigation, you will observe the structure of bone and cartilage. You will also identify the main parts of the human skeleton.

Problem

How is the human skeleton organized?

Materials *(per group)*

Prepared slide of a cross section
of hyaline cartilage
Microscope
Long bone from chicken leg
Dissecting tray
Scalpel
2 colored pencils

Safety

Put on a laboratory apron if one is available. Handle all glassware carefully. Be careful when handling sharp instruments. Always handle the microscope with extreme care. You are responsible for its proper care and use. Use caution when handling glass slides as they can break easily and cut you. Note all safety alert symbols next to the steps in the Procedure and review the meanings of each symbol by referring to the symbol guide on page 10.

Procedure

Part A. Observing the Structure of a Long Bone

1. Obtain a long bone of a chicken leg. Observe the outside covering and features of the bone. Note the smooth covering at both ends of the bone. Also notice the presence of small holes along the surface of the bone. In the appropriate place in Observations, describe the outside surface of the chicken leg bone.

2. With a scalpel, cut one end of the bone to expose the fine structure of the spongy bone and some bone marrow. **CAUTION:** *Be careful when handling a scalpel. Carefully cut in a direction away from your hands and body.*

3. Use Figure 1 as you observe the chicken bone. Observe the periosteum, which covers and protects the bone. Beneath the periosteum is a hard dense layer called compact bone. Notice the softer spongy bone near the ends of the bone. Spongy bone contains many small spaces. In long bones, the spongy bone also contains red marrow, which produces red and white blood cells. Note the internal cavity of the bone, which may contain some fat-storing yellow marrow. On the outer edge of the end of the bone, find the plate of protective cartilage. Each of the wide ends of a long bone is called the epiphysis, whereas the long shaft of the bone is called the diaphysis.

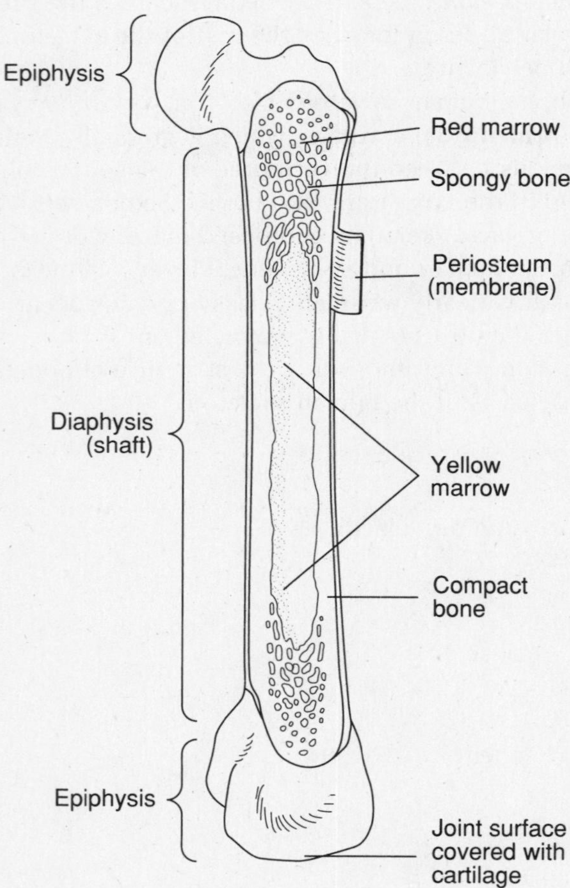

Figure 1

4. In the appropriate place in Observations, sketch the internal structure of the chicken leg bone. Label the periosteum, compact bone, spongy bone, red marrow, yellow marrow, cartilage, epiphysis, and diaphysis.

5. Follow your teacher's instructions on proper disposal of the chicken bone.

Name ______________________ Class ____________ Date ____________

Part B. Examining Cartilage

1. Cartilage is more flexible than bone; thus it can take a great deal of stress. Live cartilage cells, called chondrocytes, are found in cavities called lacunae. Unlike bone, cartilage contains no blood vessels. Materials enter and leave chondrocytes by diffusion to and from the blood vessels in adjacent layers of tissue.

2. Observe a prepared slide of the cross section of hyaline cartilage under the low-power objective of a microscope. Locate some chondrocytes. In the appropriate place in Observations, sketch the cartilage tissue under low power. Label the chondrocytes. Record the magnification of the microscope.

3. Switch to high power. **CAUTION:** *When switching to the high-power objective, always look at the objective from the side of the microscope so that the objective does not hit or damage the slide.* Observe the chondrocytes under high power. Notice that the chondrocytes usually occur in groups of two or four cells. In the appropriate place in Observations, sketch and label some chondrocytes as seen under high power. Record the magnification of the microscope.

Part C. Examining the Human Skeleton

1. With a colored pencil, color the axial skeleton in Figure 2 in Observations.

2. With the other colored pencil, color the appendicular skeleton.

3. In the appropriate place in Observations, match the names of the bones of the skeleton with their corresponding numbers as shown in Figure 2.

Observations

1. Describe the outside surface of the chicken leg bone. ______________________

____

____

Internal Structure of Chicken Leg Bone

Low-power objective

Magnification ____________

Cartilage

High-power objective

Magnification ____________

Cartilage

2. What color did you color the axial skeleton? ______________________________

The appendicular skeleton? ______________________________

3. Write the numbers of the bones in Figure 2 that correspond to the names of the bones given below.

Carpals ______

Scapula ______

Fibula ______

Clavicle ______

Radius ______

Tibia ______

Ribs ______

Metacarpals ______

Cranium ______

Tarsals ______

Phalanges ______

Ulna ______

Humerus ______

Sternum ______

Femur ______

Pelvis ______

Metatarsals ______

Vertebral column ______

Patella ______

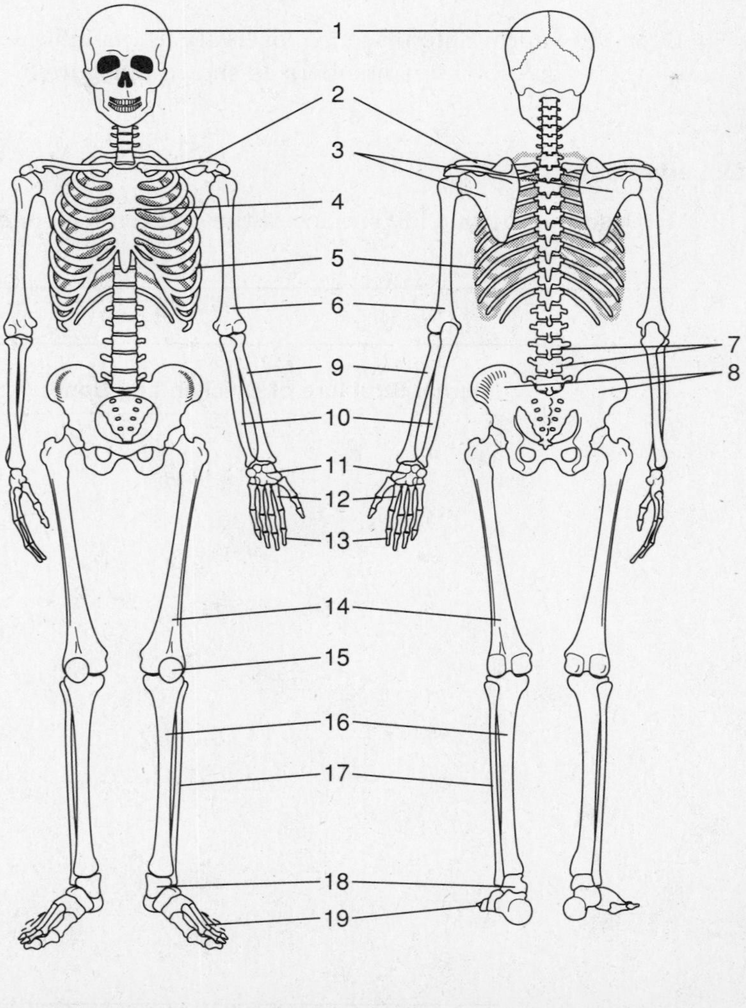

Figure 2 **Human Skeleton**

Name ______________________ Class __________ Date __________

Analysis and Conclusions

1. Why is it necessary that the ends of bones be smooth? ____________________

2. What is the function of the small holes in the bone surface? ____________________

3. What is the importance of red bone marrow? ____________________

4. How is cartilage similar to bone? ____________________

5. How do cartilage and bone differ? ____________________

6. Which part of the human skeleton, axial or appendicular, appears to have the greater number of bones? On what observations do you base this conclusion? ____________________

Critical Thinking and Application

1. The skeleton of an unborn baby consists of a large amount of cartilage, which will later change to bone. Of what advantage to the unborn child is a skeleton made of cartilage?

2. Leukemia is a disease in which white blood cells grow in an uncontrolled manner. Explain why a bone-marrow transplant is sometimes used as a treatment for this disease.

3. Of what advantage is the layer of spongy bone found on the inside of a long bone?

4. How is the large number of bones found in the feet and hands of human beings both an advantage and a disadvantage?

Going Further

1. Electricity can be used to speed up the healing of broken bones. Use reference materials to do research on this topic. Report your findings to the class.
2. Using reference materials, research the so-called soft spots, or fontanels, of a newborn baby's skull. What are fontanels and what is their function?

Name ______________________ Class __________ Date __________

75

The Muscular System

Pre-Lab Discussion

Muscle tissue consists of groups of cells that are specialized for contraction. In the body, muscle tissue is usually organized into organs called muscles. Muscles contain connective tissue, nerves, and blood vessels in addition to muscle tissue. Although the term muscle often refers to an entire organ, it is also frequently used to refer to muscle tissue alone. There are three different types of muscle tissue in the bodies of humans and other vertebrates: skeletal, smooth, and cardiac.

Skeletal muscles are attached to the bones, cover the skeleton, give shape to the body, and make movement possible. The individual cells in skeletal muscle are often called fibers rather than cells because they are long and thin and quite different in structure from typical cells. Skeletal muscle fibers are larger than most cells (as much as 40 centimeters in length!), contain many nuclei, and have many transverse stripes that are visible when the fibers are viewed under a microscope. Because of these stripes, or striations, skeletal muscle is called striated muscle. Skeletal muscle is also known as voluntary muscle because it is under conscious control.

Smooth muscle is very prominent in the walls of the stomach, intestine, and urinary bladder. Smooth muscle also occurs in the walls of blood vessels, in glands, and in the skin. The cells of smooth muscle are spindle-shaped, have individual nuclei, and are not striated. Smooth muscle is also called involuntary muscle because it is not under conscious control.

Cardiac muscle is found only within the walls of the heart. Like smooth muscle, cardiac muscle is not under conscious control—it is a type of involuntary muscle. Like skeletal muscle, cardiac muscle is striated—the cells in cardiac muscle are striped. Cardiac muscle cells contain one nucleus and form branching fibers with adjacent cardiac muscle cells.

In this investigation, you will observe prepared slides of three types of muscle cells. You will also examine a chicken wing to observe its muscle structure.

Problem

How is the human muscular system organized?

Materials *(per group)*

- Prepared slides of skeletal, smooth, and cardiac muscle
- Microscope
- Whole chicken wing
- Hand lens
- Dissecting tray
- Scalpel
- Scissors
- Probe
- Forceps
- Paper towels

Safety

Put on a laboratory apron if one is available. Handle all glassware carefully. Be careful when handling sharp instruments. Always handle the microscope with extreme care. You are responsible for its proper care and use. Use caution when handling glass slides as they can break easily and cut you. Note all safety alert symbols next to the steps in the Procedure and review the meanings of each symbol by referring to the symbol guide on page 10.

Procedure

Part A. Examining Three Types of Muscle Cells

1. Observe a prepared slide of skeletal muscle under the low-power objective of the microscope.
2. Switch to high power. **CAUTION:** *When switching to the high-power objective, always look at the objective from the side of the microscope so that the objective does not hit or damage the slide.*
3. Count the number of nuclei that are contained within one skeletal muscle fiber. Note whether this cell has striations (stripes). Observe the cell shape. Classify skeletal muscle as voluntary or involuntary. Record the information in the appropriate place in the Data Table.
4. In the appropriate place in Observations, sketch a few of the skeletal muscle fibers that you observed under high power. Label the nucleus, cell membrane, cytoplasm, and striations. Record the magnification of the microscope.
5. Observe a prepared slide of smooth muscle under low power. Switch to high power and observe the number of nuclei within one cell, the cell shape, and the presence or absence of striations. Classify smooth muscle as voluntary or involuntary. Record the information in the appropriate place in the Data Table.
6. In the appropriate place in Observations, sketch a few of the smooth muscle cells that you observed under high power. Label the nucleus, cell membrane, and cytoplasm. Record the magnification of the microscope.
7. Observe a prepared slide of cardiac muscle under low power. Switch to high power and observe the number of nuclei within one cell, the cell shape, and the presence or absence of striations. Classify smooth muscle as voluntary or involuntary. Record the information in the appropriate place in the Data Table.
8. In the appropriate place in Observations, sketch a few of the smooth muscle cells that you observed under high power. Label the nucleus, cell membrane, cytoplasm, and striations. Record the magnification of the microscope.

Part B. Examining the Muscles in a Chicken Wing

1. Rinse the chicken wing under running water. Dry it thoroughly with paper towels and place it in the dissecting tray.
2. Remove the skin from the chicken wing. With the forceps, grasp the skin near the shoulder and pull the skin away from the underlying muscles. Use scissors to cut the skin along the entire length of the chicken wing. **CAUTION:** *Be careful when handling sharp instruments. Cut in a direction away from your hands and body.* Make sure you cut the skin exactly as shown in Figure 1. **Note:** *Be careful not to cut through the underlying muscles with the scissors.*

Name ______________________ Class ____________ Date ____________

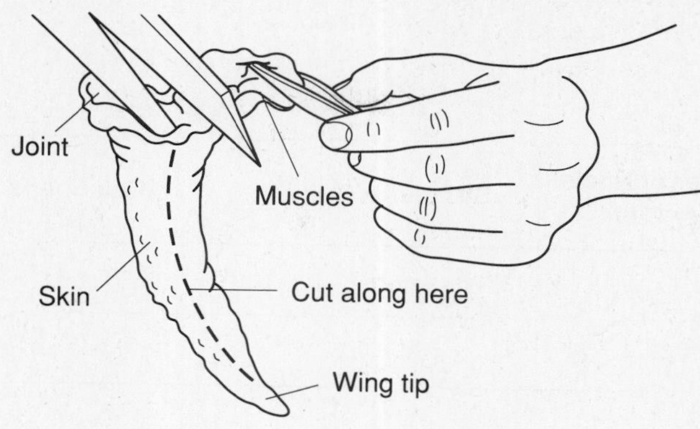

Figure 1

3. Carefully pull off the skin. Observe the fascia, the transparent connective tissue that surrounds the muscles and holds the muscles to the skin. With a probe or scalpel, gently separate the fascia from the skin. **Note:** *The skin covering the joints is difficult to remove. Work carefully, using scissors or a scalpel to remove this skin.* Remove the skin up to the second joint. It is not necessary to remove the skin from the wing tip. Cut away the excess skin around the second joint. Discard the skin according to your teacher's instructions. **Note:** *You may want to wash your hands and rinse the chicken wing at this time.*

4. Observe the muscles in the chicken wing. Observe the tendons, the shiny white cords of connective tissue at the ends of the muscles. Tendons attach muscles to bones.

5. Grasp the wing by the shoulder and the wing tip. As you bend and straighten the wing, notice how the muscles contract as they move. Locate and identify the antagonistic muscles, which are pairs of muscles that work in opposition to each other. Thoroughly wash your hands with soap and water.

6. In the appropriate place in Observations, sketch the muscles and bones of the chicken wing. Label the bones, the muscles, the tendons, and an antagonistic muscle pair.

7. Examine Figure 2, which shows the muscles and bones of the human arm. The elbow joint, which is a hinge joint, is similar to the joint in the chicken wing. On Figure 2, locate the tendons that attach the muscles to the bones. As a muscle contracts, the attachments either remain stationary or move. The attachment end that moves a bone is called the insertion. The attachment end that remains stationary, anchoring the muscle, is called the origin. Locate the origins and insertions in the chicken wing.

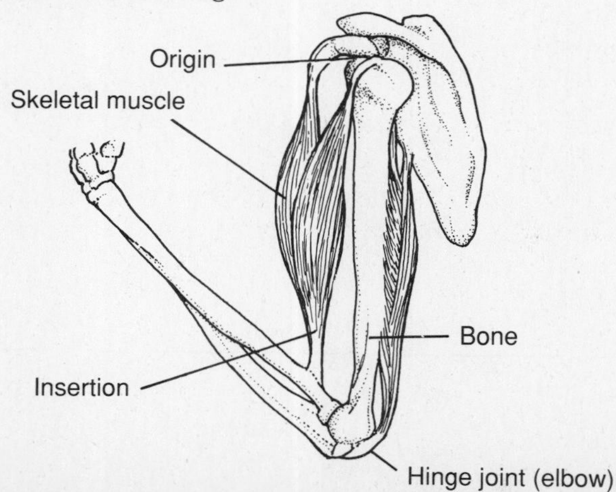

Figure 2

8. Dispose of the dissected chicken wing according to your teacher's instructions. Wash, dry, and put away the dissecting tools you used in this part of the investigation.

Observations

Data Table

Type of Muscle Cell	Number of Nuclei per Cell	Striations	Cell Shape	Voluntary or Involuntary?
Skeletal				
Smooth				
Cardiac				

Magnification ________

Skeletal Muscle

Magnification ________

Smooth Muscle

Magnification ________

Cardiac Muscle

Chicken Wing

Name ______________________ Class __________ Date __________

Analysis and Conclusions

1. a. How are skeletal muscle tissue and cardiac muscle tissue similar?

   ____

   ____

   b. How are they different? ______________________

   ____

   ____

   ____

2. How does smooth muscle tissue differ from the other two types of muscle tissue?

   ____

   ____

3. Where in the body is each of the three types of muscle tissue found?

   Skeletal muscle tissue ______________________

   Smooth muscle tissue ______________________

   Cardiac muscle tissue ______________________

4. What enables the chicken to move its wing? ______________________

   ____

5. How might injury to one of the muscles in an antagonistic muscle pair affect movement?

   ____

   ____

Critical Thinking and Application

1. What effect will the tearing of a tendon have on its corresponding muscle? How could this situation be repaired? ______________________

   ____

   ____

   ____

   ____

2. Suppose smooth muscles rather than skeletal muscles could be attached to the skeleton and given voluntary control. How might the movement of the skeleton be affected?

____

____

____

____

3. Why would a bird be unable to fly if there was some damage to the nerve in the wing?

____

____

____

Going Further

1. To prepare your own slide of muscle tissue, place a small piece of raw pork or raw beef in a solution containing 35 g of potassium hydroxide and 100 mL of distilled water. Allow the meat to soak in the solution for about 30 minutes. Remove the meat from the solution. Place the meat on a clean glass slide and with a dissecting needle tease the muscle fibers apart. Add a coverslip and observe the muscle fibers under a microscope.
2. Using reference materials, learn how muscles, tendons, ligaments, cartilage, and bones are arranged to allow movement in a knee joint. Why are knee injuries common in athletes?

Name ____________________ Class __________ Date __________

Mechanical and Chemical Digestion

Pre-Lab Discussion

When an animal eats, the food must be digested, or broken down into nutrients that the animal's cells can use. In vertebrates, unlike some other animals, digestion is extracellular and takes place in a digestive tube. Within this tube, food is mechanically and chemically digested. *Mechanical digestion* involves mixing, grinding, or crushing large pieces of food into small pieces. *Chemical digestion* occurs when digestive enzymes break down complex molecules, such as carbohydrates, into simple molecules, such as glucose. Although chemical digestion can occur without digestive enzymes, these enzymes catalyze, or speed up, the reaction. For this reason, digestive enzymes are called *catalysts.*

In this investigation, you will examine the processes of mechanical and chemical digestion. You will also observe how enzymes affect the rate of chemical digestion.

Problem

How are foods mechanically and chemically digested? What effect do digestive enzymes have on the rate of digestion?

Materials *(per group)*

Egg white from half of a boiled egg
6 test tubes
Test tube rack
5% hydrochloric acid solution
Mixture of 5% hydrochloric acid solution and 1% pepsin solution
Triple-beam balance
Scalpel
Dissecting tray
10-mL graduated cylinder
Glass-marking pencil
Olive oil
1% pancreatin solution
5% soap or bile salt solution
pH paper
Medicine dropper

Safety

Put on a laboratory apron if one is available. Put on safety goggles. Handle all glassware carefully. Always use special caution when working with laboratory chemicals, as they may irritate the skin or cause staining of the skin or clothing. Never touch or taste any chemical unless instructed to do so. Be careful when handling sharp instruments. Note all safety alert symbols next to the steps in the Procedure and review the meanings of each symbol by referring to the symbol guide on page 10.

Procedure

Part A. Chemical Digestion of Fat

1. Place three test tubes in the test tube rack. With the glass-marking pencil, label the test tubes 1, 2, and 3.
2. Add 10 mL of water and 2 drops of olive oil to each test tube. Gently swirl the test tubes to mix the contents of each.
3. Dip a piece of pH paper into each mixture. Record the pH for each mixture in the appropriate place in Data Table 1.
4. Add 5 mL of pancreatin solution to test tube 2. Add 5 mL of pancreatin solution and 3 mL of soap or bile salt solution to test tube 3. Gently swirl the test tubes to mix the contents. **CAUTION:** *Wear safety goggles when working with laboratory chemicals. Be careful not to get the solutions in your eyes or spill them on your hands or clothing.*
5. After 5 minutes, measure the pH of each mixture. Record this information in the appropriate places in Data Table 1.
6. After 5 more minutes, repeat step 5.

Part B. Mechanical and Chemical Digestion of Protein

1. Place three test tubes in a test tube rack. With a glass-marking pencil, label the test tubes 1, 2, and 3.
2. Using a scalpel, cut the egg white into three equal pieces. **CAUTION:** *Be careful when handling sharp instruments. Cut in a direction away from your hands and body.*
3. Find the mass of each piece of egg white. **Note:** *Try to make each piece have approximately the same mass.*
4. Place one piece of egg white in test tube 1.
5. Use a scalpel to chop the second piece of egg white into small pieces. Place the small pieces of egg white in test tube 2.
6. Use a scalpel to chop the third piece of egg white into small pieces. Place the small pieces of egg white in test tube 3.
7. To test tube 1, add 10 mL of hydrochloric acid solution. To test tube 2, add 10 mL of hydrochloric acid solution. To test tube 3, add 10 mL of the mixture of hydrochloric acid solution and pepsin solution. See Figure 1. Gently swirl each test tube so that the liquids mix well with the egg white.
8. Set the test tubes aside and allow them to remain undisturbed for 24 hours. After 24 hours, observe each test tube. Record your observations in the appropriate places in Data Table 2.

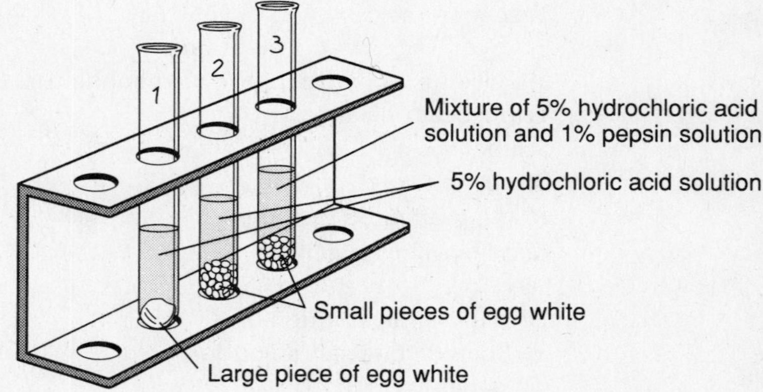

Figure 1

Name ______________________ Class __________ Date __________

Observations

Data Table 1

	pH of Substances		
Time	Test Tube 1 (water + oil)	Test Tube 2 (water + oil + pancreatin)	Test Tube 3 (water + oil + pancreatin + soap)
Start			
After 5 minutes			
After 10 minutes			

Data Table 2

Substance	Observations
Test tube 1 (large egg white + hydrochloric acid)	
Test tube 2 (small egg white + hydrochloric acid)	
Test tube 3 (small egg white + hydrochloric acid + pepsin)	

Analysis and Conclusions

1. Which test tube showed the greatest degree of fat digestion? ______________________

2. How were you able to determine experimentally that fat digestion had occurred?

____

____

3. In which test tube did the smallest degree of protein digestion occur? How do you explain this?

____

____

____

4. In which test tube did the greatest degree of protein digestion occur? How do you explain this? ______

______

______

Critical Thinking and Application

1. What roles do the teeth and tongue play in the process of digestion?

______

______

2. Bile is produced in the liver but stored in the gallbladder, where it becomes concentrated before being released into the digestive tract. How would surgery involving the removal of the gallbladder affect the process of digestion? ______

______

______

3. A person who has had a small section of small intestine removed surgically loses weight. Explain why weight loss might follow such surgery. ______

______

______

______

4. Bulimia is an eating disorder in which a person forces the body to eject the contents of the stomach, usually through the mouth. Explain why most bulimics eventually develop dental problems. ______

______

______

Going Further

Label eight test tubes 1 through 8. In each tube, place 10 mL of the mixture of 1% pepsin solution and 5% hydrochloric acid solution. Immerse test tubes 1 and 2 in ice water, keep test tubes 3 and 4 at room temperature, and place test tubes 5 and 6 in a water bath at 40°C. Boil test tubes 7 and 8 for a few minutes, allow them to cool, and then place them in the water bath. Add equal amounts of finely chopped, hard-boiled egg white to each test tube. What influence does temperature have on the effectiveness of pepsin in protein digestion? Construct a data table showing your observations.

Name ______________________ Class __________ Date __________

Measuring Food Energy

Pre-Lab Discussion

All living things need energy to carry out metabolic activities. Animals—unlike many plants, protists, and bacteria—do not have the means to get energy directly from sunlight or simple inorganic chemicals. The energy requirements of animals must be met by taking in food.

The energy content of food can be determined by burning a sample of food in a device called a calorimeter. Heat energy released by combustion is absorbed by a container of water. Any rise in water temperature is measured and then used to determine the value of the heat energy released by the burning food sample. Heat energy is expressed in units called calories. One calorie is the amount of heat needed to raise the temperature of 1 gram of water by 1 degree Celsius. This unit, however, is too small for evaluating food energy. A Calorie, which is equal to 1000 calories, is used to measure food energy.

In this investigation, you will construct a simple calorimeter and use it to measure the amount of heat energy contained in certain foods.

Problem

How is the energy in food measured?

Materials *(per group)*

Ring stand
Test tube clamp
Test tube
Paper clip
Cork stopper
Fireproof pad
Metric ruler
Four food samples
Heat-resistant gloves
Triple-beam balance
100-mL graduated cylinder
Matches
Thermometer

Safety

Put on a laboratory apron if one is available. Put on safety goggles. Handle all glassware carefully. Use extreme care when working with heated equipment or materials to avoid burns. Note all safety alert symbols next to the steps in the Procedure and review the meanings of each symbol by referring to the symbol guide on page 10.

Procedure

1. To assemble a calorimeter, set up a ring stand, test tube clamp, test tube, and fireproof pad as shown in Figure 1.
2. To make a food platform for the calorimeter, bend the outer end of a paper clip straight down so that it is at a right angle to the rest of the clip. Insert the free end of the clip into the middle of the narrow end of the cork stopper. See Figure 2.
3. Place the food platform on the fireproof pad. Adjust the height of the test tube so that the space between the food platform and the bottom of the test tube is 2 cm.
4. Use a graduated cylinder to measure exactly 15 mL of water into the test tube. Record the mass of the water in the appropriate place in the Data Table. **Note:** *Remember that 1 mL of water has a mass of 1 g.*
5. Measure the temperature of the water in the test tube. Record this number in the appropriate place in the Data Table. **Note:** *Be sure to remove the thermometer from the test tube after you record the temperature.*
6. Select a food sample and find its mass using the triple-beam balance. Record the mass in the appropriate place in the Data Table. Also record the name of the food sample used in the appropriate place in the Data Table.
7. Place the food sample on the paper clip platform. Ignite the food sample with a match, and quickly place the platform under the test tube. **CAUTION:** *Wear safety goggles when doing this part of the investigation. Be careful when using matches.* Allow the food to burn completely. Reignite the sample if necessary.

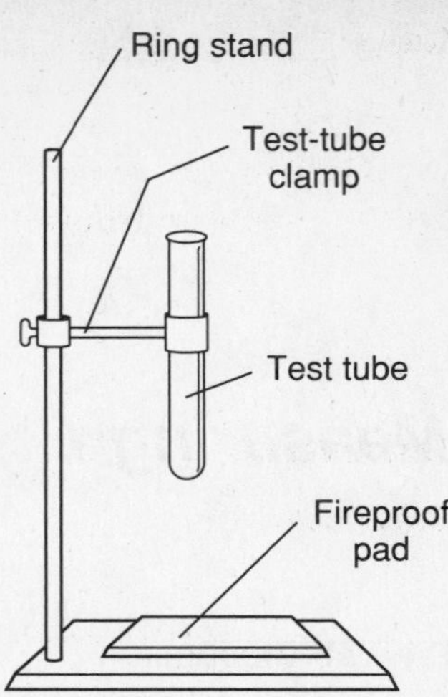

Figure 1

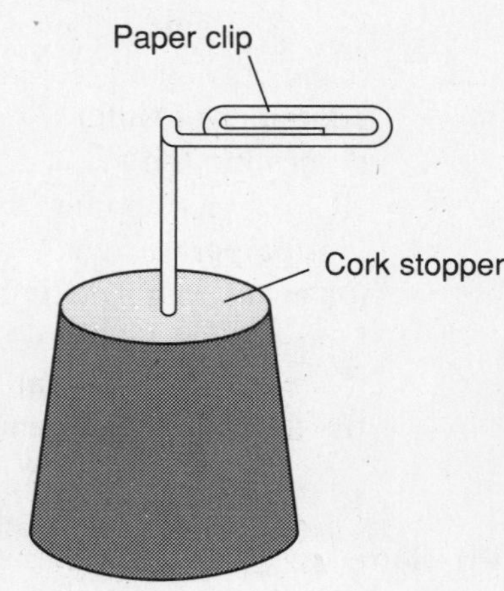

Figure 2

8. After the sample has burned completely, measure the temperature of the water in the test tube. **CAUTION:** *Do not touch the test tube; it may be hot.* Record the temperature of the water.
9. Find the mass of the remainder of the burned food sample. Record the mass.
10. Determine the change in mass of the food sample. Record the result.
11. Determine the change in the temperature of the water in the test tube. Record the result.
12. Repeat steps 3 through 11 using three other food samples. **Note:** *Remember to empty the water out of the test tube and to use cool water for each sample.*
13. Use the formula below to find the energy value, or Calories, per food sample. Record the results in the appropriate place in the Data Table. **Note:** *The specific heat of water is 1 Calorie per kilogram degree Celsius.*

$$\text{Calories per food sample} = \text{Change in water temperature} \times \text{Mass of water} \times \text{Specific heat of water} \times \frac{1\ \text{kg}}{1000\ \text{g}}$$

Name ______________________ Class __________ Date __________

14. Use the formula below to find the Calories per gram of food sample. Record the results in the appropriate place in the Data Table.

Calories per gram = Calories per food sample/Change in mass of food sample

Observations

Data Table

Variable	Food Sample			
Mass of food sample before burning (g)				
Mass of food sample after burning (g)				
Change in mass of food sample (g)				
Mass of water (g) (1 mL = 1 g)				
Temperature of water before heating (°C)				
Temperature of water after heating (°C)				
Change in water temperature (°C)				
Calories per food sample				
Calories per gram				

Analysis and Conclusions

1. What is the difference between a calorie and a Calorie? ______________________

____

2. Why must the food sample be ignited before placing the platform under the test tube?

____

____

3. Why must the thermometer be removed from the test tube when the food sample is burning?

____

____

4. How do your results compare to those of other student groups in your class? Give reasons for any variations. ______________________________

____

5. Fats yield more food energy than proteins or carbohydrates. Which of your food samples most likely contained the greatest amount of fat? ______________________________

____

Critical Thinking and Application

1. Swimming for one hour burns up 600 Calories. For each food sample you tested, calculate how many grams of food you would have to eat to get this energy. ______________________________

____

2. Fad diets, which have become popular in the past two decades, involve the consumption of large amounts of a limited variety of foods. Explain why some fad diets may be an unhealthful way to lose weight. ______________________________

____

____

3. Although fiber is not officially classified as a nutrient, it is an important component of the American diet today. What is the role of fiber in the human body?

____

____

____

4. Contrast the snacks for a person who is trying to lose weight with those for a person who is growing very rapidly. ______________________________

____

____

____

Going Further

Using the procedure from this investigation, determine the Caloric value of various diet foods and their counterparts. Is there a difference in their Caloric values?

Name ______________________ Class ____________ Date ____________

Measuring Lung Capacity

Pre-Lab Discussion

The amount of air that you move in and out of your lungs depends on how quickly you are breathing. The amount of air that is moved in and out of the lungs when a person is breathing normally is called the *tidal volume*. This amount of air provides enough oxygen for the body when the person is resting. It is possible to inhale more deeply and exhale more forcefully than usual. The maximum amount of air moved in and out of the lungs when the deepest possible inspiration is followed by the strongest possible expiration is called the *vital capacity*.

In this investigation, you will determine the tidal volume and vital capacity of your lungs.

Problem

How are the tidal volume and vital capacity of the human lungs measured?

Materials *(per pair of students)*

2 round balloons (1 for each student in the pair)
Metric ruler
Meterstick
Bathroom scale (1 per class is adequate)

Safety

Do not participate in this investigation if you have any breathing difficulties.

Procedure

Part A. Measuring Tidal Volume

1. Stretch a round balloon lengthwise several times.
2. Inhale normally and then exhale normally into the balloon. **Note:** *Do not force your breathing.*

3. Immediately pinch the end of the balloon shut so that no air escapes. Place the balloon on a flat surface. Have your partner use the metric ruler to measure the diameter of the balloon at its widest point. See Figure 1. Record this measurement in Data Table 1.
4. Deflate the balloon and repeat steps 2 and 3 two more times. Use your three measurements to calculate an average tidal volume. Record this measurement in Data Table 1.

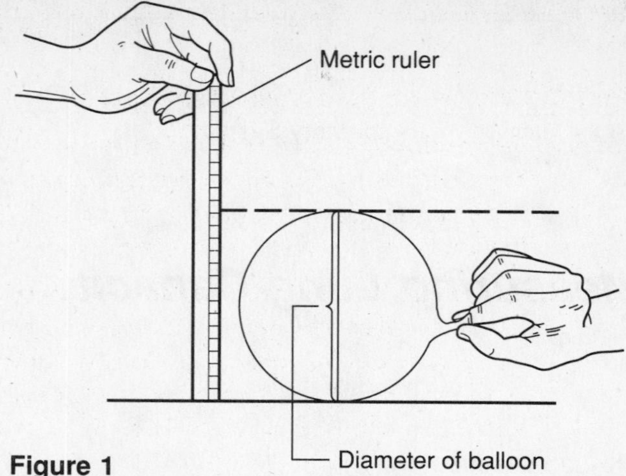

Figure 1

Part B. Measuring Vital Capacity

1. After breathing normally, inhale as much air into your lungs as possible. Exhale as much air as you can from your lungs into the balloon.
2. Immediately pinch the end of the balloon shut so that no air escapes. Place the balloon on a flat surface. Have your partner use the metric ruler to measure the diameter of the balloon at its widest point. Record this measurement in Data Table 1.
3. Deflate the balloon and repeat steps 1 and 2 two more times. Use your three measurements to calculate an average vital capacity. Record this measurement in Data Table 1.
4. Use Figure 2 to convert the balloon diameters in Data Table 1 into lung volumes. On the horizontal (x) axis, locate the diameter of the balloon in centimeters and follow the number up until it meets the curved line. Then move across in a straight line to the vertical (y) axis and approximate the lung volume. Record this measurement in Data Table 2. Repeat this procedure for all of the balloon diameters in Data Table 1.

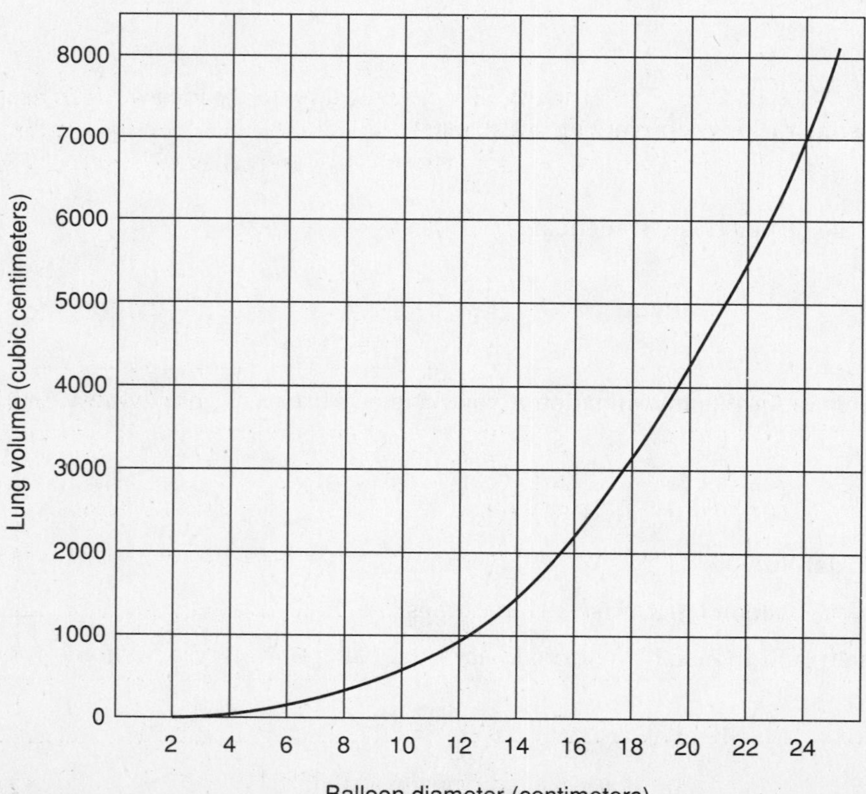

Figure 2

Name ______________________ Class __________ Date __________

Part C. Calculating Estimated Vital Capacity

1. Research has shown that the capacity of a person's lungs is proportional to the surface area of his or her body. To find the surface area of your body, you will need to know your height in centimeters and your mass in kilograms. Use a meterstick to find your height and the bathroom scale to find your mass. Record these measurements in Data Table 3.

2. Use Figure 3 to estimate the surface area of your body. Find your height in centimeters on the left scale. Mark this point. Find your mass in kilograms on the right scale. Mark this point. Use a metric ruler to draw a straight line connecting these two points. Now look at the center scale. The point at which your line crosses this scale gives your surface area in square meters. Record this number in Data Table 3.

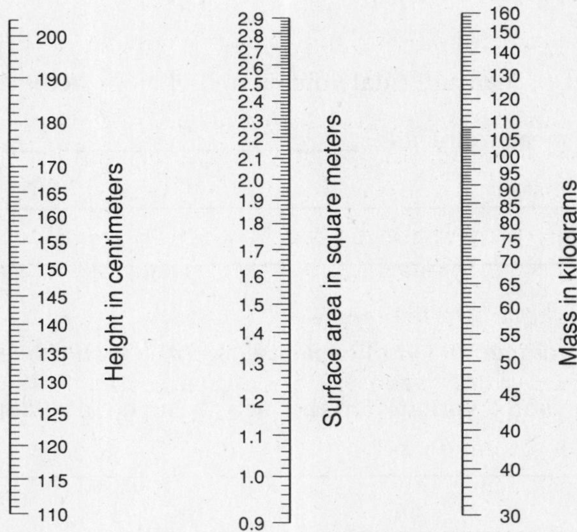

Figure 3

3. To calculate the estimated vital capacity of your lungs, multiply your surface area by the ratio of vital capacity to surface area. For females this ratio is 2000 mL per square meter. For males this ratio is 2500 mL per square meter. Record the estimated vital capacity of your lungs in Data Table 3.

Observations

Data Table 1

Balloon Diameter (cm)		
Trial	**Tidal Volume**	**Vital Capacity**
1		
2		
3		
Average		

Data Table 2

Lung Volume (cm^3)		
Trial	**Tidal Volume**	**Vital Capacity**
1		
2		
3		
Average		

Data Table 3

Body Component	Data
Height (cm)	
Mass (kg)	
Surface area (m^2)	
Vital capacity (cm^3)	

Analysis and Conclusions

1. Why is it important to measure tidal volume and vital capacity three times and calculate averages for these measurements? ____________________

2. How do your tidal volume and vital capacity compare with those of other class members? Why might there be some variation in the measurements of different people?

3. How does your estimated vital capacity compare to your measured vital capacity?

4. If there is a difference, suggest a reason for this. ____________________

5. Why might it be important to know a person's tidal volume or vital capacity?

Name ______________________ Class ____________ Date ____________

6. Even when you exhale as forcefully as possible, some air remains in the lungs. Why is this an important occurrence? ______________________

____

____

Critical Thinking and Application

1. Figure 4 shows measurements of a jogger's vital capacity. The measurements were taken using a method similar to the one you used in this investigation. The vital capacity was measured at the beginning of the jogger's training period and then every five days after that. Use the data in Figure 4 to construct a line graph on the graph provided.

Day of Training Period	Vital Capacity (cm^3)
0	4800
5	4830
10	4890
15	4910
20	4960
25	5040
30	5130

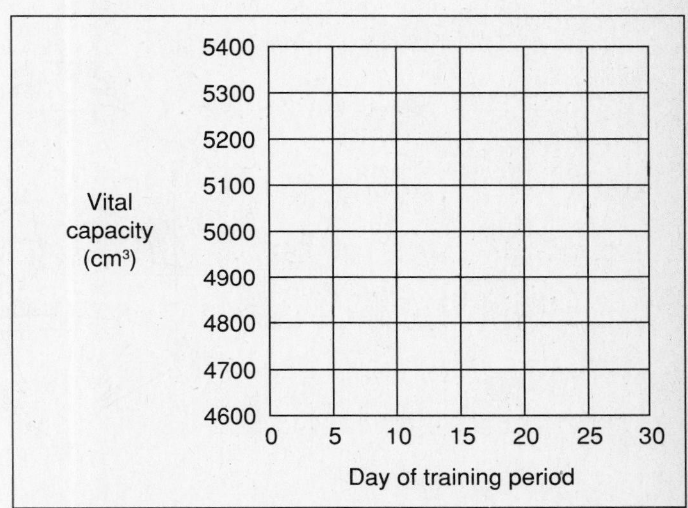

Figure 4

2. What happened to the jogger's vital capacity as the training period progressed?

____

____

3. What probably caused the change in the jogger's vital capacity? ______________________

____

____

4. How might vital capacity be important to some musicians? ______________________

____

____

5. How do you think smoking would affect vital capacity? ______________________________

____

____

Going Further

1. The vital capacity of the lungs is affected by the anatomical build of a person, the position of the person during the vital capacity measurement, and the stretching capability of the lungs and chest cavity. Measure the vital capacity of your lungs while sitting up and then while lying down. In each position, inhale as much air into your lungs as possible and exhale as much air as you can into a balloon. Measure the diameter of the balloon each time. Compare your vital capacity in each position.
2. If a bell jar is available, construct a model of human lungs as shown in Figure 5. To make the model "breathe," pull down and then push up on the rubber sheet. Describe what happens. How does the working of this model relate to the working of your diaphragm and lungs?

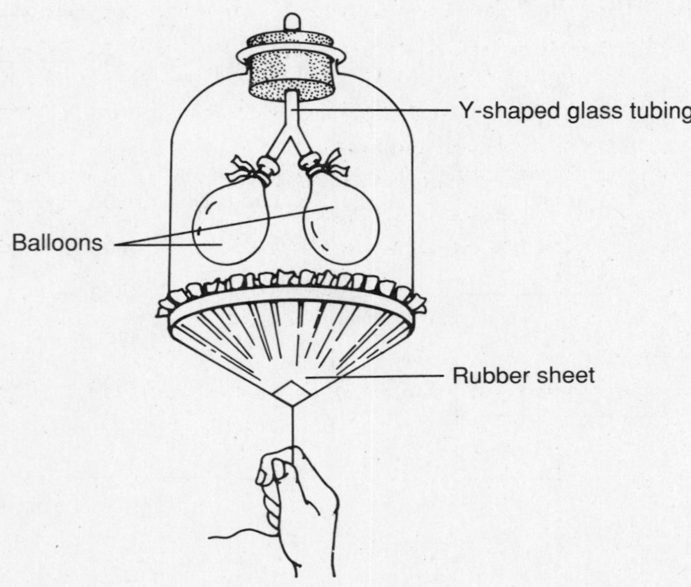

Figure 5

Name ______________________ Class __________ Date __________

The Effect of Exercise on Respiration

Pre-Lab Discussion

In multicellular organisms, specialized organs and organ systems are used to exchange the gases involved in cellular respiration with the external environment. The function of the human respiratory system is to take oxygen into the body and to release carbon dioxide from the body. As a result of changes in the environment or the amount of physical activity, levels of carbon dioxide and oxygen in the blood change. The body attempts to maintain the proper balance of carbon dioxide and oxygen levels in the blood by increasing or decreasing the rate of breathing. The presence of carbon dioxide can be detected with the testing solution bromthymol blue. A change of color in this solution indicates the presence of carbon dioxide.

In this investigation, you will determine the effect of exercise on the breathing rate. You will also use bromthymol blue to determine the presence of carbon dioxide in exhaled air, and the effect of exercise on the amount of carbon dioxide produced.

Problem

How are breathing rate and production of carbon dioxide affected by exercise?

Materials *(per group)*

200 mL of bromthymol blue solution
2 500-mL jars or flasks
Glass stirring rod
8 drinking straws
100-mL graduated cylinder
65 mL of 4% ammonia solution
Medicine dropper
Clock or watch with second hand

Safety

Put on a laboratory apron if one is available. Put on safety goggles. Handle all glassware carefully. Always use special caution when working with laboratory chemicals, as they may irritate the skin or cause staining of the skin or clothing. You will be exercising during this investigation. If at any time you feel faint or dizzy, sit down and immediately call your teacher. Note all safety alert symbols next to the steps in the Procedure and review the meanings of each symbol by referring to the symbol guide on page 10.

Procedure

Part A. Breathing Rate at Rest

After each part of this investigation, rest and allow your heart rate and breathing rate to return to normal. **CAUTION:** *If at any time during this investigation you feel faint or dizzy, sit down and immediately call your teacher.*

1. Sit down and relax for several minutes before doing this part of the investigation. Remain seated throughout this part of the investigation.
2. Count the number of times you inhale at rest in one minute. You may wish to have one of your group members watch the clock as you count your number of breaths. Record this measurement in Data Table 1.
3. Repeat step 2 two more times and calculate your average breathing rate at rest. Record this number in Data Table 1.

Part B. Breathing Rate After Exercise

1. Run in place for one minute.
2. Return to your chair and record the number of times you inhale in one minute. Record this measurement in Data Table 1.
3. Repeat steps 1 and 2 two more times and calculate your average breathing rate after exercise. Record this number in Data Table 1.

Part C. Amount of Carbon Dioxide Exhaled at Rest

1. Pour 100 mL of bromthymol blue solution into a jar or flask.
2. Have one member of your group use a drinking straw to breathe out normally into the bromthymol blue solution for exactly one minute. **CAUTION:** *Be careful not to suck the solution into your mouth.* The bromthymol blue solution should turn a pale yellow color at the end of one minute.
3. With a medicine dropper, add 1 drop of ammonia to the jar and stir once with a glass stirring rod.
4. Continue to add ammonia one drop at a time. Count each drop and stir once between drops until the solution turns blue. In Data Table 2, record the total number of drops of ammonia needed to turn the solution blue.
5. Repeat steps 2 through 4 two more times using the same bromthymol blue solution. Find the average number of ammonia drops needed for the three trials and record this number in Data Table 2.

Part D. Amount of Carbon Dioxide Exhaled After Exercise

1. Pour 100 mL of bromthymol blue solution into a second jar or flask.
2. Have the same group member who was the test subject in Part C stand up and run in place for one minute.
3. Have the group member use another drinking straw to exhale into the bromthymol blue solution for exactly one minute. **CAUTION:** *Be careful not to suck the solution into your mouth.*
4. With a medicine dropper, add 1 drop of ammonia to the jar and stir once with a glass stirring rod.
5. Continue to add ammonia one drop at a time. Count each drop and stir once between drops until the solution turns blue. In Data Table 2, record the total number of drops of ammonia needed to turn the solution blue.

Name ______________________ Class __________ Date __________

6. Repeat steps 2 through 5 two more times using the same bromthymol blue solution. Find the average number of ammonia drops needed for the three trials and record this number in Data Table 2.

Observations

Data Table 1

Breathing Rate (times per minute)		
Trial	**At Rest**	**After Exercise**
1		
2		
3		
Average		

Data Table 2

Number of Ammonia Drops		
Trial	**At Rest**	**After Exercise**
1		
2		
3		
Average		

Analysis and Conclusion

1. Compared with your normal breathing rate at rest, what happened to your breathing rate after exercise? ______________________

____

2. Why would your breathing rate change after exercise? ______________________

____

3. Were your results the same as the results of all of your classmates? What might account for any differences? ______________________

____

4. In Parts C and D, what does the number of ammonia drops represent?

____

____

5. Are the average numbers of ammonia drops needed to restore the blue color to the solutions in Parts C and D the same? If not, why are they different? ______________________________

____

____

6. What effect does exercise have on the amount of carbon dioxide released from the lungs?

____

____

Critical Thinking and Application

1. What factors would an athlete have to consider if he or she were training or competing at a high-altitude location such as Denver, Colorado? ______________________________

____

2. Why do some athletes inhale pure oxygen after strenuous activity? ______________________________

____

____

3. List three benefits of exercise to your body. ______________________________

____

4. The nicotine in cigarette tobacco enters the bloodstream, where it constricts the blood vessels. How would this affect the breathing rate? ______________________________

____

____

Going Further

1. Find out if germinating seeds give off carbon dioxide during respiration. Obtain about 10 to 15 germinating lima bean seeds. Place them in a flask, or small jar, and add enough bromothymol blue solution to cover about half of the seeds. Stopper the flask or, if a jar is used, place a lid on it. Place the jar aside for 30 minutes and then observe the results. What color should the bromthymol blue solution turn if carbon dioxide is given off by the seeds?
2. Collect breathing rates for members of your family and other people of different ages. Categorize the data by 10-year increments and illustrate the results with a bar graph.

Name ______________________ Class ____________ Date ______________

80

Simulating Blood Typing

Pre-Lab Discussion

Human blood may be classified according to the presence or absence of certain *antigens*, or factors, that are attached to the surface of the red blood cells, or erythrocytes. Two of the antigens used in blood typing are known as A and B. A person whose red blood cells have only antigen A has type A blood, whereas a person whose red blood cells have only antigen B has type B blood. People who have both A and B antigens on their red blood cells have type AB blood. Those whose blood cells have neither A nor B antigens have type O blood.

The plasma of each blood group contains a certain type or combination of *antibodies*. Antibodies are substances that attack antigens. Blood type A plasma contains anti-B antibodies, whereas blood type B plasma has anti-A antibodies. Anti-A antibodies attack red blood cells that have A antigens; anti-B antibodies attack those that have B antigens. The attacking antibodies bind to the red blood cells, causing them to *agglutinate,* or clump together. Type AB plasma has both A and B antigens and has neither type of antibody. Type O blood has neither A nor B antigens and contains both anti-A and anti-B antibodies. In transfusions, the blood types of the donor and recipient must be carefully matched because transfusion of the wrong type of blood can be fatal to the recipient.

In this investigation, you will simulate human blood typing.

Problem

How is a person's blood type determined?

Materials *(per group)*

Glass slide with two depressions
Simulated anti-A serum
Simulated anti-B serum
Simulated blood—types A, B, AB, O

Safety

Put on a laboratory apron if one is available. Put on safety goggles. Handle all glassware carefully. Always use special caution when working with laboratory chemicals, as they may irritate the skin or cause staining of the skin or clothing. Never touch or taste any chemical unless instructed to do so. Note all safety alert symbols next to the steps in the Procedure and review the meanings of each symbol by referring to the symbol guide on page 10.

Procedure

1. Put on safety goggles. Place 2 drops of the solution in the dropper bottle labeled Bottle 1 in each of the two depressions in the glass slide. To the left depression, add 2 drops of the solution in the bottle labeled anti-A serum. To the right depression, add 2 drops of the solution in the bottle labeled anti-B serum. **CAUTION:** *Use caution when working with laboratory chemicals. If a laboratory chemical comes into contact with your skin, wash the area with water immediately.*

2. Examine the substances in the two depressions for signs of clumping. If clumping occurs only on the left side of the depression slide, this simulates the presence of type A blood. If clumping occurs only on the right side of the depression slide, this simulates the presence of type B blood. If clumping occurs on both sides of the depression slide, this simulates the presence of type AB blood. If no clumping occurs on either side of the depression slide, this simulates the presence of type O blood. See Figure 1. Record your observations in the appropriate places in the Data Table.

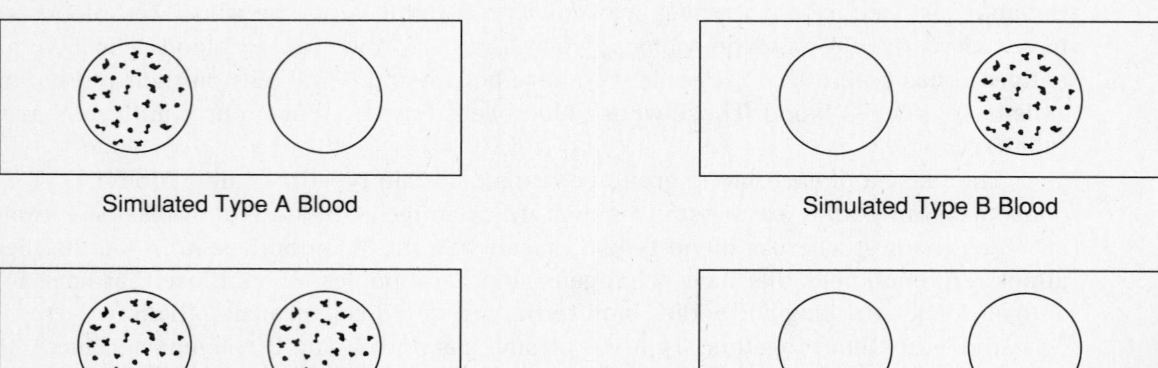

Figure 1

3. Based on your results, determine the type of simulated blood that is contained in Bottle 1. Record this information in the appropriate place in the Data Table.
4. Carefully wash and dry the glass slide thoroughly.
5. Repeat steps 1 through 4 using the solutions in Bottles 2, 3, and 4.

Observations

Data Table

Bottle	Clumping in the Left Side of the Depression Slide?	Clumping in the Right Side of the Depression Slide?	Simulated Blood Type
1			
2			
3			
4			

Name ______________________ Class ____________ Date ____________

Analysis and Conclusions

1. a. Which simulated blood type(s) showed clumping when simuiated anti-A serum was added?

____

b. Which simulated blood type(s) showed clumping when simulated anti-B serum was added?

____

2. a. If clumping occurs when both anti-A serum and anti-B serum are added, what is the blood type? ____

b. If clumping does not occur when either anti-A serum or anti-B serum is added, what is the blood type? ____

3. Which blood type(s) must people have in order to safely receive a transfusion of type A blood? ____

4. Which blood type(s) must people have in order to safely receive a transfusion of type B blood?

____

5. Which blood type(s) must people have in order to safely receive a transfusion of type AB blood? ____

6. How is this simulation similar to actual human blood typing? ______________________

____

____

____

____

____

Critical Thinking and Application

1. Why is it advisable for you to know your own blood type? ______________________

____

____

2. How could you tell whether two blood samples are compatible for a transfusion if no typing serum were available? ______________________________

____

____

3. Why is a person with type AB blood sometimes called a "universal recipient"?

____

____

____

4. Why is a person with type O blood sometimes called a "universal donor"?

____

____

____

5. Why is a person with type O blood unable to receive blood from any type other than O?

____

____

____

Going Further

1. Successful organ transplant surgery depends on advances made in the area of tissue typing and compatibility. Using reference materials, write a report comparing blood typing with the relatively new science of tissue typing.
2. Use reference materials to research the topic of Rh factors. How were they discovered? How is a person's blood tested for Rh factor? Why is this knowledge important?

Name ______________________ Class ____________ Date ______________

81

Simulating Urinalysis

Pre-Lab Discussion

All organisms produce wastes. These waste materials must be removed so that the organism is not poisoned by its own metabolism. In humans, *urine* is the fluid produced by the kidneys as they remove waste chemicals from the blood. Urine is made up primarily of water, with some salts and organic materials dissolved in it. The concentration of each of these substances varies with a person's health, diet, and degree of activity.

By testing the chemical composition of urine, doctors can learn much about the general health of an individual. Urinary tract infections, kidney malfunction, diabetes, and liver disease are just some of the medical problems that can be diagnosed through *urinalysis*. Urinalysis involves the physical, chemical, and visual examination of a urine sample.

In this investigation, you will perform several tests to detect substances in a sample of artificial urine. You will also determine the contents of an artificial urine sample of unknown composition.

Problem

What chemical substances are found in a sample of artificial urine?

Materials *(per group)*

12 test tubes
Glass-marking pencil
Test tube rack
Test tube holder
10-mL graduated cylinder
Hot plate
Bunsen burner
400-mL beaker
Matches
Urine sample with glucose
Urine sample without glucose
Urine sample with phosphate
Urine sample without phosphate
Urine sample with albumin
Urine sample without albumin
Urine sample with chloride
Urine sample without chloride
Benedict's solution
Silver nitrate solution
10% acetic acid solution

Safety

Put on a laboratory apron if one is available. Put on safety goggles. Handle all glassware carefully. Use extreme care when working with heated equipment or materials to avoid burns. Always use special caution when working with laboratory chemicals, as they may irritate the skin or cause staining of the skin or clothing. Never touch or taste any chemical unless instructed to do so. Observe proper laboratory procedures when using electrical appliances. Note all safety alert symbols next to the steps in the Procedure and review the meanings of each symbol by referring to the symbol guide on page 10.

Procedure

Part A. Test for Glucose

1. Place two test tubes in a test tube rack. With a glass-marking pencil, label one test tube "G" for glucose. Allow the other test tube to remain unlabeled as it will act as the control.
2. Use the 400-mL beaker to prepare a hot water bath. **CAUTION:** *Use extreme care when working with hot water. Do not let the water splash onto your body.*
3. Add 3 mL of Benedict's solution to both test tubes. **CAUTION:** *Use extreme care when handling Benedict's solution to avoid staining of the skin and clothing.*
4. Add 3 mL of the urine sample with glucose to the test tube labeled "G." Add 3 mL of the urine sample without glucose to the unlabeled test tube. Note the appearance of the substance in each test tube. Record this information in Data Table 1.
5. Place both test tubes in the hot water bath for 2 minutes.
6. After 2 minutes, remove the test tubes from the hot water bath with a test tube holder. Place the test tubes in the test tube rack. **CAUTION:** *Be careful when working with heated equipment or materials to avoid burns.* Note any color changes in the test tubes. Record your observations in Data Table 1.

Part B. Test for Chloride

1. Place two test tubes in a test tube rack. With a glass-marking pencil, label one test tube "C" for chloride. Allow the other test tube to remain unlabeled as it will act as the control.
2. Add 5 mL of the urine sample with chloride to the test tube labeled "C." Add 5 mL of the urine sample without chloride to the unlabeled test tube. Note the appearance of the substance in each test tube. Record this information in Data Table 2.
3. Carefully add 3 drops of silver nitrate solution to each test tube. **CAUTION:** *Use caution when working with silver nitrate solution because it can stain the skin and clothing.* Observe the top surface of the liquid in each test tube. Record its appearance in Data Table 2.

Part C. Test for Albumin

1. Place two test tubes in a test tube rack. With a glass-marking pencil, label one test tube "A" for albumin. Allow the other test tube to remain unlabeled as it will act as the control.
2. Half fill the test tube labeled "A" with the urine sample with albumin. Half fill the unlabeled test tube with the urine sample without albumin.
3. Using a test tube holder, pass the top surface of each test tube over the flame of a Bunsen burner for 15 to 20 seconds. See Figure 1. **CAUTION:** *Secure all loose clothing and hair when using a Bunsen burner. When heating a test tube, always point it away from yourself and other students.* After heating each test tube, place it in a test tube rack. Note the appearance of each substance. Record your observations in Data Table 3.

Name ______________________ Class __________ Date __________

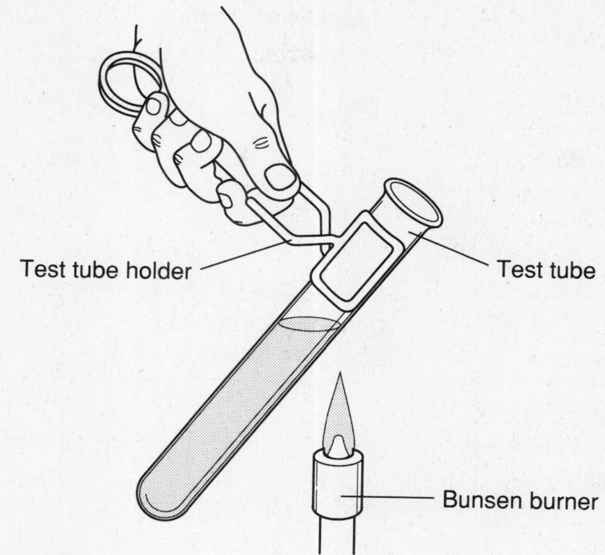

Figure 1

4. Add 5 drops of acetic acid to each test tube. **CAUTION:** *Be careful when using an acid.* Note the appearance of each substance. Record your observations in Data Table 3.

Part D. Test for Phosphate

1. Place two test tubes in a test tube rack. With a glass-marking pencil label one test tube "P" for phosphate. Allow the other test tube to remain unlabeled as it will act as the control.
2. Half fill the test tube labeled "P" with the urine sample with phosphate. Half fill the unlabeled test tube with the urine sample without phosphate.
3. Using a test tube holder, pass the top surface of each test tube over the flame of a Bunsen burner for 15 to 20 seconds. **CAUTION:** *Secure all loose clothing and hair when using a Bunsen burner. When heating a test tube, always point it away from yourself and other students.* After heating each test tube, place it in a test tube rack. Note the appearance of each substance. Record this information in Data Table 4.
4. Add 5 drops of acetic acid to each test tube. Record your observations in Data Table 4. **CAUTION:** *Be careful when using an acid.*

Part E. Testing an Unknown Urine Sample

1. Obtain a sample of artificial urine marked "unknown." Record the identification number of the sample in Data Table 5.
2. Using four clean test tubes, perform the glucose, chloride, albumin, and phosphate tests on the unknown sample. Follow the Procedures in Parts A through D of this investigation.
3. Record your results in Data Table 5.

Observations

Data Table 1

Substance	Color Before Heating	Color After Heating
Urine sample with glucose		
Urine sample without glucose		

Data Table 2

Substance	Color Before Adding Silver Nitrate	Color After Adding Silver Nitrate
Urine sample with chloride		
Urine sample without chloride		

Data Table 3

Substance	Appearance After Heating	Appearance After Adding Acetic Acid
Urine sample with albumin		
Urine sample without albumin		

Data Table 4

Substance	Appearance After Heating	Appearance After Adding Acetic Acid
Urine sample with phosphate		
Urine sample without phosphate		

Data Table 5

Composition of Unknown Urine Sample # ________	
Test	**Present or Absent?**
Glucose	
Chloride	
Albumin	
Phosphate	

Name ______________________ Class __________ Date __________

Analysis and Conclusions

1. Why is it necessary to perform tests on the urine samples that do not contain any chemical substances? ______________________

2. If the top portion of a urine sample is heated and no haze forms, what conclusion can be drawn about the sample? ______________________

3. If you add Benedict's solution to a urine sample from a person who has diabetes and heat it, what color would you expect the heated sample to be? Explain your answer.

4. Which substances did you find in your unknown urine sample? Do you think this sample might have come from a healthy person? Explain your answer. ______________________

Critical Thinking and Application

1. Why does the chemical content of urine change throughout the day?

2. If a doctor finds high levels of protein in a patient's urine sample, the doctor will probably test the patient's urine several times over a week before drawing any conclusions. What might be the reason for this? ______________________

3. If a white haze forms at the top of a heated urine sample, how can you determine whether it is due to the presence of phosphate or albumin? ____

____

____

4. While a blood sample contains glucose, phosphate, albumin, and chloride molecules, a normal urine sample contains only phosphate and chloride molecules. What does this indicate about one of the functions of the kidneys? ____

____

____

____

Going Further

1. People with diseased kidneys must frequently have waste products removed from their blood by a procedure called dialysis. Use reference materials to research this procedure. You may also wish to interview a hospital dialysis technician.
2. Aquatic animals get rid of waste products by continuously excreting them directly to the outside watery environment. Design an experiment to show this using a goldfish. How is pH an important factor in maintaining tropical fish?

Name ______________________ Class ____________ Date ____________

82

Observing Human Growth

Pre-Lab Discussion

The human *endocrine system* is made up of a number of ductless glands that secrete chemicals that control certain body activities. The chemicals released by endocrine glands are called *hormones.* A hormone is a chemical that is produced in one part of an organism and is transported to other parts of the organism where it controls certain activities of tissues and organs. The *pituitary gland,* found near the base of the brain, is a small gland that releases many hormones. One of these hormones is *growth hormone,* or GH. GH stimulates the growth of bone, muscle, and many other tissues.

Humans grow at varying rates from birth through adolescence. These rates include a growth spurt when the body grows at a rapid rate. Although both males and females go through a growth spurt, they do not grow at the same rates and they do not have their growth spurts at the same ages.

In this investigation, you will observe and interpret male and female growth statistics.

Problem

How do the growth rates of males and females compare?

Materials *(per pair of students)*

Tape measure or meterstick
2 colored pencils

Procedure

1. Have your partner use the tape measure or meterstick to measure your height in centimeters. Record this number and your age in the appropriate place in Observations. Also record this information in the data table provided by your teacher on the chalkboard.

2. After all students in your class have recorded their heights and ages on the chalkboard, calculate the average height for males and the average height for females in each age group. Record these averages in the Data Table.

3. On Graph 1, construct line graphs of the information shown in the Data Table. Use one of the colored pencils to draw the line graph for females and the other colored pencil to draw the line graph for males.

4. Figure 1 shows the average heights for human females and males from birth to 18 years of age. On Graph 2, construct line graphs of the information shown in Figure 1. Use one of the colored pencils to draw the line graph for females and the other colored pencil to draw the line graph for males.

Average Height (cm)		
Age Group (yr)	**Female**	**Male**
At birth	50	51
2	87	88
4	103	104
6	117	118
8	128	128
10	139	139
12	152	149
14	160	162
16	163	172
18	163	174

Figure 1

Observations

Your height = ____________ centimeters Your age = ______ years

Data Table

Average Height (cm)		
Age Group (yr)	**Female**	**Male**
13		
14		
15		
16		
17		

Name ______________________ Class ____________ Date ____________

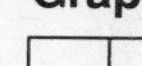

Graph 1

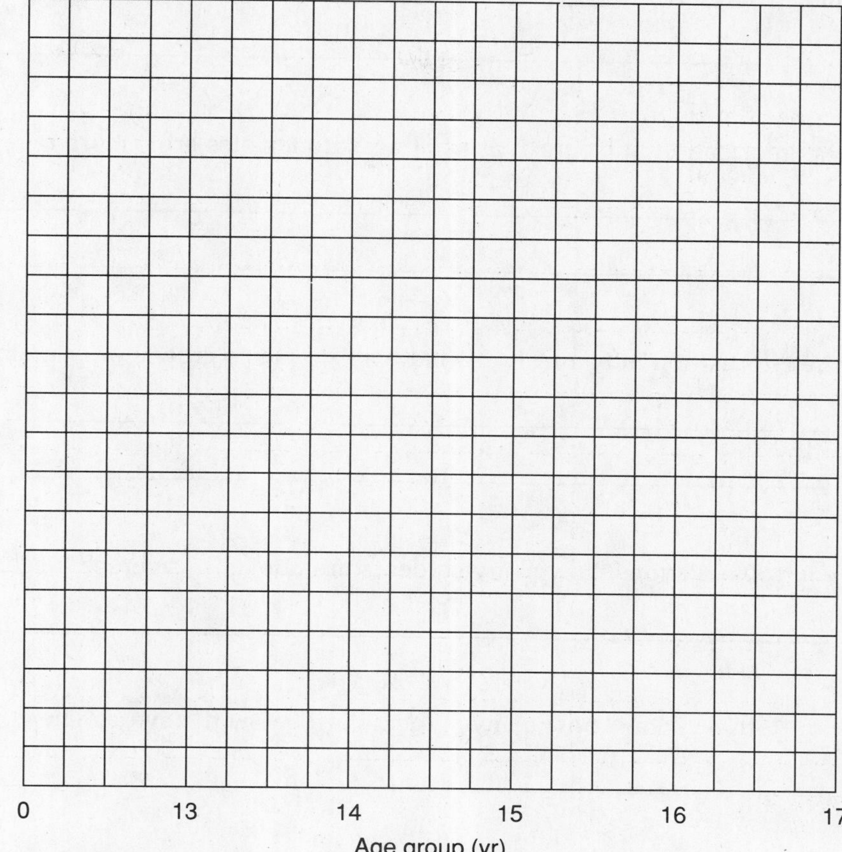

Graph 2

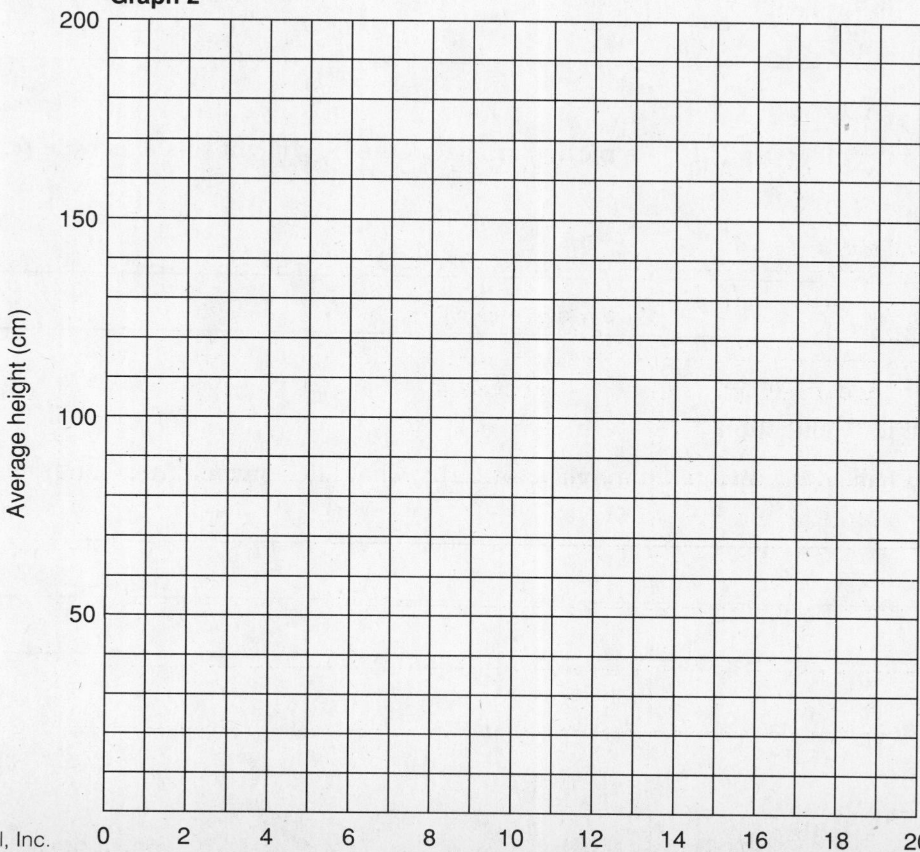

Analysis and Conclusions

1. Based on the information in Figure 1, at approximately what age does the height of males and females double from birth? ______

2. Based on the information in Figure 1, at what ages do females grow more rapidly than males?

______

3. Based on the information in Figure 1, at what ages do males grow more rapidly than females?

______

4. Based on your observations, how many students are above the average height for their age?

______

5. Based on your observations, how many students are below the average height for their age?

______

6. As a group, are the students in your class above or below the average height?

______

7. If there are very few students in a particular age group, how might this affect the reliability of your graph? ______

Critical Thinking and Application

1. If a person has an inactive pituitary gland at birth, what effects would this condition have on the body? ______

Name ______________________ Class ____________ Date ____________

2. Why might damage to the pituitary gland be far more serious than damage to other endocrine glands? ______________________

____

____

3. A condition known as acromegaly is the sudden growth of certain body parts, such as the hands and face, after normal body growth has been completed. How might this condition occur? ______________________

____

____

4. Some hormones prescribed by doctors are taken orally while others are injected directly into the body. Why might these hormones be administered in different ways?

____

____

Going Further

1. Collect and record the data from other classes that have completed this investigation. Calculate the average height for females and the average height for males in each age group for all classes. Construct a line graph for female heights and a line graph for male heights. Do these averages more closely resemble the averages reported in Figure 1 of this investigation?
2. Anabolic steroids are hormones that stimulate muscle tissue growth. Using reference materials, research the health hazards associated with the use of anabolic steroids by weight lifters and other athletes. How can it be determined whether an athlete is using anabolic steroids?

Name ______________________ Class ____________ Date ______________

Ovaries and Testes

Pre-Lab Discussion

Reproduction is the process by which offspring are produced. The most important function of reproduction is to continue the species. Reproduction may also serve to increase the number of individuals in a species.

Humans reproduce sexually. Gametes are produced in specialized sex organs, or *gonads*. The gonad of the human male is the *testis* (plural, *testes*). The two functions of this organ are the production of *sperm cells,* which are the male gametes, and the production of the male hormone *testosterone*. The gonad of the human female is the *ovary*. The two ovaries of the human female produce *egg cells,* which are the female gametes, and also secrete female sex hormones.

In this investigation, you will examine prepared slides of mammalian ovaries and testes. You will also investigate the process of egg and sperm production.

Problem

What structures are found in a mammalian ovary and testis? How are eggs and sperm produced?

Materials *(per group)*

Microscope
Prepared slides of:
- Cat ovary, transverse cross section
- Rat testis, longitudinal cross section
- Rat epididymis, transverse cross section

Safety

Handle all glassware carefully. Always handle the microscope with extreme care. You are responsible for its proper care and use. Use caution when handling glass slides as they can break easily and cut you. Note all safety alert symbols next to the steps in the Procedure and review the meanings of each symbol by referring to the symbol guide on page 10.

Procedure

Part A. The Mammalian Ovary

1. Obtain a stained slide of a transverse cross section of a mature ovary of a nonpregnant cat. Using the low-power objective of a microscope, focus on a part of the ovary where you can see both the outer edge and the interior of the ovary. The outermost layer of cells of the ovary consists of the germinal epithelium, a single layer of epithelial cells. Immediately below the germinal epithelium is the connective tissue called the stroma. Within the stroma you will observe numerous round structures of various sizes. These structures are the follicles, where the ovarian eggs develop and mature. In some of the follicles you can observe the developing egg, a large, circular cell with a darkly stained nucleus.
2. Examine different areas of the slide under low power to find follicles containing eggs in various stages of development. Using Figure 1 as a guide, trace the development of one developing egg cell. A developing egg is surrounded by a covering, or corona, of follicle cells. As the follicle matures, a cavity filled with the female hormone estrogen develops. Estrogen is essential to the growth and development of the maturing egg. The mature follicle gradually moves to the surface of the ovary, where it ruptures the surface and releases the egg into the Fallopian tube.

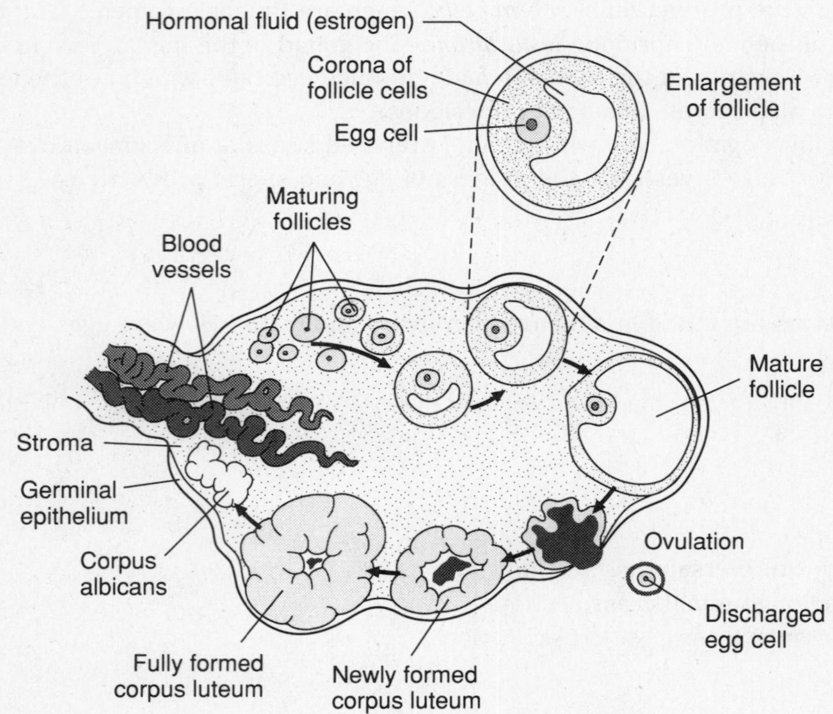

Figure 1

3. After a follicle ruptures and releases an egg from the ovary, the follicle undergoes a series of changes and becomes an endocrine structure called the corpus luteum. In this process, the follicle cells enlarge and a yellowish substance called lutein accumulates in their cytoplasm. The lutein cells secrete the hormones estrogen and progesterone, which are responsible for building up the lining of the uterus in preparation for pregnancy. Examine a corpus luteum on your slide. If pregnancy does not occur, the corpus luteum continues to grow for about 10 to 12 days and then shrinks, eventually becoming a small, white ovarian scar called a corpus albicans. Locate a corpus albicans on your slide.

Name ______________________ Class __________ Date __________

Part B. The Mammalian Testis

1. Obtain a stained slide of the testis of a mature rat for examination under the low-power objective of a microscope. The thick outer covering of the testis is called the tunica albuginea. The testis is separated into several wedge-shaped compartments by partitions called septa (singular, *septum*). Within each compartment examine the cluster of small circles. These are the cut surfaces of the tiny, coiled seminiferous tubules, which are involved in the production of sperm. The seminiferous tubules lead into the epididymis, a long, narrow, flattened structure attached to the posterior surface of the testis. Sperm complete their maturation while passing through the epididymis. The lower portion of the epididymis uncoils and widens into a long duct called the vas deferens.
2. In the appropriate place in Observations, label the following structures of the testis: tunica albuginea, septa, seminiferous tubules, epididymis, and vas deferens.
3. Carefully switch to the high-power objective and adjust your slide so that you are focusing on one of the seminiferous tubules. **CAUTION:** *When turning to the high-power objective, always look at the objective from the side of the microscope so that the objective does not hit or damage the slide.*
4. Notice that the walls of the seminiferous tubule consists of many layers of cells. The cytoplasm of these cells will most likely be stained red and the chromosomes in the nuclei will most likely be stained blue. Closely examine these cells in order to trace the various stages of spermatogenesis, or sperm cell production. The cells nearest to the outer surface of the seminiferous tubule are called spermatogonia. These are the sperm-producing cells. The spermatogonia divide by mitosis. Half of the daughter cells remain as spermatogonia, while the other half undergo changes and become cells called primary spermatocytes. The primary spermatocytes make up the layer next to the spermatogonia. Try to observe the chromosome arrangement in the primary spermatocytes.
5. The diploid primary spermatocyte undergoes the first stage of meiosis, producing two haploid cells called secondary spermatocytes. The secondary spermatocytes are found in a layer next to the primary spermatocytes. The two secondary spermatocytes undergo the second meiotic division, producing four haploid spermatid cells. The spermatid cells make up the innermost layer of the seminiferous tubule. A spermatid cell develops into a sperm with an oval head and a long, whiplike tail.
6. Observe how the sperm cells are clustered around elongated cells that are evenly spaced around the circumference of the seminiferous tubule. These cells, called Sertoli cells, probably provide nutrients for the developing sperm cells. Observe how the sperm cells have their tails facing out into the central opening, or lumen, of the seminiferous tubule.
7. In the appropriate place in Observations, label the following parts of the seminiferous tubule: spermatogonia, primary spermatocytes, secondary spermatocytes, spermatids, sperm cells, Sertoli cells, and lumen.
8. Obtain a prepared slide of a cross section of the epididymis of a mature rat. Observe the epididymis under low and high power. Note the sperm cells clustered within the lumen of the epididymis. Observe that the cells lining the lumen are lined with cilia. These cilia help propel the sperm cells through the epididymis. Observe the smooth muscle cells in the walls of the epididymis. As these muscles contract, the sperm cells are pushed through the epididymis, toward the vas deferens.
9. In the appropriate place in Observations, label the following parts of the epididymis: lumen, sperm cells, cilia, and smooth muscle cells.

Observations

Rat Testis

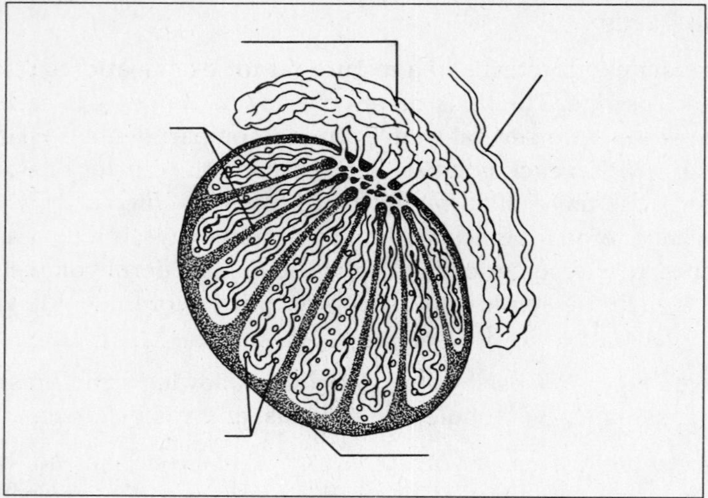

Rat Seminiferous Tubule

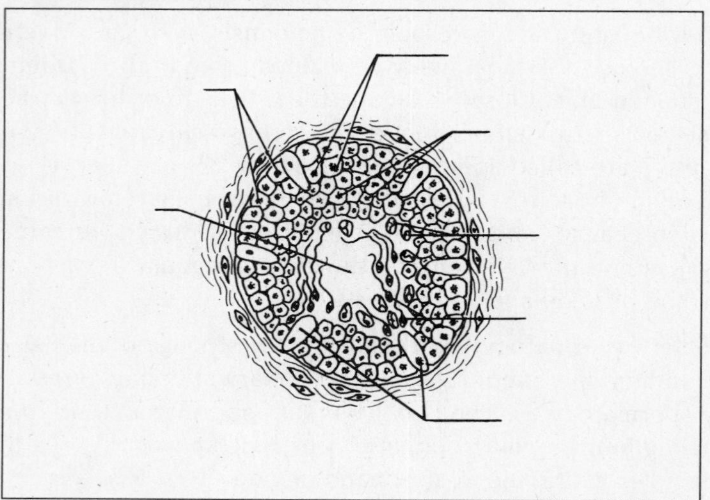

Rat Epididymis

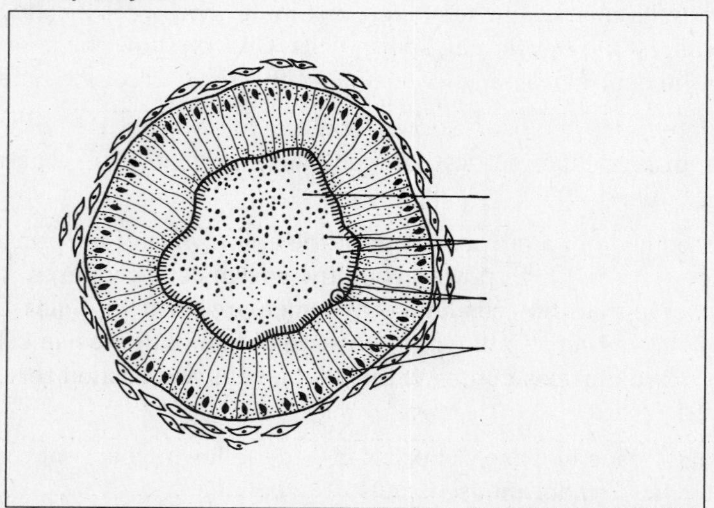

Name ______________________ Class __________ Date __________

Analysis and Conclusions

1. In what way are mature sperm and egg cells different from all other types of body cells?

2. What is the function of the corpus luteum in the ovary?

3. What is the adaptive advantage of the tail on the sperm cell?

4. What are two functions that are common to both ovaries and testes?

5. Describe three ways in which sperm-cell development is different from egg-cell development.

Critical Thinking and Application

1. The middle section of a sperm cell is packed with mitochondria. Use your knowledge of cell organelle function to determine the purpose the mitochondria serve in the sperm cell.

2. What might happen if more than one egg were released at the same time from the ovaries?

3. A mature egg cell contains a great deal more cytoplasm than a mature sperm cell. What is the possible function of the added cytoplasm found within the egg cell?

____

____

Going Further

1. Examine prepared slides of the sperm cells of different animals. Compare their similarities and differences.
2. Infertility is a problem faced by some couples who desire to have children. Identify possible causes of this problem. Is it simply the inability to produce sperm or eggs? What corrective measures can be taken for males and females?
3. Do library research on *in vitro* fertilization. Be sure to include information on why the technique is used, what procedures are involved, and how successful the technique is.

The Human Menstrual Cycle

Pre-Lab Discussion

The *menstrual cycle* is the hormone-controlled reproductive cycle of human females. It involves the periodic development and release of an egg and the periodic shedding of the uterine lining. The menstrual cycle involves interactions of the hypothalamus, pituitary, ovary, and uterus. The cycles usually occur between puberty and menopause. The cycle may be interrupted by pregnancy, illness, or other factors. The menstrual cycle is about 28 days long, but it varies from one female to another.

Each egg produced by the female matures inside a *follicle,* or egg sac, near the surface of the ovary. When the egg is fully mature, the follicle bursts and the egg is released into a Fallopian tube, which leads to the uterus. An unfertilized egg will pass from the female's body within a short time. If this occurs, the lining of the uterus, which has been prepared for the implantation of a fertilized egg, deteriorates and also passes out of the body. The process by which blood and uterine cells are shed through the vagina is called *menstruation.* Menstruation is one phase of the four-phase menstrual cycle.

In this investigation, you will examine the changes that occur during the different phases of the menstrual cycle.

Problem

What changes occur within the body of a human female during the menstrual cycle?

Materials *(per student)*

Metric ruler
Colored pencils

Procedure

Part A. The Follicle Phase

1. Many egg cells are located within the ovary of a human female. Each egg is contained within a structure called a follicle. Under the influence of follicle-stimulating hormone, or FSH, the follicle matures within the ovary. The amount of FSH in the bloodstream influences the growth and development of the maturing follicle. Figure 1 shows the various stages of a follicle's maturation. Notice that an immature follicle is much smaller than a mature follicle.

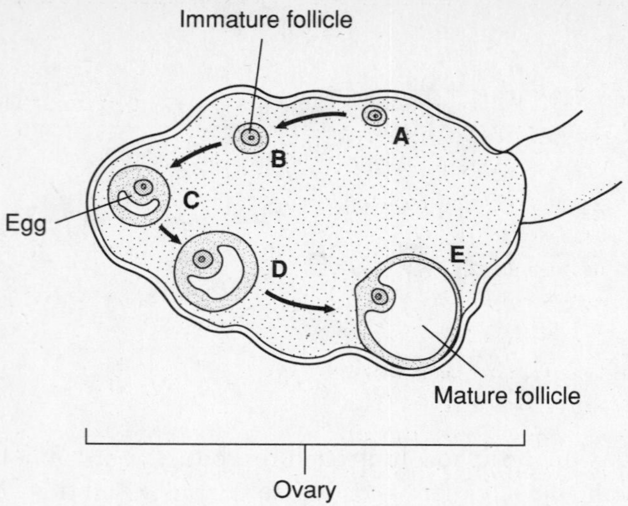

Figure 1

2. Figure 2 shows the concentration of FSH in the bloodstream of an average human female throughout the 28-day menstrual cycle. On Graph 1, prepare a line graph of the data in Figure 2. In the boxes above Graph 1, make drawings of the various stages of follicle maturation. The drawings should match the lettered stages shown in Figure 1.

Amount of FSH in Bloodstream (units per milliliter)			
Day	**FSH**	**Day**	**FSH**
1	9	15	9
2	11	16	8
3	13	17	8
4	14	18	8
5	15	19	8
6	14	20	7
7	14	21	7
8	15	22	6
9	13	23	5
10	11	24	5
11	9	25	6
12	18	26	7
13	13	27	7
14	9	28	8

Figure 2

3. Answer questions 1 through 4 in Observations.

Name ______________________ Class ____________ Date ____________

Part B. The Luteal Phase

1. Once a follicle is mature, it bursts open and the egg is released. This process is called ovulation. The egg passes into the Fallopian tube where it may or may not be fertilized. Once the mature follicle loses its egg, it forms a body within the ovary called the corpus luteum. Figure 3 shows the changes that occur within the corpus luteum during the menstrual cycle. After maturation, the corpus luteum begins to break apart and disappear. A hormone called luteinizing hormone, or LH, is responsible for causing changes in the corpus luteum.

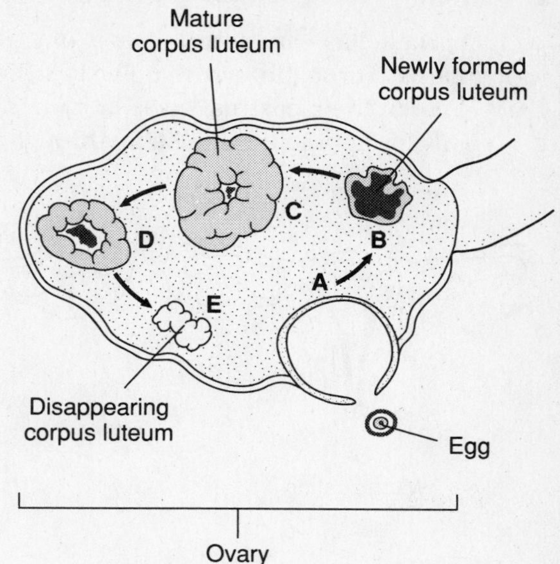

Figure 3

2. Figure 4 shows the concentration of LH in the bloodstream of an average human female during the 28-day menstrual cycle. On Graph 2, prepare a line graph of the data shown in Figure 4. In the boxes above Graph 2, make drawings of the various stages of corpus luteum maturation. The drawings should match the lettered stages shown in Figure 3.

Amount of LH in Bloodstream (units per milliliter)			
Day	**LH**	**Day**	**LH**
1	9	15	30
2	12	16	14
3	16	17	10
4	18	18	9
5	19	19	7
6	16	20	5
7	12	21	3
8	19	22	3
9	15	23	2
10	16	24	3
11	20	25	3
12	30	26	4
13	75	27	4
14	58	28	4

Figure 4

3. Answer questions 5 through 8 in Observations.

Part C. Uterine Changes During the Menstrual Cycle

1. As the follicle and luteal phases of the menstrual cycle occur, a series of changes occurs in the uterus. Through rapid cell division, the lining of the uterine walls becomes very thick. At one point in the menstrual cycle, the uterus lining ceases to thicken and begins to break apart. This loss of the uterine lining through the vagina is called menstruation. See Figure 5.

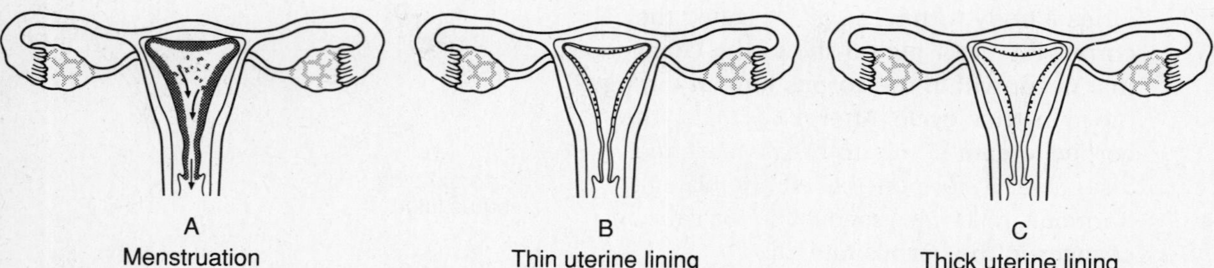

Figure 5

2. Two hormones that are responsible for the building up of the uterine lining are estrogen and progesterone. Figure 6 shows the amounts of estrogen and progesterone found in the bloodstream of an average human female during the 28-day menstrual cycle. On Graph 3, prepare line graphs of the data in Figure 6. Use a different colored pencil to construct each of the line graphs. In the squares shown above Graph 3, make drawings of the various stages of uterine thickness. The drawings should match the lettered stages shown in Figure 5.

Amount of Estrogen and Progesterone in Bloodstream (units per milliliter)					
Day	**Estrogen**	**Progesterone**	**Day**	**Estrogen**	**Progesterone**
1	20	6	15	180	23
2	20	8	16	150	37
3	25	10	17	120	58
4	25	10	18	100	83
5	30	10	19	50	104
6	80	10	20	30	120
7	130	12	21	25	120
8	140	12	22	25	118
9	180	13	23	25	103
10	200	15	24	25	72
11	220	15	25	30	40
12	230	16	26	30	30
13	220	18	27	25	20
14	200	20	28	25	18

Figure 6

Name ____________________ Class __________ Date __________

3. Answer questions 9 through 11 in Observations.

Observations

Graph 1

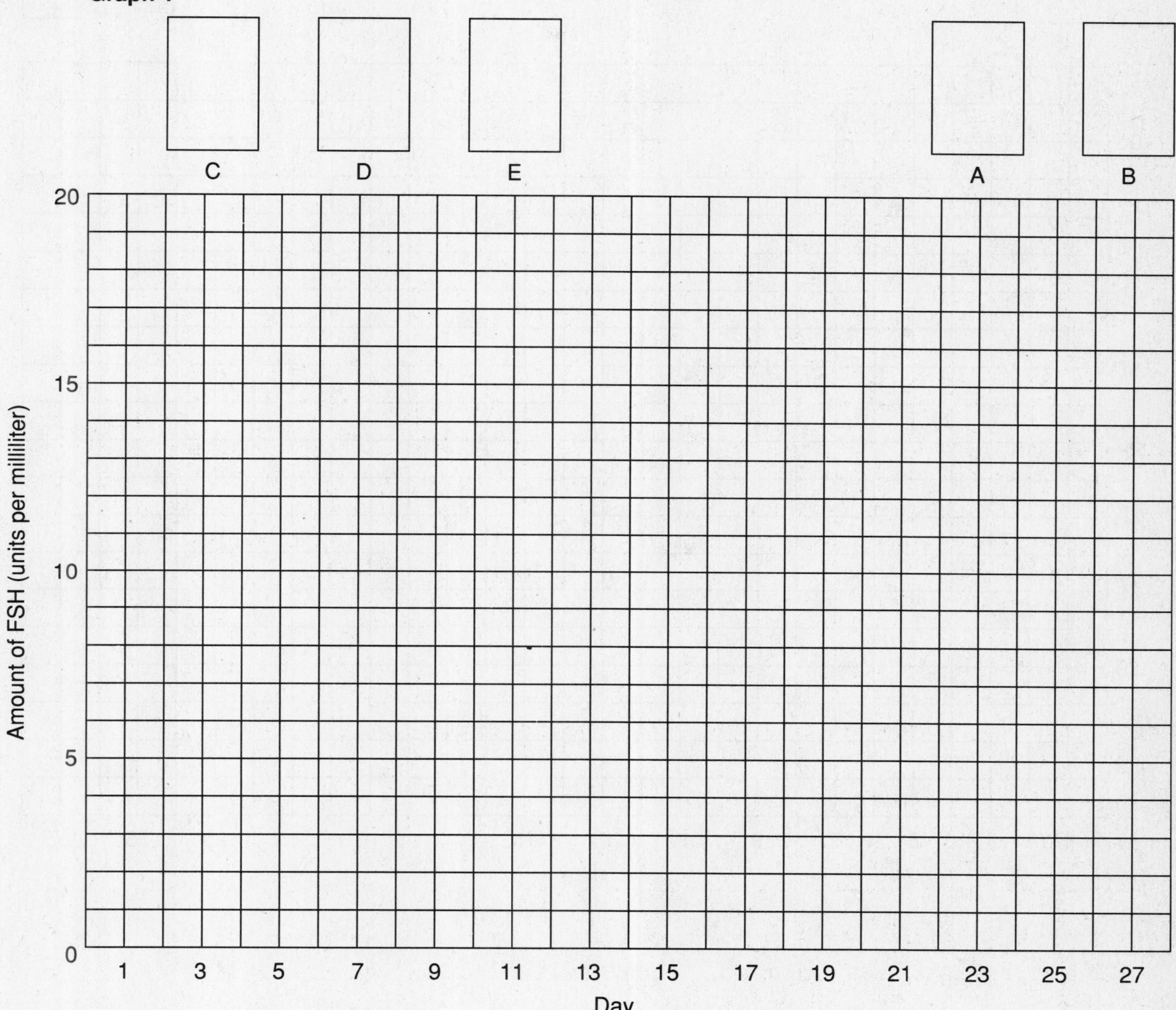

Graph 2

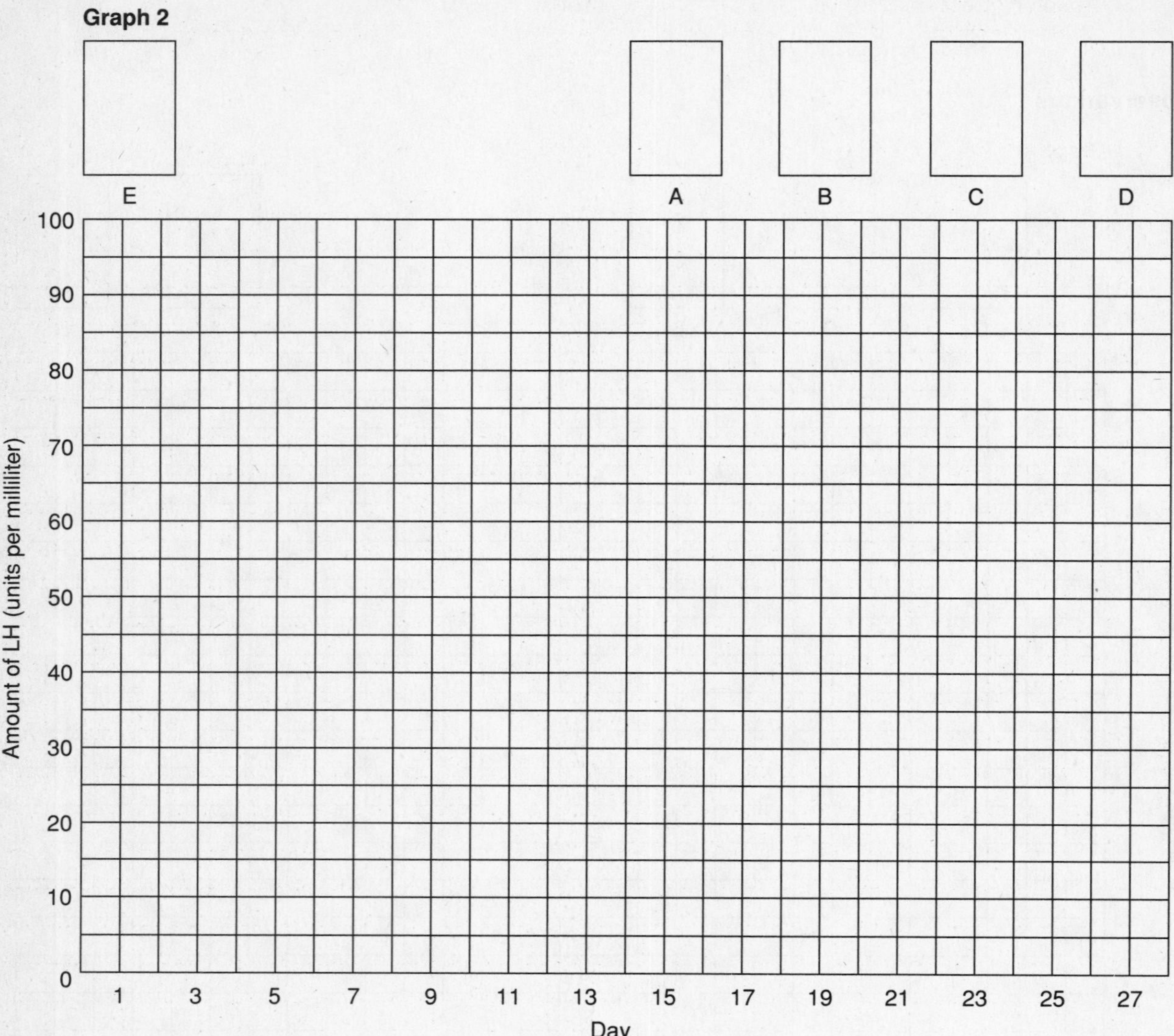

E
A
B
C
D
100
90
80
70
60
50
40
30
20
10
0
Amount of LH (units per milliliter)
1
3
5
7
9
11
13
15
17
19
21
23
25
27
Day

Name ______________________ Class ____________ Date ____________

Graph 3

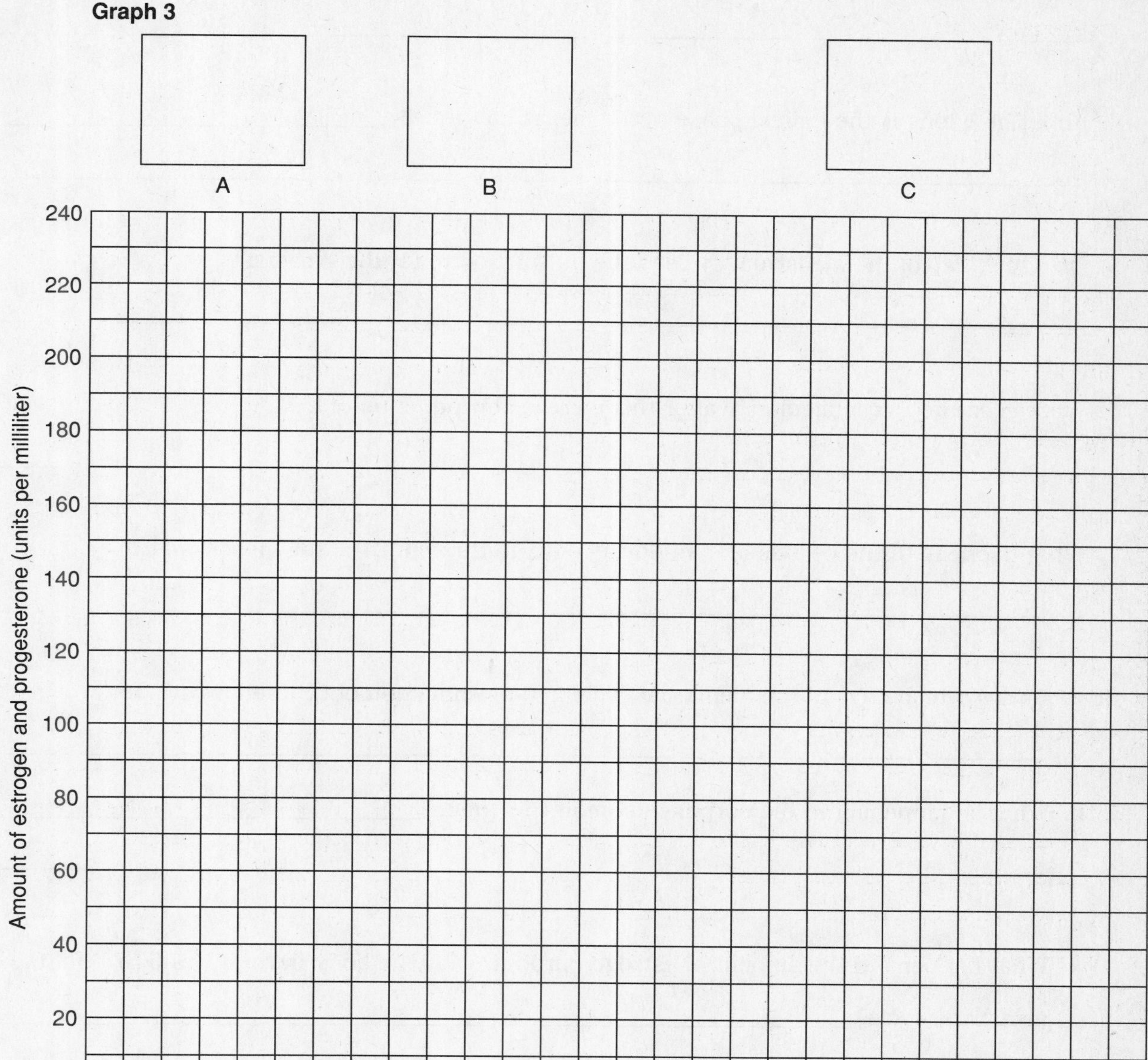

1. Between days 25 and 12 of the menstrual cycle, what happens to the amount of FSH produced by the body of an average human female? ______________________

____

2. What happens to the follicle between days 25 and 12 of the menstrual cycle?

____

3. During which days of the menstrual cycle is the level of FSH at its lowest in the bloodstream?

______

______

4. About how long is the follicle phase of the menstrual cycle? ______

______

5. On which day of the menstrual cycle is the production of LH the greatest?

______

6. What event occurs immediately after this increased production of LH?

______

7. What happens to the corpus luteum during days 15 through 24 of the menstrual cycle?

______

8. a. During which days of the menstrual cycle is the production of LH the lowest?

______

b. What is happening to the corpus luteum at this time? ______

______

9. a. What happens to the amount of estrogen produced by the body during days 6 to 12 of the menstrual cycle? ______

______

b. What is occurring to the uterus during this time? ______

______

10. a. What happens to the amount of progesterone produced by the body during days 13 to 23 of the menstrual cycle? ______

______

b. What is occurring to the uterus during this time? ______

______

Name ______________________ Class __________ Date __________

11. **a.** During which days of the menstrual cycle are the levels of both estrogen and progesterone at their lowest? ______________________

____

b. What event is occurring at this time? ______________________

____

Analysis and Conclusions

1. How is the name follicle-stimulating hormone appropriate for its function?

____

____

2. How is the name luteinizing hormone appropriate for its function?

____

____

3. Based on your observations, do you think estrogen and progesterone both cause similar changes in the uterus? Explain your answer. ______________________

____

____

4. What events occur during the follicle phase of the menstrual cycle?

____

____

5. What events occur during the luteal phase of the menstrual cycle?

____

____

Critical Thinking and Application

1. Why are the events described in this investigation referred to as a cycle?

2. The word "menstrual" comes from the Latin word *mensis,* meaning "month." How is the name appropriate for this cycle of the human female reproductive system?

3. If a female did not produce sufficient quantities of FSH and LH, how would her ability to have children be affected?

Going Further

1. A comparable cycle involving a periodic growth and loss of uterine cells occurs in mammals other than primates, but little or no blood is lost. The shedding of cells is called estrus, and the cycle is called the estrous cycle. Use reference materials to investigate the estrous cycles of common mammals such as dogs and cats.
2. Women commonly experience a variety of pains, discomforts, and/or emotional disturbances before or during menstruation. In some women, the premenstrual period may be characterized by irritability, sluggishness, or deep depression. Use reference materials to research other symptoms of Premenstrual Syndrome (PMS) and some of the ways in which these symptoms can be medically treated.

Name ______________________ Class __________ Date __________

Examining an Unfertilized Chicken Egg

Pre-Lab Discussion

The embryonic development of an organism begins soon after the fertilization of an egg cell by a sperm. During this period of development, the embryo is protected and nourished by a set of surrounding membranes. As development proceeds, all of the complex organs and tissues appear in a definite order.

It is difficult to study human embryological development because the embryo develops internally in a placenta. For this reason embryos that develop externally have been widely used in developmental studies. Because of their availability, easily observed features, and similarity to mammalian embryo development, chick embryos are used extensively.

In order for a chick egg to develop, the egg must first be fertilized by a sperm from a male chicken. The egg must then be kept warm by the female chicken or placed in an incubator for 21 days. Eggs purchased in the grocery store usually have not been fertilized or incubated.

In this investigation, you will examine the internal structures of an unfertilized chicken egg.

Problem

How is the chicken egg adapted to support an embryo?

Materials *(per group)*

Unfertilized chicken egg
Hand lens
Petri dish
Metric ruler
Forceps

Safety

Put on a laboratory apron if one is available. Handle all glassware carefully. Note all safety alert symbols next to the steps in the Procedure and review the meanings of each symbol by referring to the symbol guide on page 10.

Procedure

1. Obtain an unfertilized chicken egg from your teacher. Gently crack the egg crosswise on the side of a petri dish. Separate the two halves of the egg shell as you would to put an egg in a frying pan. Allow the contents of the egg to drop into the petri dish. **Note:** *Be careful not to break the yolk.*
2. Examine the outside of the egg shell with a hand lens. Place the egg shell to the side for later examination. Answer question 1 in Observations.
3. Using Figure 1 as a guide, examine the internal structures of the chicken egg. On the surface of the yolk locate a small, white, round dot. This dot is the germinal disk, from which an embryo would form after fertilization. Use a metric ruler to find the diameter of the germinal disk in millimeters. Answer question 2 in Observations.
4. Find two dense, white, cordlike structures. These structures are called the chalazae (singular, chalaza). Using forceps, gently pull on one of the chalazae. Answer question 3 in Observations.

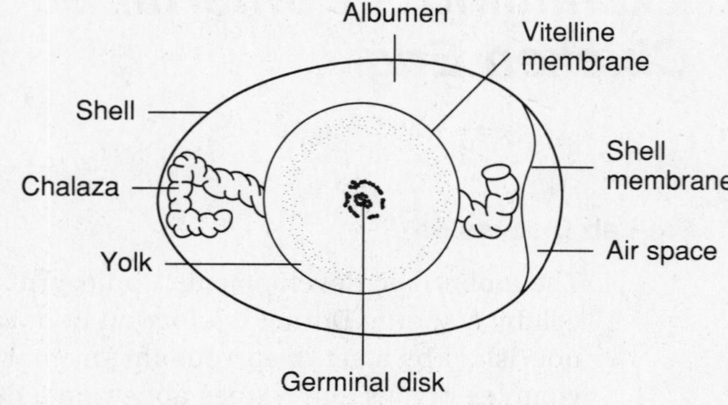

Figure 1

5. The white of the egg contains water and a protein called albumen. Notice that the egg white surrounds the yolk and is fairly thick.
6. Observe the yolk. The yolk of the egg contains fats, minerals, and vitamins. These materials, which come from the liver of the hen, pass to the ovary and become part of the egg. Touch the yolk gently with the forceps. Note that it is surrounded by a membrane called the vitelline membrane.
7. Examine the inside of the shell and the membranes attached to it. Try to locate an air space between the membranes and shell. The chick uses the air in the air space only when it is ready to hatch. Answer question 4 in Observations.
8. In the space provided in Observations, draw the egg you have examined. Label the albumen, yolk, chalazae, germinal disk, air space, shell membrane, vitelline membrane, and shell.
9. Follow your teacher's instructions for the proper disposal of the chicken egg and shell.

Observations

Chicken Egg

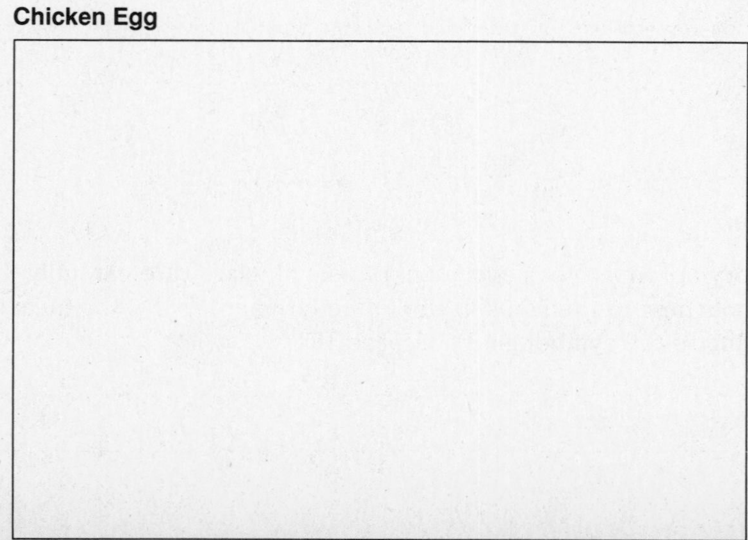

Name ______________________ Class __________ Date __________

1. Describe the texture of the chicken egg shell. ______________________

2. What is the approximate diameter of the germinal disk? ______________________

3. To what are the chalazae attached? ______________________

4. At which end of the egg, the blunt or the pointed end, is the air space located?

____

Analysis and Conclusions

1. What functions does the shell provide for the embryo? ______________________

2. If the shell membranes are removed from the shell and water is added, the water will seep out of the shell. What feature of the shell would allow this to occur? ______________________

3. If the membranes remain in a shell, very little water will seep out. What function do the membranes probably perform for the developing embryo? ______________________

4. What is the likely function of the chalazae? ______________________

5. What might the egg white provide for the developing embryo? ______________________

6. What might the egg yolk provide for the developing embryo? ______

______

______

Critical Thinking and Application

1. In the developing human, there is no albumen or yolk serving as a source of nutrients. How does the human embryo obtain nutrients for growth and development?

______

______

______

2. A chick embryo develops in 21 days, but a human embryo develops in about 266 days. Why are longer periods of development possible in humans and other mammals?

______

______

______

3. "Candling" is a process by which egg processors look at an egg through a light source to determine whether or not it has been fertilized. Why is the "candling" of chicken eggs important to egg processors? ______

______

______

______

Going Further

1. Examine prepared glass slides of developing chicken embryos. Identify and describe the different stages of development. Compare the development of a chick embryo to the development of a human embryo. What similarities do you find? What differences?
2. Construct your own chicken egg incubator, obtain some fertilized eggs, and care for the eggs for 20 days. On day 21, watch as the chicks hatch from their eggs.

Name ______________________ Class ____________ Date ____________

86

A Model for Disease Transmission

Pre-Lab Discussion

A *disease* is a condition that interferes with the normal functioning of a living thing but is not the result of an injury. Generally, a disease has certain symptoms. The disease can affect the entire body or only parts of the body. People can be born with certain diseases, such as genetic disorders. Other diseases can develop during a person's lifetime.

An *infectious disease* is one that is caused by an organism or virus that enters the body. Most disease-causing microorganisms do not move from one person to another on their own. Instead, the microorganisms are transmitted through contact with an infected person or a contaminated object or substance.

In this investigation, you will observe how easily a harmless microorganism is transmitted from person to person.

Problem

How easily can infectious disease be spread?

Materials *(per group of four students)*

4 petri dishes containing sterile nutrient agar
Glass-marking pencil
Sterile cotton swabs
Sterile distilled water
Yeast culture
Wire inoculating loop
Bunsen burner
Flint striker or matches

Safety

Put on a laboratory apron if one is available. Put on safety goggles. Handle all glassware carefully. Note all safety alert symbols next to the steps in the Procedure and review the meanings of each symbol by referring to the symbol guide on page 10.

Procedure

1. Assign a number from 1 to 4 to each member of your group.
2. Using a glass-marking pencil, draw a line on the bottom of an agar plate to divide it in half. Label one side of the plate "Control" and the other side "Experimental." Also write your assigned number on the bottom of the plate.
3. Swab your left hand with a sterile cotton swab moistened in sterile distilled water. Then swab one corner of the control side of the agar with the same swab.
4. Sterilize a wire loop by passing it through the flame of a Bunsen burner until the entire length of the wire has been heated to a red glow. **CAUTION:** *Put on safety goggles whenever you use a Bunsen burner.*
5. Streak your plate with the sterilized wire loop as shown in Figure 1. The streak should begin from the point at which the plate was touched with the swab. **Note:** *The wire loop should be sterilized after each use by heating as before.*

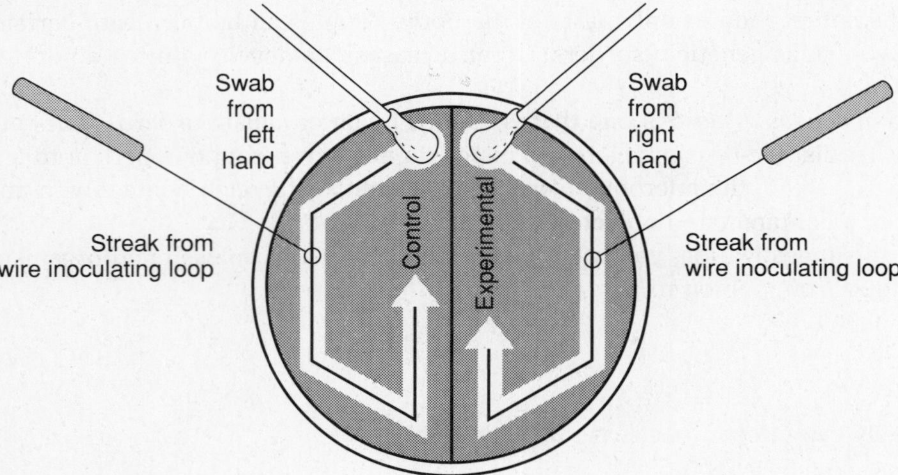

Figure 1

6. Your teacher should swab the right hand of group member 1 with a culture of yeast. Group member 1 should then shake hands with group member 2. Group member 2 should then shake hands with group member 3. Finally, group member 3 should shake hands with group member 4.
7. You and your group members should swab the right hand with sterile cotton swabs moistened with sterile distilled water. Then using the same swab, you all should swab one corner of the experimental side of your individual agar plate and repeat steps 4 and 5 of the Procedure.
8. Wash your hands thoroughly with soap and water after swabbing.
9. You and your group members should cover all the agar plates and incubate them in an inverted position for 48 hours, after which they should be examined.

Observations

1. Compare the control and experimental sides of the plates. Describe the differences between them. ______________________________

____

____

____

Name ______________________ Class __________ Date __________

2. Why is it necessary to compare your control plates with the yeast culture?

____

____

Analysis and Conclusions

1. How are most disease-causing microorganisms transmitted? ______________________

____

____

2. How were the yeast colonies transmitted from plate 1 to plates 2, 3, and 4?

____

____

3. Which plate should have contained the greatest number of yeast? Which plate should have contained the least? Explain your answer. ______________________

____

____

____

Critical Thinking and Application

1. What do you think is the best way to prevent the transmission of infectious diseases?

____

____

____

2. How would population density influence the transmission of diseases?

____

____

____

3. List three infectious diseases that are commonly spread among human populations.

____

____

____

4. List three ways in which diseases can be directly and indirectly spread.

____

____

____

5. Children must be immunized for certain diseases before they enter school. Why is immunization required by law? _______________

____

____

Going Further

1. Test the sensitivity of the bacteria in your plates to antibiotics. Using forceps, place a pretreated antibiotic disk in the center of each of your petri dishes. Turn the petri dishes upside down and incubate them at 37°C for 24 to 48 hours. When incubation is complete, some areas on the surface of the agar should look cloudy or white. These areas have bacteria growing on them. Check for clear or less dense circular regions around each disk. A clear region around the disk indicates that the antibiotic has either killed bacteria or inhibited their growth.
2. Carry out an experiment to show that microbes are sprayed into the air by talking and laughing. Talk for 30 seconds facing a sterile nutrient agar plate held 15 cm from your mouth. Repeat this procedure with a second agar plate held at arm's length from your mouth. Repeat the same procedure for two more agar plates, but this time cough onto each. Repeat the procedure for two more agar plates, but this time laugh in the direction of each plate. Cover each plate and incubate them at room temperature for 24 to 48 hours. Observe the amount of microbial growth on each plate. Write a brief report explaining your results.

Name ______________________ Class ____________ Date ____________

87

Relating Chronic Diseases and Nutrition

Pre-Lab Discussion

During your teens, you will grow at a faster rate than at any other time in your life except infancy. This growth involves more than just height and weight. Your bones increase in density, and your muscles develop in size and strength. Your endocrine glands also grow and develop. Good eating habits are especially important during this period of rapid growth.

A well-balanced diet definitely contributes to a healthy body. Yet many teenagers do not have good eating habits. They may skip breakfast, choose snacks that are rich in fats and sugars, go on crash diets, and neglect foods that contain important nutrients. For most Americans, improper nutritional habits cause health problems.

Good eating habits during the teenage years usually mean fewer problems during later years. Scientists have begun to take a closer look at the relationship between nutrient intake and chronic, life-threatening diseases. Heart disease, diabetes, high blood pressure, kidney disease, various digestive disorders, and even certain types of cancer have all been found to be connected with nutrition. It is becoming more and more evident that "you are what you eat."

In this investigation, you will plan nutritionally balanced menus and calculate the number of Calories you consume and burn in an average day.

Problem

Can you plan a menu that meets the Recommended Dietary Allowances (RDA) published by the Food and Nutrition Board of the National Academy of Sciences? How closely does your Calorie consumption equal the Calories you burn?

Materials *(per group)*

Calculator (optional)
Pencil
Additional tables of nutrition information (optional)

Procedure

1. From the foods listed in the Table of Nutrition Information, plan a day's menu. Choose those items you would like to eat for breakfast, lunch, dinner, and snacks. List them in the chart entitled "My Menu" in Observations. Also list the nutrition information for each item.
2. Total each column in your menu chart. Compare the totals with the data in the Recommended Dietary Allowances Table.

Recommended Dietary Allowances

Sex	Age	Calories	Protein (g)	Calcium (mg)	Iron (mg)	Vitamins A (i.u.)	Vitamins B$_1$ (mg)	Vitamins C (mg)
Males	12–16	2700–3000	46–54	1200	18	5000	1.4	50–60
Females	12–16	2100–2400	44–48	1200	18	4000	1.1	50–60

g = grams, mg = milligrams, i.u. = international units

Table of Nutrition Information

Food	Amount	Calories	Protein (g)	Calcium (mg)	Iron (mg)	Vitamins A (i.u.)	Vitamins B$_1$ (mg)	Vitamins C (mg)
Hamburger	113 g	300	20.2	11	3.1	40	.09	0
Hot dog or hamburger bun	1	90	2.5	22	0.6	trace	.08	trace
T-bone steak	227 g	800	30.0	16	4.4	150	.12	0
Fried chicken	1/4	230	22.4	18	1.8	230	.07	0
Egg	11 g	80	6.5	27	1.2	590	.15	0
Flounder	100 g	70	14.9	61	0.8	—	.06	—
Bacon	1 slice	50	1.8	1	0.2	0	.04	0
Hot dog	1	125	7.0	3	0.6	0	.08	0
Shrimp	0.2 kg	145	28.3	26	1.8	—	—	—
American cheese	28 g	105	7.0	198	0.3	350	.01	0
Milk	1 cup	160	9.0	288	0.1	350	.07	2
Ice cream	1 cup	255	6.0	194	0.1	590	.05	1
Lima beans	1/2 cup	130	8.0	28	2.9	—	.12	—
Green beans	1/2 cup	15	1.0	31	0.4	340	.05	8
Broccoli	1 stalk	25	3.1	88	0.8	2500	.09	90
Corn	1 ear	70	3.0	2	0.5	310	.09	7
Blackeyed peas	1/2 cup	90	6.5	20	1.7	280	.02	14
Baked potato	1 med.	145	4.0	14	1.1	trace	.15	31
French fries	20	310	4.0	18	1.4	trace	.14	24
Potato chips	20	230	2.0	16	0.8	trace	.08	6
Apple	1 med.	80	0.3	10	0.4	120	.04	6
Banana	1 med.	100	1.0	10	0.8	230	.06	12
Fresh strawberries	1/2 cup	35	0.5	14	0.6	80	.02	16
Orange juice	1 cup	120	2.0	25	0.2	550	.22	124
White bread	1 slice	65	2.0	22	0.6	trace	.06	trace
Chocolate chip cookie	1	50	0.5	4	0.2	10	.01	trace
Chocolate cake	1 piece	235	3.0	41	0.6	100	.02	trace
Corn flakes	1 cup	95	2.0	6	0.5	0	.10	0
Pancake	1	105	3.2	27	0.6	.54	.08	trace
Syrup	1 Tbsp.	50	0	33	0.6	0	—	0

Name ________________ Class ________ Date ________

3. The number of Calories you burn depends in part on the activities you perform. Fill in the chart entitled "My Activities" with the time you spend during an average day on each activity listed.

4. To calculate the total Calories burned in each activity category, multiply A x B x C. Keep the time units the same within each category. For example, if time is spent in hours, use the figure for Calories burned per hour. Add the categories to find the total Calories burned in 24 hours.

Observations

My Menu

Meal	Calories	Protein (g)	Calcium (mg)	Iron (mg)	Vitamins		
					A (i.u.)	B_1 (mg)	C (mg)
Breakfast							
Lunch							
Dinner							
Snacks							
TOTALS							

My Activities

Activity	A Calories Burned (per hr or min)	B Minutes or Hours Spent in Activity	C Your Mass	Total Calories
Sleeping Napping	.0075/min .45/hr			
Reading Watching TV Eating Sitting in class	.0108/min .64/hr			
Dressing Showering Driving car	.015/min .90/hr			
Light activity Walking Lab work	.0308/min 1.84/hr			
Moderate activity Gym class Bicycling Dancing Easy jogging	.0395/min 2.37/hr			
Heavy work Swimming Tennis Basketball Wrestling Climbing stairs	.0483/min 2.9/hr			

1. Compare your total Calories burned in 24 hours with the total Calories you would consume according to the menu you planned in step 2 of the Procedure.

Calories burned: ______________________________

Calories consumed: ______________________________

Analysis and Conclusions

1. a. How do the totals from your menu chart compare with the Recommended Dietary Allowances? ______________________________

____

b. Does your menu provide too many, too few, or the right number of Calories?

____

Name ______________________ Class __________ Date __________

c. In what areas, if any, is your menu deficient?

2. Not all of the recommended nutrients have been included in the charts for this investigation. Name four other nutrients, including minerals and vitamins, that should be included in your diet.

3. What portion of a gram is 1 milligram?

4. Compare your total Calories burned in 24 hours with your total Calories consumed. If this is your normal pattern, what conclusion can you draw regarding your Calorie intake?

5. List three diseases and disorders that have been associated with poor nutrition.

Critical Thinking and Application

1. What two things can you do to safely lose weight?

2. Of the two weight-loss methods you listed in question 1, which do you think is the better method and why?

3. Many companies now advertise breakfast cereals that are lower in sugar. Why do you think it is better to eat a breakfast cereal that is lower in sugar? ______________________________

____

____

____

4. Explain how it is possible for a person to appear overweight but suffer from malnutrition.

____

____

____

Going Further

1. Keep a record of all foods you eat every day for one week. Compute your Calorie intake each day. Compute your average daily Calorie intake for the week. How well does your list of foods eaten meet the Recommended Dietary Allowances? Is your daily food intake well balanced in terms of the four basic food groups? What conclusions can you draw regarding your eating habits?
2. Many food products have nutritional information listed on the package. Look at this information for some of your favorite foods. How well do these foods meet the requirements for Recommended Dietary Allowances?
3. Obtain nutrition information for the major items sold at your favorite fast-food restaurant. Make a chart or booklet of this information and share it with your classmates.

Name ______________________ Class ____________ Date ____________

Lysis by a Bacteriophage

Pre-Lab Discussion

Some diseases that affect people are caused by viruses. Viruses also cause plant diseases and infect bacteria. A virus consists of a core of nucleic acid surrounded by a protein coat. Although unable to grow or independently carry out the processes characteristic of living things, a virus has the ability to reproduce inside living cells. Electron microscope examination shows that when a virus infects a cell, it first attaches itself to the wall or cell membrane of the cell. The virus then injects its nucleic acid into the interior of the cell through an opening it has made in the host-cell wall or membrane. Once inside, the nuclear material of the virus uses the ribosomes and enzymes of the cell to produce more viral nucleic acids and protein coats. The viral nucleic acids and protein coats are assembled into new viruses. By the action of enzymes produced by these new viruses, the host cell bursts, or *lyses.* The viruses leave the cell and infect other cells.

Lysis is an important tool used by virologists to detect the presence of viruses in a cell population. If a culture of bacteria is inoculated with bacteriophages, or viruses that infect bacteria, the lysis of the bacterial cells makes the culture appear clear. This clear area is called a *plaque*. The clearer a phage-infected culture of bacteria is, the greater the number of phages present in the culture.

The culturing of bacteria and bacteriophages requires the use of *aseptic,* or sterile, techniques. Aseptic techniques prevent contamination of existing cultures by other microorganisms. The heat produced by autoclaving or passing certain materials through the flame of a Bunsen burner is sufficient to destroy microorganisms and keep the environment sterile.

In this investigation, you will detect the presence of bacteriophages in a bacterial culture. You will also acquire skill in preparing viral and bacterial cultures using aseptic techniques.

Problem

How is the presence of bacteriophages in a culture of bacteria demonstrated?

Materials *(per group)*

Bunsen burner
Culture of T-4 bacteriophage
Inoculating loop
6 sterile test tubes each containing
  nutrient broth
Glass-marking pencil
Test tube rack
Nutrient agar plate
Culture of *Escherichia coli*
Sterile cotton swab
Incubator

Safety

Put on a laboratory apron if one is available. Handle all glassware carefully. Put on safety goggles. When using the Bunsen burner, be sure that hair or loose clothing does not come into contact with the flame. Follow your teacher's directions and all appropriate safety procedures when handling microorganisms. Wash your hands thoroughly after each part of this investigation. Note all safety alert symbols next to the steps in the Procedure and review the meanings of each symbol by referring to the symbol guide on page 10.

Part A. Preparing a Bacteriophage Culture

1. Obtain a culture of T-4 bacteriophage from your teacher. **CAUTION:** *Be very careful when working with microorganisms. Wash your hands thoroughly after each part of this investigation.*
2. Sterilize the inoculating loop by passing it through a Bunsen burner flame as shown in A of Figure 1. Allow the loop to remain in the flame until the entire length of the metal part has turned red. **CAUTION:** *When using a Bunsen burner, wear safety goggles and be sure that hair and loose clothing do not come into contact with the flame. Do not touch the end of the inoculating loop that is placed in the flame or allow it to come into contact with any flammable objects.*
3. While holding the sterile inoculating loop in one hand, pick up in the other hand the test tube containing the T-4 bacteriophage culture and one sterile test tube containing the nutrient broth.
4. Remove the cotton plugs from each test tube by grasping the plugs between the fingers of the hand holding the loop. This technique is shown in B of Figure 1. **Note:** *Transfer the cotton plugs from in between the last two fingers to between the second and third fingers as shown in D of Figure 1. Do not touch the parts of the cotton plugs that are placed inside the test tubes.*
5. Sterilize the mouths of the test tubes by quickly passing them through the flame two or three times as shown in C of Figure 1.
6. Carefully insert the inoculating loop into the test tube containing the T-4 bacteriophage culture and remove one loopful of culture. **Note:** *Allow the inoculating loop to cool before inserting it into the test tube.*

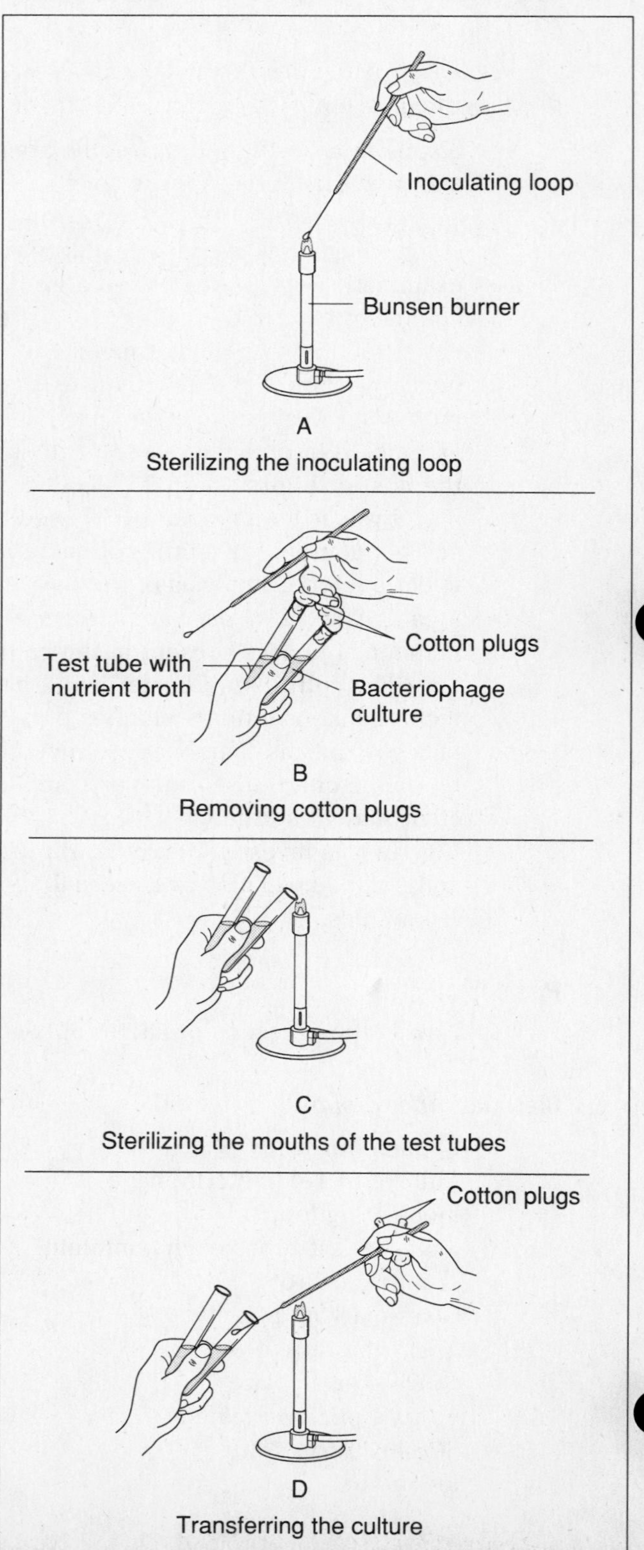

Figure 1

Name ______________________ Class ____________ Date ____________

7. Transfer this loopful of T-4 bacteriophage culture to the test tube containing the nutrient broth.

8. Before replacing the cotton plugs on the two test tubes, sterilize the mouths of the test tubes. Also, resterilize the inoculating loop.

9. Mix the contents of the test tube containing the nutrient broth and bacteriophage culture by rotating the tube rapidly between your hands. With the glass-marking pencil, label this tube "Stock Phage Culture."

10. Place the remaining five tubes containing 1 mL each of nutrient broth in a test tube rack. With the glass-marking pencil, label them 1 through 5.

11. Transfer a loopful of culture from the stock phage culture to test tube 1 following steps 2 through 8.

12. Continue transferring loopfuls from test tube to test tube as shown in Figure 2. Stop when test tube 4 is inoculated. **Note:** *Do not inoculate test tube 5 with the bacteriophage culture.*

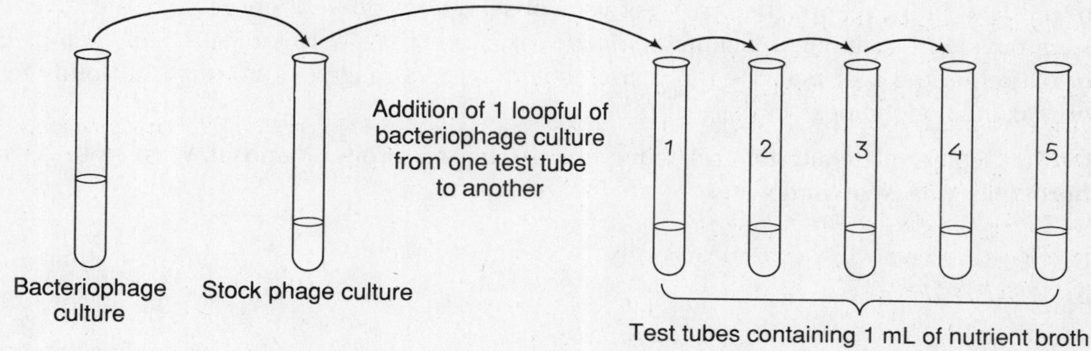

Figure 2

Part B. Growing Bacteria and Bacteriophages on Agar

1. Obtain a nutrient agar plate and turn it upside down. Across the bottom of the plate, draw five lines with the glass-marking pencil. Number each line as shown in Figure 3. Turn the petri dish upright.

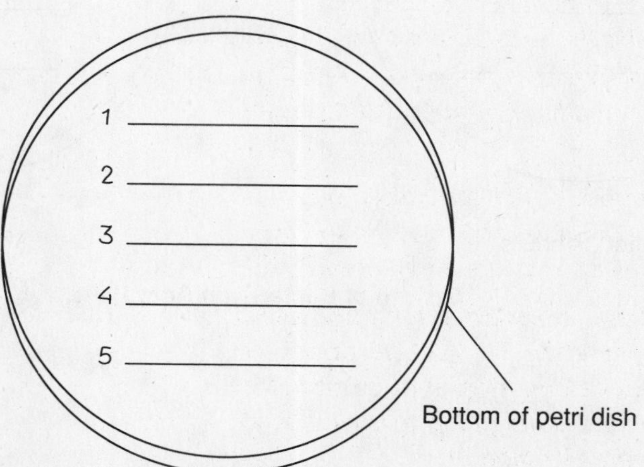

Figure 3

2. Obtain a culture of *E. coli*. Dip the sterile cotton swab into the culture of *E. coli*.
3. Raise the lid of the petri dish slightly and swab the entire surface of the agar plate with the cotton swab. Try to cover the surface of the agar plate as uniformly as possible
4. Using the sterile techniques described in steps 2 through 8 of Part A, remove a loopful of material from test tube 1 and streak this on the surface of the nutrient agar plate following line 1. Replace the cotton plug.
5. Follow the same procedure for each of the remaining four test tubes, including test tube 5. Be sure to streak one loopful along the line that corresponds to the test tube number. **Note:** *Be sure to sterilize the loop after each streaking.*
6. After completing the streaking of the bacteriophage dilutions onto the nutrient agar plate, inoculate test tubes 1 through 5 with a loopful of *E. coli* culture. Be sure to follow the sterilization procedures for each inoculation. Mix the contents of each tube well by rapid rotation.
7. Incubate the plate and broth cultures in the incubator at 35° to 37°C.
8. Examine the plate and test tubes after 24 and 48 hours. In the appropriate place in Observations, sketch what you observe on the agar plate after 24 and 48 hours.
9. In the Data Table, record the results of your observations after 48 hours. In the first column, use a range of + to indicate minimum plaque formation to ++++ to indicate maximum plaque formation on the agar plate. In the second column, use a range of + for slightly cloudy to ++++ for maximum cloudiness in the test tubes.
10. Return the agar plate and test tubes to your teacher for proper disposal. Wash your hands thoroughly with soap and water.

Observations

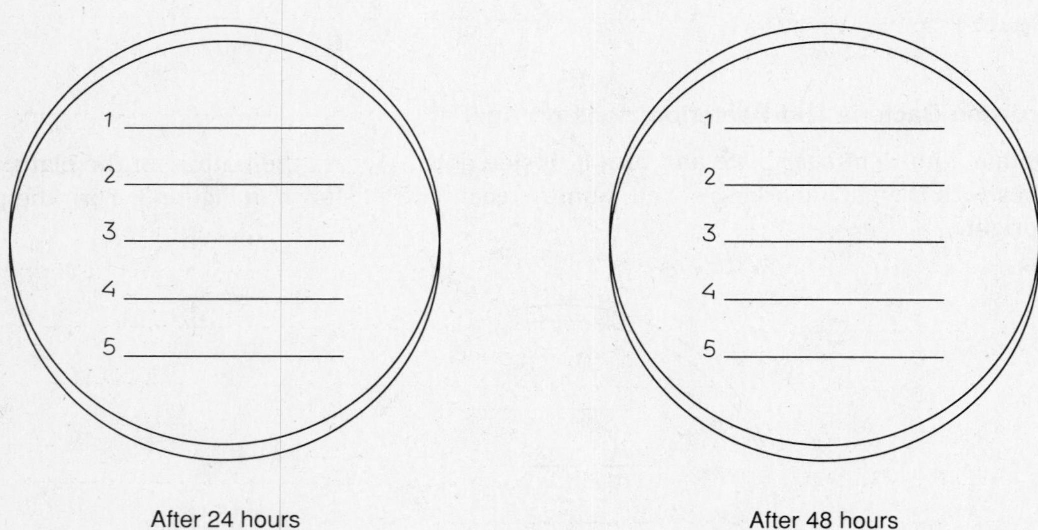

Growth of Bacteria on Agar Plate

Name ______________________ Class ______________ Date ______________

Data Table

Test Tube	Plaque formation (clearing on streaks)	Growth of *E. coli* (cloudiness in test tubes)
1		
2		
3		
4		
5		

Analysis and Conclusions

1. Why is it necessary to use aseptic techniques when culturing bacteria?

____

____

2. What is happening to the bacteriophage culture as it is transferred from one test tube to the next? ____

____

3. In which streak is the growth of *E. coli* most abundant? ______________________

____

4. In which streak are the plaques most evident? ______________________

____

____

5. **a.** In which tube was *E. coli* growing alone? ______________________

____

b. Describe the appearance of the broth in this test tube. Explain this appearance.

____

____

6. a. In which test tube was the concentration of bacteriophages the greatest?

____

b. Describe the appearance of the broth in this test tube. Explain this appearance.

____

____

Critical Thinking and Application

1. Why are viruses considered to be parasites? ______________________________

____

____

2. Could the viruses that exist today have been the first type of living thing? Explain your answer.

____

____

____

3. Viruses contain either DNA or RNA. How do you account for the fact that viruses need only one type of nucleic acid? ______________________________

____

____

____

4. Viruses can sometimes induce the formation of tumors. Formulate a hypothesis about how viruses cause cells to grow out of control. Explain your answer. ______________________________

____

____

____

____

Going Further

1. To demonstrate that viruses are filterable, obtain a microbiological filter from your teacher and filter the contents of test tube 4. Using the inoculating loop and aseptic techniques, inoculate test tube 5 with the filtered material and incubate the test tube at 35° to 37°C for 24 hours. What do you observe in test tube 5? What conclusions can you make?
2. Prepare a report on why agar is used in microbiological culturing. What properties of this medium make it suitable for the study of microorganisms?

Name ______________________ Class ____________ Date ____________

89

The Effect of Alcohol on the Growth of Microorganisms

Pre-Lab Discussion

The body's immune system sometimes cannot control an infection by itself. An *antimicrobial agent* is a chemical substance that kills or inhibits the growth of microorganisms. Such a substance may be either a synthetic chemical or a natural product. One type of antimicrobial agent is an *antiseptic.* An antiseptic is a chemical used to inhibit the growth of or kill microorganisms on living tissues. Antiseptics that are used on the skin may be swabbed on with sterile cotton. Soaps containing antiseptics leave a residue of the antiseptic on the skin after the soap is rinsed off, and this active residue may continue to kill microorganisms for an extended period of time.

In this investigation, you will observe how alcohol can be used as an antiseptic.

Problem

What is the effect of alcohol on the growth of microorganisms?

Materials *(per group)*

2 petri dishes containing sterile nutrient agar
Forceps soaking in alcohol
2 paper clips
2 thumbtacks
2 pennies
Glass-marking pencil
Masking or clear tape
50-mL graduated cylinder
70% alcohol solution
100-mL beaker

Safety

Put on a laboratory apron if one is available. Put on safety goggles. Handle all glassware carefully. Be careful when handling sharp instruments. Note all safety alert symbols next to the steps in the Procedure and review the meanings of each symbol by referring to the symbol guide on page 10.

Procedure

1. Set out the two petri dishes containing sterile nutrient agar, as well as the two thumbtacks, two pennies, and two paper clips. Each of the metal objects should be handled before beginning the procedure.

2. Use a glass-marking pencil to label the lids of the dishes. Label one dish "Control" and the other dish "Soaked in Alcohol." Also write the names of the group members and the date on the cover of each dish. **Note:** *Be sure to keep the dishes closed while labeling them.*

3. Use a graduated cylinder to pour 50 mL of 70% alcohol solution into a beaker.

4. Place one paper clip, one thumbtack, and one penny into the alcohol solution. Allow these objects to remain in the solution for 10 minutes.

5. Using a forceps that has been soaked in alcohol, place the other paper clip, other thumbtack, and other penny into the dish marked "Control." **Note:** *These objects should not be soaked in alcohol solution.* Immediately cover the dish and tape it closed.

6. Using the forceps, remove the paper clip, thumbtack, and penny from the beaker of alcohol solution. Place these objects into the dish marked "Soaked in Alcohol." Immediately cover the dish and tape it closed. See Figure 1.

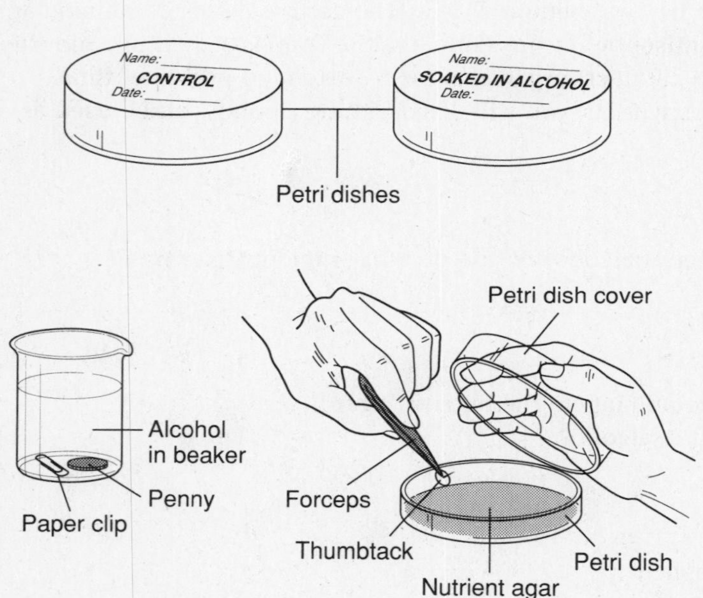

Figure 1

7. Place the petri dishes in an undisturbed location indicated by your teacher for one week.

8. After one week, examine the contents of the two petri dishes. In the appropriate places in Observations, draw what you observe in each petri dish.

9. Follow your teacher's instructions for the proper disposal of all materials.

Name ______________________ Class ____________ Date ____________

Observations

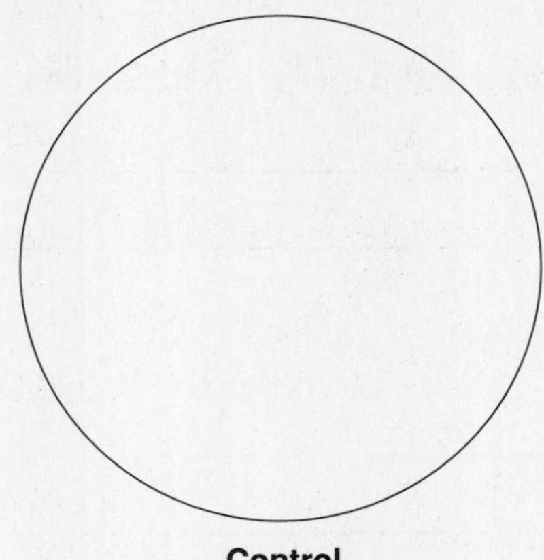
Control

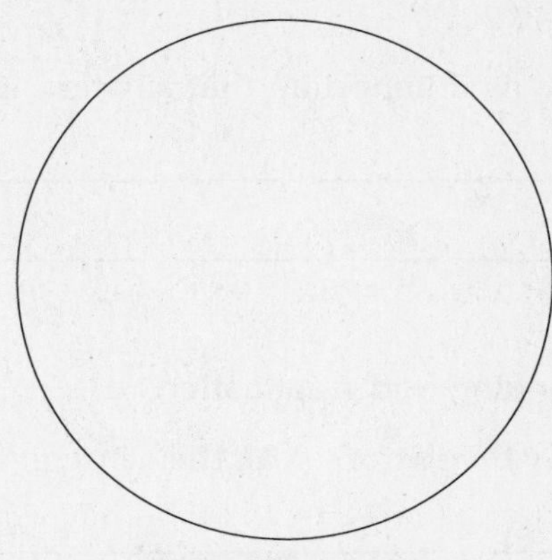
Soaked in Alcohol

1. After one week what did you observe in the dish marked "Control"?

2. After one week, what did you observe in the dish marked "Soaked in Alcohol"?

3. How did the growth of microorganisms in the two dishes compare?

Analysis and Conclusions

1. What effect did the alcohol have on the growth of microorganisms?

2. Why did you use forceps, rather than your fingers, to place the objects in the petri dishes?

3. Why was it necessary to immediately close the petri dishes after you added the objects?

4. Why is it important that all starting pennies, thumbtacks, and paper clips have been touched?

Critical Thinking and Application

1. Why do doctors soak their instruments in alcohol?

2. Why should cuts on the skin be cleaned with an antiseptic and bandaged?

3. A person who is burned over large areas of the body runs a high risk of developing a serious infection. Why is an infection likely to occur?

4. Suppose a person was born without an immune system. What are some of the precautions that would have to be taken so that the person could survive?

Going Further

Repeat this investigation with other antiseptics such as hydrogen peroxide, mouthwashes, deodorants, and antiseptic soaps to determine their effectiveness in killing or inhibiting the growth of microorganisms.

The Effects of Alcohol on Human Reactions

Pre-Lab Discussion

Ethyl alcohol is a drug that is absorbed directly into the bloodstream from the stomach and small intestine. It takes only two minutes, and even less on an empty stomach, for the alcohol to reach the bloodstream. The blood carries the alcohol to various parts of the body. The liver is the major organ that metabolizes ethyl alcohol to carbon dioxide and water and releases energy.

Alcohol acts as a depressant on the central nervous system. High blood-alcohol concentrations tend to reduce the activity of bodily functions. The result is a numbing, or anesthetic, effect on the central nervous system. Alcohol affects the areas of the brain controlling judgment, memory, emotion, speech, vision, motor skills, muscular coordination, and balance. Heart, digestive, and respiratory rates slow down under the influence of alcohol.

In this investigation, you will simulate some of the effects of alcohol without really drinking any alcohol.

Problem

What are the effects of alcohol on a person's reactions?

Materials *(per group)*

Masking tape
Watch or clock with second hand
Pencil

Safety

Do not spin around in this investigation if you are under a doctor's care, have dizzy spells or heart problems, or are unable to participate in physical education classes.

Procedure

Part A. Walking the Line

1. Using masking tape, make a straight line about 3 meters long on the floor. One member of your group should walk from the beginning of this line to the end, putting the heel of one foot right against the toe of the other. A second member of the group should time how long it takes the person to walk from one end of the line to the other. A third group member should keep track of the number of times the walker accidentally misses the line.

2. To simulate the effect of drinking an excess amount of alcohol, the walker should spin around in place for 10 seconds. Be certain that the area is cleared of furniture and other obstacles. The other group members should stand nearby and act as spotters ready to catch the spinner if he or she starts to fall. **CAUTION:** *Do not spin if you are under a doctor's care, have dizzy spells or heart problems, or are unable to participate in physical education classes.*

3. As soon as the person finishes 10 seconds of spinning, lead him or her to the beginning of the line and have him or her repeat the walk described in step 1. Be sure to record the time and number of misses in Data Table 1.
4. Repeat steps 1 through 3 with each group member having an opportunity to be the walker. Record their results in Data Table 1.

Part B. Connecting the Dots

1. The object of this test is to draw a wavy line through the dots as shown in the sample in Observations. One of your group members will time you for 10 seconds while you connect as many dots as you can. At the end of 10 seconds, give yourself one point for each dot you cross and for each time you touch the top or bottom line without crossing over it. Record this information in Data Table 2.
2. Repeat the spinning procedure described in step 2 of Part A. Then repeat the dot test and see how many points you score this time. Record this information in Data Table 2.
3. Repeat steps 1 and 2 with each member of your group having an opportunity to connect the dots. Record their results in Data Table 2.

Part C. Using Statistics to Construct Graphs on Alcoholism

1. It is estimated that there are 14 million alcoholics in the United States. Of this number, 3.5 million are teenagers, 13 to 17 years old. Among teenage alcoholics, 6.25% are females.
2. On the circle graph in Observations, indicate the portion of the total population of alcoholics who are between the ages of 13 and 17, and male. Also indicate the number of female alcoholics between the ages of 13 and 17.
3. A recent survey of teenage drinking habits revealed the following:
   - 62% drank alcoholic beverages occasionally
   - 19% tried alcohol once
   - 18% never tried alcohol
   - 1% no comment

   Use the circle graph in Observations to graph these survey results.

Observations

Data Table 1

Name	Before Spinning		After Spinning	
	Time (sec)	Mistakes	Time (sec)	Mistakes

Sample

Name ____________________ Class __________ Date __________

Data Table 2

Name	Score Before Spinning	Score After Spinning

Alcoholics in the United States

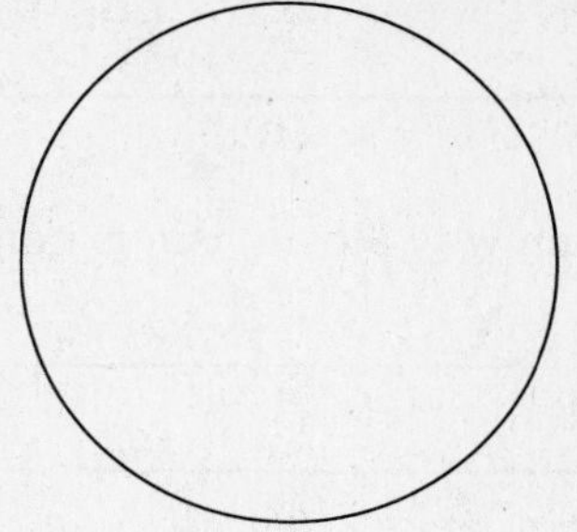

Teenage Drinking Habits

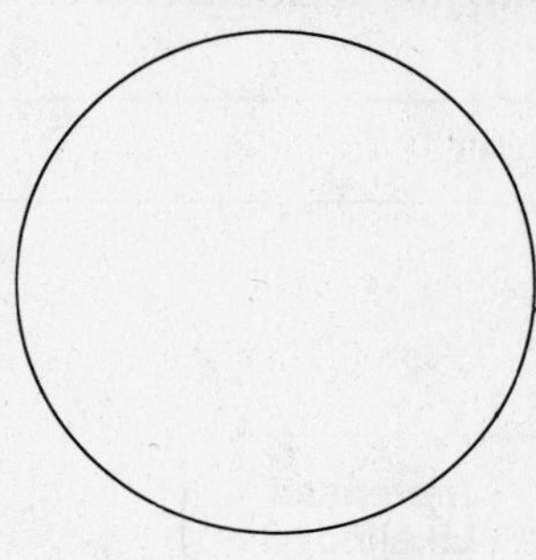

Analysis and Conclusions

1. Based on your observations in this simulation of drinking alcohol, what effect does alcohol have on a person's reactions? ____________________

2. What is the effect of alcohol on the central nervous system? ____________________

3. According to the statistics given in Part C, what percentage of the total population of alcoholics are between the ages of 13 and 17? ____________________

Critical Thinking and Application

1. Why is alcohol technically classified as a legal drug? ____________________

2. Alcohol is known to be a depressant. Why then does the consumption of alcohol sometimes result in a feeling of happiness? ______________________________

____

____

3. Explain why alcoholism is considered to be a disease. ______________________________

____

____

4. Use the information in the accompanying chart to construct a line graph. Does the likelihood of having an accident increase proportionally with the amount of alcohol in the blood?

____

____

% of Alcohol in Blood	Increased Likelihood of Having an Accident While Driving
0.00	1.0 x
0.02	1.5 x
0.04	2.0 x
0.06	2.5 x
0.08	4.0 x
0.10	6.0 x
0.12	10.0 x
0.14	19.0 x
0.16	30.0 x

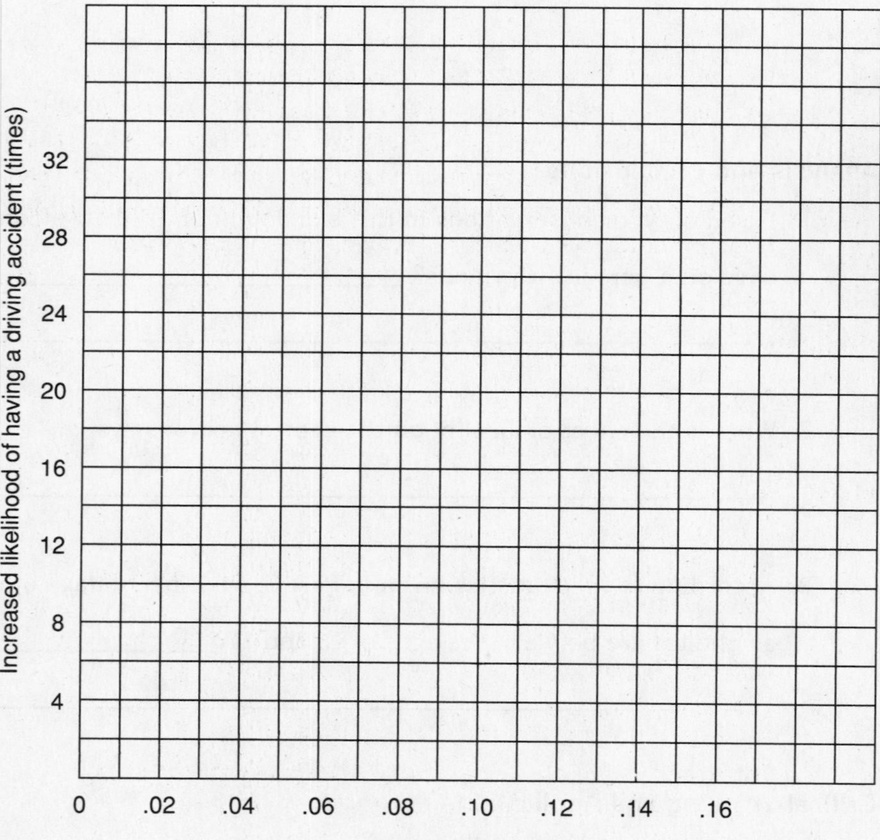

Going Further

1. Call your local chapter of Alcoholics Anonymous, Al-Anon, or Alateen and request information about the warning signs of alcoholism.
2. Using resources from within your community, prepare a panel discussion on topics related to drugs and their abuse that would interest students in your biology class. Invite outside experts from law enforcement groups, local and federal government agencies, and the medical profession to discuss these topics.

Name ______________________ Class ____________ Date ____________

91

The Effects of Tobacco and Alcohol on Seed Germination

Pre-Lab Discussion

Ethyl alcohol, or ethanol, which is an ingredient in alcoholic beverages, is a drug absorbed directly into the bloodstream from the stomach and small intestine. A *drug* is any substance that causes change in body function. Alcohol acts as a depressant on the central nervous system. Depressants decrease the action of the central nervous system by reducing the ability of nerves to transmit impulses between the brain and the rest of the body.

Tobacco leaves contain substances that have been shown by scientists to be harmful to people and animals. The tar in tobacco has been linked to cancer. Tars and other substances in tobacco may also lead to heart and respiratory illnesses.

In this investigation, you will test the effect of alcohol and tobacco smoke on seed germination. A seed that is alive and healthy will germinate under proper conditions of moisture and temperature. But if a harmful substance is added to the seed's environment, it may change the number of seeds that will germinate.

Problem

How do tobacco and alcohol affect seed germination?

Materials *(per group)*

- 100-mL graduated cylinder
- 250-mL graduated cylinder
- 250-mL Erlenmeyer flask
- 3 50-mL beakers
- Glass-marking pencil
- Faucet aspirator
- Scissors
- Forceps
- Wide rubber tubing
- Aluminum foil
- Cotton
- Filter paper
- 10 filterless cigarettes
- Matches
- 25 mL ethanol
- 150 mustard seeds
- Two-hole rubber stopper
- Glass tubing
- Narrow rubber tubing
- Metric ruler
- 4 petri dishes and covers

Safety

Put on a laboratory apron if one is available. Put on safety goggles. Handle all glassware carefully. Be careful when using matches. Always use special caution when working with laboratory chemicals, as they may irritate the skin or cause staining of the skin or clothing. Never touch or taste any chemical unless instructed to do so. Keep alcohol away from any open flame. Be careful when handling sharp instruments. Note all safety alert symbols next to the steps in the Procedure and review the meanings of each symbol by referring to the symbol guide on page 10.

Procedure

1. Construct a smoking machine similar to the one shown in Figure 1.

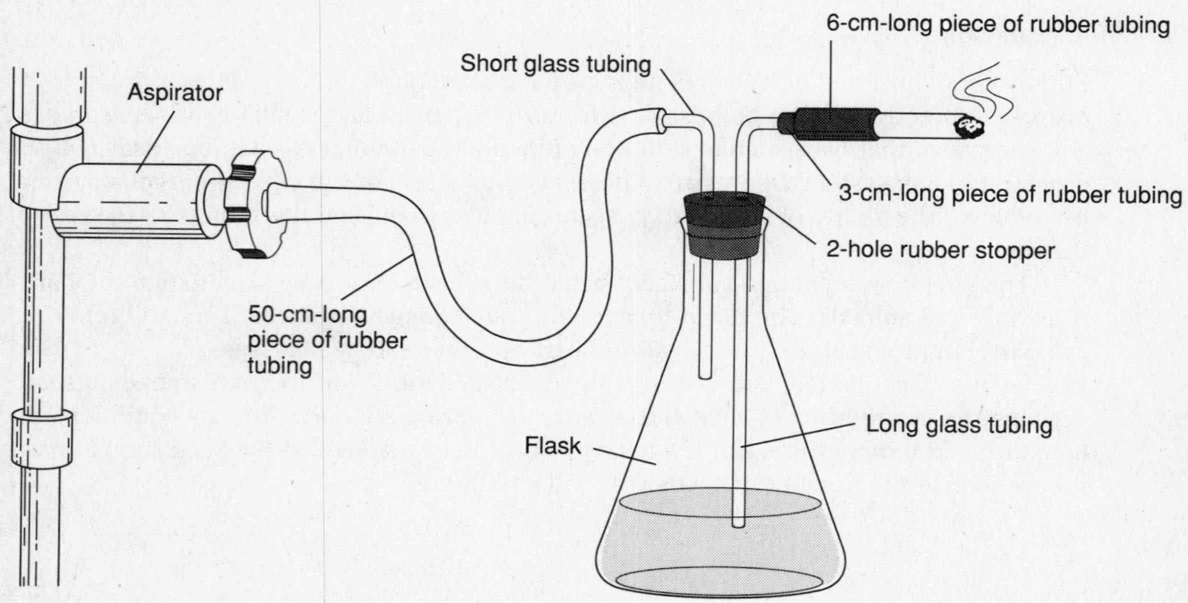

Figure 1

2. Using a graduated cylinder, pour 150 mL of water into a 250-mL flask. Obtain from your teacher a two-hole rubber stopper with glass tubing already inserted. Place the stopper in the flask. Make sure the long glass tubing goes into the water. The short glass tubing, however, should not touch the water.
3. With a gentle twisting motion, slide a 3-cm-long piece of narrow rubber tubing over the free end of the long glass tubing. Cut a 2-mm notch in each end of the 6-cm-long piece of wide rubber tubing. To construct your cigarette holder, slide the wide rubber tubing over the narrow rubber tubing.
4. Attach one end of a piece of narrow rubber tubing about 50 cm in length to the free end of the short glass tubing. Attach the other end to a faucet aspirator. When the water is turned on, it will produce a partial vacuum that will draw smoke from the cigarette through the water in the flask.
5. Insert a filterless cigarette into the cigarette holder. To catch falling ashes, put a petri dish under the cigarette.

Name ______________________ Class ______________ Date ______________

6. Light the cigarette with a match. **CAUTION:** *Be careful when using matches.* Turn on the water. Smoke should bubble through the water. Increase the flow of faucet water through the aspirator if the smoke does not bubble freely through the water.

7. When the cigarette has burned almost to the holder, turn off the water. Using forceps, remove the cigarette, run water over it, and dispose of it in the petri dish with the ashes.

8. Repeat steps 5 through 7 until the smoke from 10 cigarettes has been bubbled through the water.

9. Place 50 mustard seeds in each of three 50-mL beakers.

10. Add 25 mL of water to one of the beakers. Label the beaker accordingly. Add 25 mL of ethanol to the second beaker and label it. **CAUTION:** *Keep alcohol away from an open flame.* To the last beaker add 25 mL of smoky water and label it. Cover the beakers with aluminum foil and set them aside for 24 hours.

11. After 24 hours, spread a thin layer of cotton across the bottom of each of three petri dishes. Place a piece of filter paper on top of the cotton. Pour the contents of each 50-mL beaker into a separate petri dish. Carefully spread the seeds out over the filter paper. Cover the petri dishes and label them accordingly. Set the petri dishes aside.

12. After another 24 hours, check the petri dishes for germinated seeds. Count the number of germinated seeds in each petri dish after one day. Record the results in the Data Table. Then set the petri dishes aside for another 24 hours. Count the number of germinated seeds in each petri dish after two days. Record this information in the Data Table.

Observations

Data Table

Substance	Number of Seeds Germinated	
	Day 1	Day 2
Water		
Alcohol		
Tobacco smoke		

1. How does the number of germinated seeds in ethanol compare to the number of germinated seeds in tap water? ____

____

____

2. How does the number of germinated seeds in smoky water compare to the number of germinated seeds in tap water? ____

____

____

Analysis and Conclusions

1. How does alcohol affect the central nervous system? ______________________________

____

____

2. Based on your results, do you think alcohol is harmful to seeds? Explain your answer.

____

____

3. To what human illnesses has tobacco been linked? ______________________________

____

____

4. Based on your results, do you think tobacco is harmful to seeds? Explain your answer.

____

____

Critical Thinking and Application

1. How might using filtered instead of unfiltered cigarettes change the results of this experiment?

____

____

____

2. Cigarette smoking causes a person's blood vessels to constrict (close up). What negative effects might this have on the body? ______________________________

____

____

3. Why should a person who has consumed alcohol not drive an automobile?

____

____

____

Name ______________________ Class ____________ Date ____________

4. Cigarette smoke contains the toxic gas carbon monoxide, which has a 250 times higher ability to combine with the hemoglobin in red blood cells than atmospheric oxygen does. What effect would this have on the human body? ______________________

____

____

____

5. Caffeine is a substance found in coffee, tea, and chocolate that tends to increase a person's heart rate and breathing rate. Would you consider caffeine to be a drug? Explain your answer.

____

____

____

Going Further

1. Using reference materials, research the medical and social problems created by drug abuse. Choose one drug, give its chemical name, and list its effects on the body. Also find out some of the social problems caused by the selling, buying, and use of this drug.
2. Write to organizations such as the American Lung Association, American Heart Association, and American Cancer Society for information about the hazards of smoking. Using this information, prepare charts for a bulletin board display.

Name ______________________ Class ____________ Date ____________

92

The Oxygen Cycle

Pre-Lab Discussion

Oxygen, which is needed by most living things for respiration, makes up about 21 percent of the atmosphere. It is also dissolved in water and is a part of water molecules themselves. The movement of oxygen through an ecosystem is called the *oxygen cycle.*

One phase of the oxygen cycle occurs during the process of photosynthesis. During photosynthesis, molecules of water are split, releasing oxygen into the atmosphere. Oxygen in the atmosphere is used by most organisms for respiration. One of the waste products of respiration is water. Water is released into the atmosphere and may later fall to the earth as precipitation. This water can be absorbed by plants, which break down the water and release oxygen, thus continuing the oxygen cycle.

In this investigation, you will examine the operation of the oxygen cycle in a closed environment.

Problem

How does the oxygen cycle work?

Materials *(per group)*

2 snails
4 sprigs of *Elodea*
Pond water
Masking tape
Bright light source
Bromthymol blue solution in dropper bottle
4 culture tubes with tops
Test tube rack

Safety

Put on a laboratory apron if one is available. Put on safety goggles. Handle all glassware carefully. Always use special caution when working with laboratory chemicals, as they may irritate the skin or cause staining of the skin or clothing. Never touch or taste any chemical unless instructed to do so. Follow your teacher's directions and all appropriate safety procedures when handling live animals. Note all safety alert symbols next to the steps in the Procedure and review the meanings of each symbol by referring to the symbol guide on page 10.

Procedure

1. Obtain four culture tubes with tops. Use the masking tape to prepare four labels as shown in Figure 1. Place one label on each culture tube.

Tube 1	Tube 2	Tube 3	Tube 4
2 sprigs *Elodea* and 1 snail	1 snail	2 sprigs *Elodea*	No organisms
Name	Name	Name	Name
Date	Date	Date	Date

Figure 1

2. Into tube 1 place two springs of *Elodea* and one snail. Into tube 2 place one snail. Into tube 3 place two sprigs of *Elodea*.
3. Fill all four tubes to the top with pond water. Add four drops of bromthymol blue solution to each tube. Seal each tube tightly. **CAUTION:** *Handle the bromthymol blue solution with care because it stains the skin and clothing.*
4. Set the tubes in a test tube rack and place them near a bright light.
5. After 24 hours, observe the tubes. Notice if the organisms are still alive. Note any color change in the water. Bromthymol blue solution is an indicator. In the presence of carbon dioxide, it changes color from blue to yellow. Record your observations in the Data Table.
6. Observe the tubes every day for seven days. Record your observations in the Data Table.
7. Empty all the tubes and dispose of the organisms according to your teacher's directions.

Observations

Data Table

	Observations			
Day	**Tube 1**	**Tube 2**	**Tube 3**	**Tube 4**
1				
2				
3				
4				
5				
6				
7				

Name ______ Class ______ Date ______

Analysis and Conclusions

1. What is the purpose of tube 4? ______

2. What changes occurred in the bromthymol blue solution in each tube?

   Tube 1 ______

   Tube 2 ______

   Tube 3 ______

   Tube 4 ______

3. At the end of seven days, what happened to the organism in tube 2? Why?

4. Why did the organisms in tube 1 remain alive? ______

5. Explain why the plants alone in tube 3 remained alive. ______

6. How does this investigation demonstrate the oxygen cycle? ______

Critical Thinking and Application

1. What do you predict would happen if all of the tubes in the investigation were placed in the dark? Explain your prediction. ____________________

2. Explain the following statement: "Some of the oxygen you breathed yesterday might have been breathed by George Washington." ____________________

3. Is it true that plants could exist on Earth without animals, but animals could not exist without plants? Explain your answer. ____________________

Going Further

Repeat this investigation with different organisms or different combinations of organisms in each tube. You may also want to alter the size of the culture tubes or other containers. Be sure that all containers remain tightly sealed throughout the observation period. Report your observations and conclusions to the class.

Name ______________________ Class __________ Date __________

Adapting to the Cold

Pre-Lab Discussion

The *tundra* is a large terrestrial biome located in northern latitudes. This biome covers about 5.5 percent of the Earth's landmass. The climate of the tundra is very cold and dry. Winters are nine months long, and temperatures remain below freezing most of this time.

Picture yourself trudging through the tundra in –10°C weather, with the wind and wolves howling. A pretty chilling image, no doubt! Most people can avoid long exposure to such cold, but many tundra animals—the Arctic fox, polar bear, caribou, and moose, for example—spend winters entirely outdoors. Fortunately, they show adaptations to their cold biome.

In this investigation, you will discover two ways in which animals are protected from extreme temperatures in their biomes. You will also learn how to determine whether a material is a good insulator or a poor insulator of heat.

Problem

How do animals adapt to long-term exposure to cold?

Materials *(per group)*

3 250-mL containers with lids
3 750–1000-mL containers with lids
3 Celsius thermometers
100-mL graduated cylinder
Piece of foam rubber, 2.5 cm thick and approximately 21 cm x 28 cm
Scissors
Masking tape
Article of winter clothing such as a heavy sock, mitten, hat, or scarf
3 colored pencils
Hot tap water
Sand or soil
Natural materials such as leaves, dried grass, sawdust, or shredded paper

Safety

Put on a laboratory apron if one is available. Handle all glassware carefully. Use extreme care when working with heated equipment or materials to avoid burns. Be careful when handling sharp instruments. Note all safety alert symbols next to the steps in the Procedure and review the meanings of each symbol by referring to the symbol guide on page 10.

Procedure

Part A. Homegrown Insulation

1. Do layers of natural insulation such as fur, feathers, or fat help animals stay warm? You will construct a model of three warmblooded animals to find out. A 250-mL container with lid will represent each of the three animals.

2. Number the containers 1, 2, and 3.

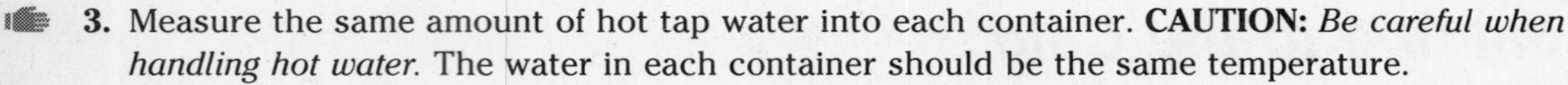

3. Measure the same amount of hot tap water into each container. **CAUTION:** *Be careful when handling hot water.* The water in each container should be the same temperature.

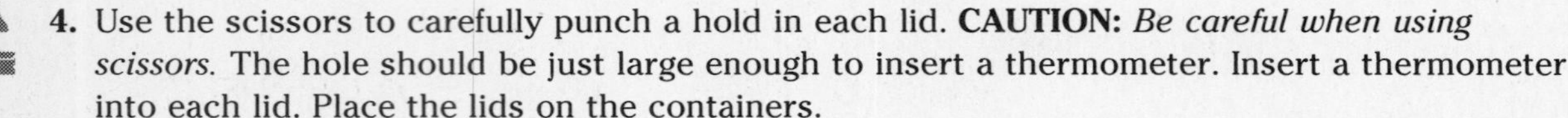

4. Use the scissors to carefully punch a hold in each lid. **CAUTION:** *Be careful when using scissors.* The hole should be just large enough to insert a thermometer. Insert a thermometer into each lid. Place the lids on the containers.

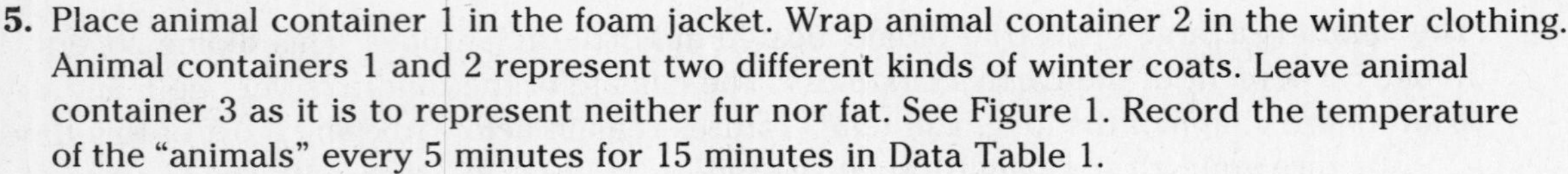

5. Place animal container 1 in the foam jacket. Wrap animal container 2 in the winter clothing. Animal containers 1 and 2 represent two different kinds of winter coats. Leave animal container 3 as it is to represent neither fur nor fat. See Figure 1. Record the temperature of the "animals" every 5 minutes for 15 minutes in Data Table 1.

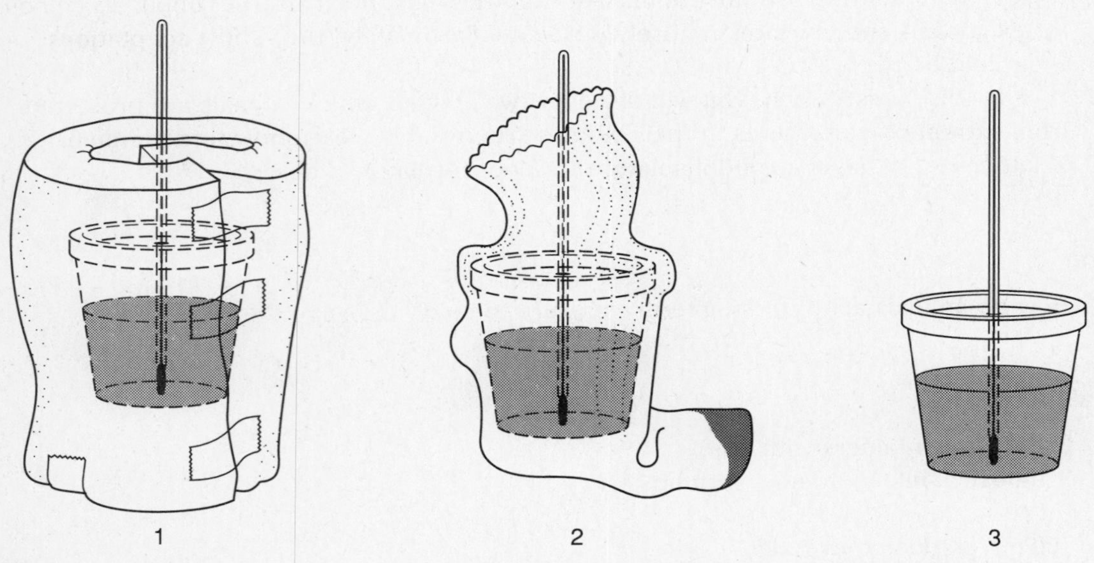

Figure 1

6. On Graph 1, plot the temperatures for each "animal." Use a different colored pencil for each "animal." Label the lines.

7. Clean the containers for use in Part B.

Part B. Finding Shelter in the Biome

1. If the outdoor temperature is –10°C, most animals need more protection than a coat of fur provides. Endothermic animals and some exothermic animals, such as snakes and salamanders, often live in underground burrows. Does the ground provide protection against the cold?

Name ______________________ Class ____________ Date ____________

2. Set up three small containers as shown in Figure 2. They are now animals X, Y, and Z. All three go into shelters (the three 750–1000-mL containers). X is simply out of the wind, surrounded by air. Y finds a snug burrow in soil or sand. Z's home is lined with natural materials of your choice, such as leaves or wood chips.

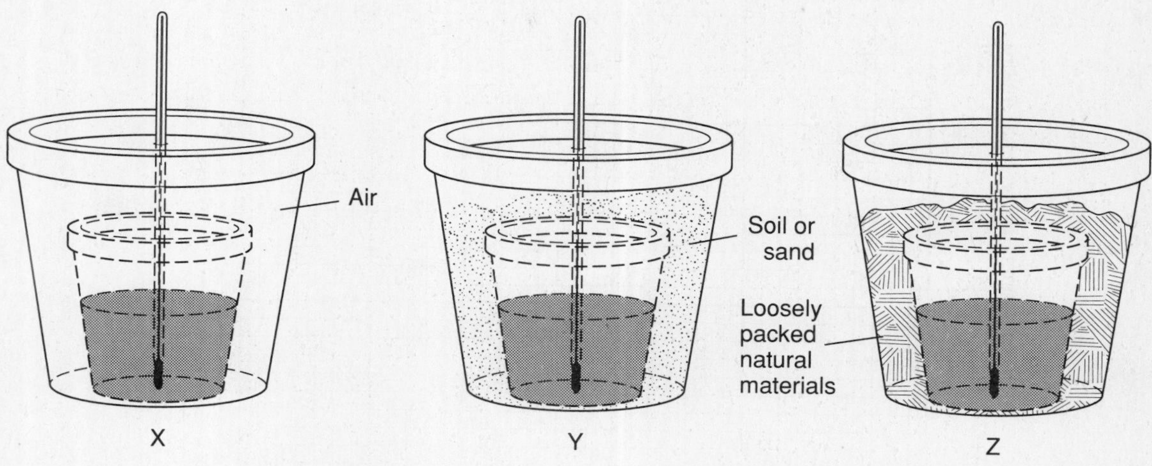

Figure 2

3. Start with the same amount of hot water at the same temperature as you used in Part A. Record the temperature every 5 minutes for 15 minutes in Data Table 2.

4. On Graph 2, plot the temperatures for each "animal." Use a different colored pencil for each "animal." Label the lines.

Observations
Part A. Homegrown Insulation

Data Table 1

	Temperature (°C)			
"Animal"	0 min	5 min	10 min	15 min
1 Foam				
2 Winter clothing				
3 No coat				

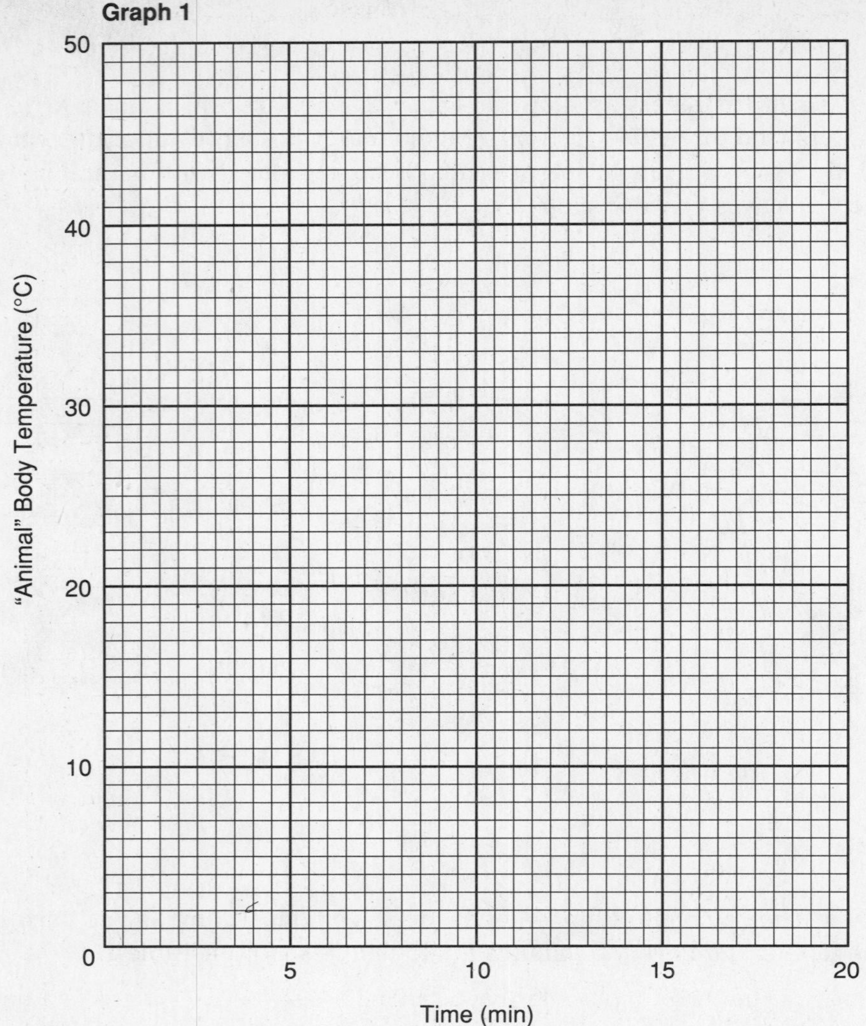

1. Which "animal" stayed warmest? ___

2. Which "animal" cooled fastest? ___

Part B. Finding Shelter in the Biome

Data Table 2

	Temperature (°C)			
"Animal"	0 min	5 min	10 min	15 min
X In air				
Y In soil				
Z In natural materials				

Name ______________________ Class __________ Date __________

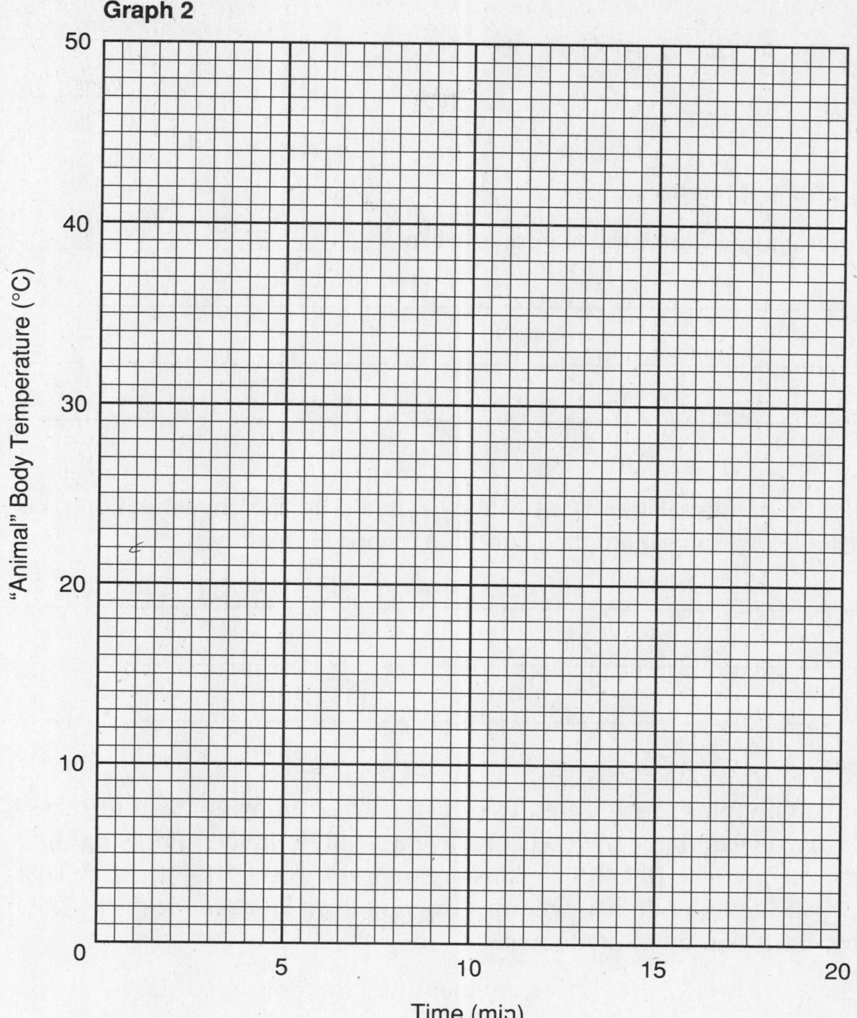

1. Which "animal" stayed warmest? ______________________

2. Which "animal" cooled fastest? ______________________

Analysis and Conclusions

1. What effect does a layer of insulation have on the temperature of an endothermic animal?

____

____

2. What is the best estimate of the temperature of "animal" X after 20 minutes?

____

____

3. How does an underground burrow protect an animal from freezing?

____

____

____

Critical Thinking and Application

1. How does a heavy jacket keep a person warm? ______________________

____

____

____

2. Eventually, the temperatures of all of the "animals" in this investigation return to room temperature. How do real animals continue to produce body heat?

____

____

____

3. Two identical puppies are adopted by cousins. One puppy is taken to Alaska. The other goes to Florida. Two years later the cousins have a family reunion and bring their dogs. Both dogs have thick, furry coats. But the cousins are amazed to see that the dog from Alaska has soft, thick fur covering its abdomen whereas the dog from Florida does not. The dog from Florida has almost bare skin on its abdomen.

   a. How does the fur covering the abdomen help the dog from Alaska?

   ____

   ____

   ____

   b. How does fur on the head, back, and legs help the dog from Florida? The dog from Alaska?

   ____

   ____

   ____

c. How does the lack of fur on the underside help the dog from Florida?

____

____

____

4. In addition to fur, what other adaptive body feature might an animal have to protect itself from the cold? ______________________________

____

____

____

Going Further

Temperatures in the desert can easily climb to 50°C. Do fur and underground burrows protect animals from intense heat? Repeat the investigations. Put the "animals" above heating vents or in direct sunlight.

Name ______________________ Class ____________ Date ____________

94

Counting a Population

Pre-Lab Discussion

A *population* is a group of individuals of the same species living in a given area. Biologists can use two different methods to determine the number of living things in a given area. The most accurate data would be obtained by counting each member of the population. However, in most situations this counting method is impractical and very time-consuming. A second method is to count a small sample of the population. The sample counted must be representative of the entire population. If the sample is not representative of the entire population, then the data collected are biased and therefore inaccurate. A random selection of areas in which to count organisms helps to eliminate bias. Every member of the population stands an equal chance of being counted.

In this investigation, you will demonstrate the technique of counting a random sample of population, from which you can estimate the size of an actual population.

Problem

How can the number of a population be estimated from a sample?

Materials *(per group)*

Plant and weed identification charts or books
3 sheets of notebook paper
Metric ruler
2-meter length of string
4 stakes or large nails
Forceps or garden gloves

Safety

To avoid possible contact with poisonous or prickly plants, use forceps or wear gloves when collecting plant specimens. Use a field guide to identify dangerous plants. Be careful when handling sharp instruments. Note all safety alert symbols next to the steps in the Procedure and review the meanings of each symbol by referring to the symbol guide on page 10.

Procedure

1. Work with two partners.
2. On three separate sheets of notebook paper, make a copy of the chart shown in Figure 1. **Note:** *You will have three copies of the chart when you are finished.* Label the first chart "Sunlight," the second "Shade," and the third "Neglected."

(Location)	
Plant Type	Number

Figure 1

3. Select three different locations in the lawn around your school. One location should be in an area that receives constant sunlight. The second location should be in an area that is mostly shady. The third location should be a neglected area where weeds grow well.
4. At the sunny location, designate one person in the group to toss an object such as a pencil, a small rock, or a coin into the area. The person should close his or her eyes when tossing the object. Mark the spot where the object lands.
5. With the string and four stakes or nails, outline a 30 cm x 30 cm square around the marked spot. The marker should be in the center of the 30 cm x 30 cm square, as shown in Figure 2.

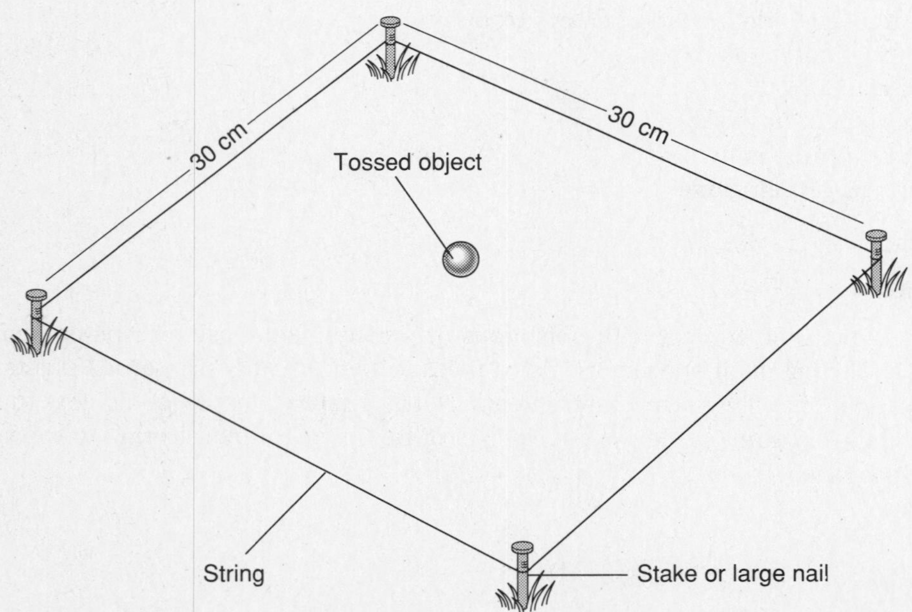

Figure 2

Name ______________________ Class __________ Date __________

6. On the chart labeled "Sunlight," record the type and number of all plants in the square. If a plant cannot be identified, carefully remove a single leaf, flower, or individual plant for later identification. **CAUTION:** *Be able to identify poisonous or prickly plants. Use forceps or wear gloves when removing the plant specimens to avoid possible contact with poisonous or prickly plants.* Use plant identification books and charts in the classroom or library to identify unknown plants.

7. When you have finished counting population numbers and collecting specimens in the area, remove the stakes and string.

8. Repeat steps 4 through 7 for the shaded and neglected locations.

Observations

1. What was the total number of plants recorded at each location?

   Sunlight ______________________

   Shade ______________________

   Neglected ______________________

2. What type of plant had the largest population number at each location?

   Sunlight ______________________

   Shade ______________________

   Neglected ______________________

3. What type of plant had the smallest population number at each location?

   Sunlight ______________________

   Shade ______________________

   Neglected ______________________

Analysis and Conclusions

1. Why was the person tossing the marker directed to close his or her eyes?

   ____

   ____

   ____

2. What characteristic or adaptation of the dominant plants enabled them to be so successful?

3. What environmental factors at each of the three locations might have limited the various plant populations?

4. Would you expect the estimated total population number to be closer to the actual population number if an average of your classmates' results were taken? Explain your answer.

5. Based on the results of this investigation, do you think that counting samples is an accurate method for counting a population? Explain your answer.

Critical Thinking and Application

1. How would you estimate the total population of any one type of plant in the entire school lawn?

Name ______________________ Class ____________ Date ____________

2. What considerations must be made when a marketing research team is polling a sample population for the sale of a new product? ______________________

____

____

____

3. How might you count the number of individuals in a sample of a mobile population, such as mice on a prairie or fish in a pond? ______________________

____

____

____

4. List two factors that might cause the number of individuals in a population sample to change from month to month. ______________________

____

____

____

Going Further

1. Estimate the number of the following organisms in an area. Use the method of counting described in this investigation.
   - **a.** tree seedlings in an area of dense shade versus an area with full sunlight
   - **b.** earthworms in a garden
   - **c.** insects in a field
2. Design an experiment to investigate the effect of the availability of space on populations. Plant seeds so that the density of the population varies, then compare their survival and growth rates.

Name ______________________ Class __________ Date __________

Relationships in an Ecosystem

Pre-Lab Discussion

An *ecosystem* consists of groups of organisms and their nonliving environment. Each ecosystem contains *biotic factors,* or living organisms, and *abiotic factors,* or physical (nonliving) components. Abiotic factors include soil, water, temperature, and light.

If you examine an outdoor ecosystem, you will find producers, consumers, and decomposers that form a variety of food webs. You may also find evidence of competition for food, space, and shelter; predator-prey relationships; and various types of symbiosis.

In this investigation, you will examine an outdoor ecosystem. In any outdoor activity, it is important that you disturb the environment as little as possible. When doing this investigation, remember that humans share this environment with wildlife.

Problem

What effect does protective coloration have on a predator-prey relationship? How many different interactions among organisms can you observe, directly or indirectly, in the environment around your school?

Materials *(per group)*

500 colored toothpicks (100 each of 5 different colors, including green)
Pencil and notepad

Safety

Handle toothpicks carefully when picking them up from the ground. As you observe the ecosystem around your school, do not pick up or touch any of the organisms you see. You will want to disturb organisms as little as possible to observe how they interact with other organisms in the environment. Note all safety alert symbols next to the steps in the Procedure and review the meanings of each symbol by referring to the symbol guide on page 10.

Procedure
Part A. Effects of Protective Coloration

1. Your teacher will scatter colored toothpicks over a grassy area. These toothpicks represent the food (prey) that you (predator) are hunting.
2. After the signal to begin is given, see how many prey you can capture in the allotted time.
3. Count the number of prey of each color that you captured. Include your results with the rest of the class. Record the class totals in the Data Table.
4. Repeat steps 2 and 3 and record the results for this trial in the Data Table.

Part B. Ecological Scavenger Hunt

1. In most scavenger hunts, you are asked to collect various items. However, in order to disrupt the wildlife in the area as little as possible, you will do your collecting with a pencil and notepad.
2. Search your assigned area for evidence that:
   a. Plants support all forms of animal life, either directly or indirectly.
   b. Humans and wildlife share environments.
   c. People have changed the environment.
   d. Wildlife exists in many different colors.
   e. Animals live in or have passed through this area.
   f. Humans and wildlife are subject to similar environmental problems.

   Record your observations.
3. What evidence can you find of the following?
   a. predator-prey relationships
   b. plant disease
   c. insect damage
   d. food webs
   e. symbiotic relationships

   Record your observations.

Observations
Part A. Effects of Protective Coloration

Data Table

Color of Toothpick	Trial 1	Trial 2
1.		
2.		
3.		
4.		
5.		
Total prey captured		

Name ______________________ Class ____________ Date ____________

Part B. Ecological Scavenger Hunt

1. Summarize the observations you made in step 2 of Part B.

2. Summarize the observations you made in step 3 of Part B.

Analysis and Conclusions

1. What color prey were captured least often? Explain why this is so.

2. What kinds of producers did you find in the area you studied? ______

______

______

3. What kinds of consumers did you find in the area you studied? ______

______

______

Critical Thinking and Application

1. List three animals that have protective coloration and can camouflage themselves in their environment. ______

______

2. Do all ecosystems have consumers and producers? Explain your answer.

______

______

______

3. How might a change in season affect an ecosystem? ______

______

______

______

4. How might the community around your school be affected if a new wing were added to your school building? ______

______

______

Going Further

1. Design variations for this predator-prey investigation. For example, you might shorten or lengthen the time allowed for searching, or assign each student one color of prey as the only kind he or she can capture. Study the results of each variation that you carry out. Draw conclusions showing the probable effect of each variation on a real-life situation.
2. Learn about local wild or native plant species that are used for food. See how many of these edible plants you can find. Prepare a wild edible plant book for your area.

Name ______________________ Class ____________ Date ____________

Ecological Succession

Pre-Lab Discussion

Over time, an ecosystem goes through a series of changes known as *ecological succession.* The changes happen within the structure of the community and usually are caused by the community itself. During ecological succession, organisms often change their environment in such a way as to make conditions less favorable for themselves and more favorable for other organisms.

Abiotic factors also play a role in ecological succession. Temperature, amounts of rainfall and light, geological features, and other abiotic factors may change. These changes may produce biotic changes. As ecological succession occurs, a stable *climax community* that is resistant to change is eventually established.

Most examples of ecological succession must be studied over a period of many years because it takes a long time for ecological succession to occur in the environment. However, a pond water culture can be used as a small-scale model of ecological succession that shows results in a relatively short period of time.

In this investigation, you will study ecological succession in a pond water culture.

Problem

How can a pond water culture be used to study a succession of populations in a community?

Materials *(per group)*

- Pond water culture
- Large jar with lid
- Medicine dropper
- Aged tap water
- Glass-marking pencil
- 3 glass slides
- 3 coverslips
- Microscope
- pH paper

Safety

Put on a laboratory apron if one is available. Handle all glassware carefully. Always handle the microscope with extreme care. You are responsible for its proper care and use. Use caution when handling glass slides as they can break easily and cut you. Note all safety alert symbols next to the steps in the Procedure and review the meanings of each symbol by referring to the symbol guide on page 10.

Procedure

1. Obtain a pond water sample from your teacher. Use the glass-marking pencil to put your group members' names on the outside of the jar. Carefully examine the culture. Observe the color, cloudiness, odor, and any layering of materials that may occur. Record the date and your observations in the Data Table.

2. Remove the lid from the culture jar and make three wet-mount slides of the pond water culture. Take the sample for each slide from a different part of the culture jar—top, middle, and bottom.

3. Examine each slide under a microscope, first under low power, then under high power. **CAUTION:** *When turning to the high-power objective, you should always look at the objective from the side of your microscope so that the objective does not hit or damage the slide.* Observe any organisms in the culture. Use the drawings in Figure 1 to identify the organisms you might observe in your pond water culture. In the appropriate place in the Data Table, record the type and approximate number of each organism you find on each slide.

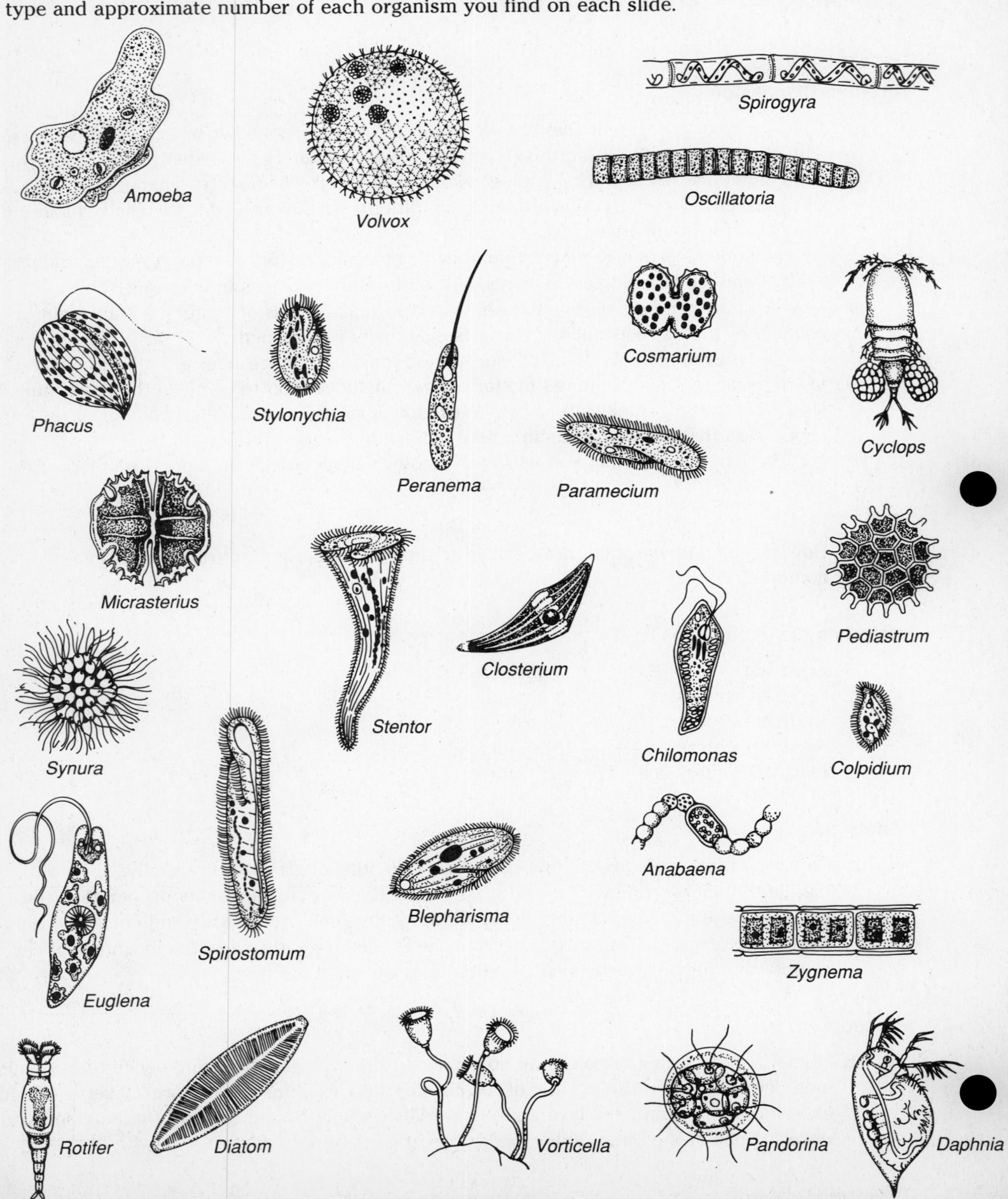

Figure 1

Name ______________________________ Class ______________ Date ______________

4. Use pH paper to test the pH of the pond water culture in the jar. Record this information in the Data Table.
5. Repeat steps 1 through 4 every two days for a period of two weeks.
6. Use the graph provided in Observations to construct a line graph of the growth of any organism of your choice.

Observations

Data Table

Date	Appearance of Culture Water	pH of Culture Water	Species Present in Culture	Approximate Number of Individuals in Each Population

Graph

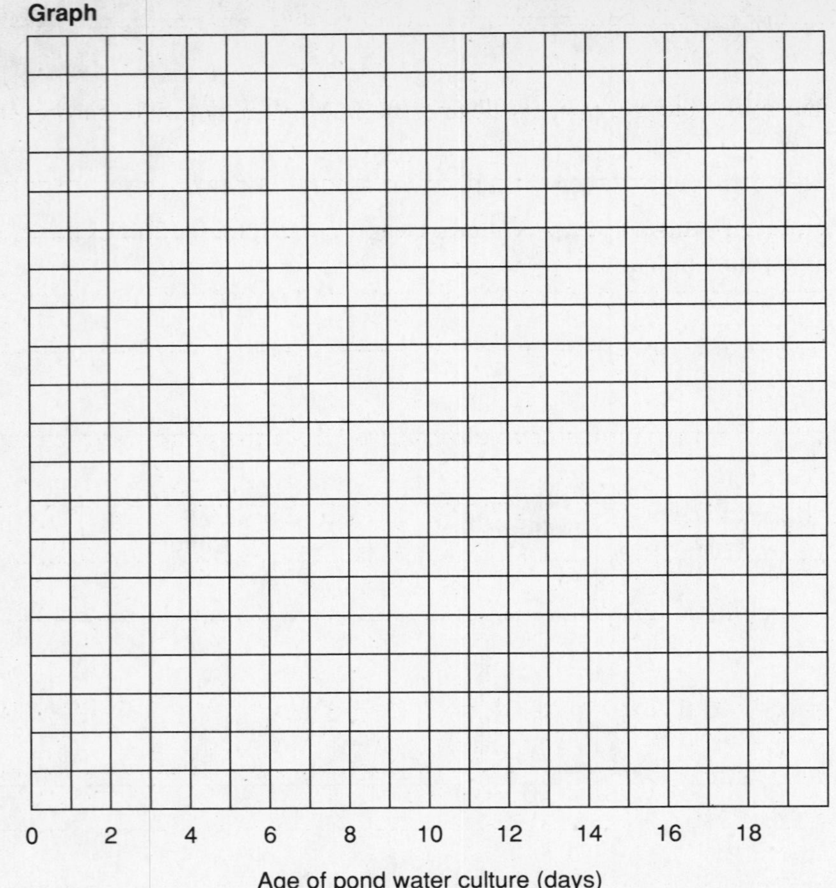

Analysis and Conclusions

1. Why should the pond water jar lid have holes punched in it? ______________________

____

____

____

2. What organisms were the first to become abundant in your pond water culture?

____

____

____

3. What population of organisms would you consider to be the climax community in the succession of the pond water culture? Explain your answer. ______________________

____

____

____

____

Name ______________________ Class __________ Date __________

4. What is the relationship between pH and the number of populations present in the pond water culture? ______________________

5. Why might some populations have disappeared from the pond water culture?

6. Describe any evidence of ecological succession you observed in your pond water culture.

Critical Thinking and Application

1. A stable population suddenly experiences a tremendous growth in size. List two factors that might be responsible for this growth. ______________________

2. How is the competition between species related to the process of succession?

3. Explain Figure 2 in terms of ecological succession. ______________________________

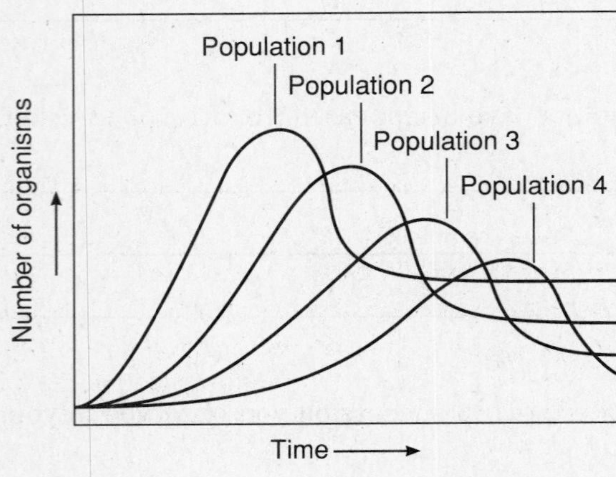

Figure 2

Going Further

1. Design an experiment to determine the effect of changes in the physical conditions (temperature, light, volume of container, etc.) on the ecological succession that takes place in the pond water culture.
2. Design an experiment to discover how the introduction of a foreign population into a pond water culture affects a normal succession.
3. Isolate a particular type of organism in your pond water culture. Set up an artificial environment with ideal conditions for the growth of the isolated species.

Name ______________________ Class ____________ Date ____________

Investigating Air and Water Pollution

Pre-Lab Discussion

Although life on Earth depends on air and water, we are endangering these important resources by polluting them with harmful substances. The air we breathe contains the oxygen our cells need, but it also contains many other chemicals that can damage our bodies. Our atmosphere is polluted with smog, acid rain, carbon dioxide, and a variety of other chemicals that create problems for all organisms dependent on the atmosphere.

Despite the fact that the Earth's surface is 75 percent water, only a small fraction of that water can be used by living things. The pollution of our freshwater supply by chemicals, sewage, oil, and heated waste water interferes with many food chains. It also requires costly treatments to ensure the safety of our water supply.

In this investigation, you will conduct tests to determine the types of pollutants present in air and water samples taken from the area in which you live.

Problem

How can air and water pollution be detected?

Materials *(per group)*

4 glass slides
Petroleum jelly
Tongue depressors
4 petri dishes
Glass-marking pencil
Microscope
Water samples A, B, and C
3 stirring rods
pH test paper
3 test tubes with stoppers
3 medicine droppers

Safety

Put on a laboratory apron if one is available. Put on safety goggles. Handle all glassware carefully. Always use special caution when working with laboratory chemicals, as they may irritate the skin or cause staining of the skin or clothing. Never touch or taste any chemical unless instructed to do so. Always handle the microscope with extreme care. You are responsible for its proper care and use. Use caution when handling glass slides as they can break easily and cut you. Note all safety alert symbols next to the steps in the Procedure and review the meanings of each symbol by referring to the symbol guide on page 10.

Procedure

Part A. Air Pollution

1. Make four particle traps by using a tongue depressor to smear the center of four glass slides with petroleum jelly.
2. Place each particle trap in the bottom half of a petri dish. Cover each petri dish immediately.
3. Select one location on the school grounds for your particle traps. After obtaining your teacher's approval, place the four petri dishes containing the particle traps in the selected location. Make sure that the petri dishes are side by side.
4. Remove the lids of the petri dishes and expose the traps to the air for 20 minutes.
5. At the end of the exposure time, replace the lids. Use the glass-marking pencil to record the test location on the lid of each petri dish. Return to your classroom.
6. Carefully remove the particle trap from the first petri dish and place it on the stage of a microscope. Use the low-power objective to examine the slide. Count the number of trapped particles and record this number in Data Table 1.
7. Repeat step 6 for each of the remaining particle traps.
8. Calculate the average number of particles trapped at your selected location. Record this average in Data Table 1. To find the average number, add the number of particles counted on each trap and divide by 4.
9. Report your location and the average number of particles you trapped to the class. Using class data, complete Data Table 2.
10. Clean your laboratory equipment before proceeding with Part B of the investigation.

Part B. Water Pollution

1. Stir water sample A with a clean stirring rod.
2. Describe the appearance of the water in sample A. Record your observations in Data Table 3.
3. Using proper laboratory technique, determine if the water has an odor. Record your observations in Data Table 3.
4. To determine the pH of the water sample, use a clean medicine dropper to place a drop of the water sample on the pH test paper. Immediately compare the color with the chart on the pH paper container. Record the pH in Data Table 3.
5. Pour water from the sample into a test tube until it is half full. To check for the presence of detergents, place a stopper in the test tube and shake for 10 seconds. Foam on the top of the sample indicates the presence of detergents. Record your observations in Data Table 3.
6. Repeat steps 1 through 5 with water samples B and C.

Name ______________________ Class __________ Date __________

Observations

Data Table 1

Particle Trap	Number of Particles
1	
2	
3	
4	
Average	

Data Table 2

Particle Trap Location	Average Number of Trapped Particles

Data Table 3

Test For:	Sample A	Sample B	Sample C
Appearance			
Odor			
pH			
Detergents			

Analysis and Conclusions

1. Why did you cover the particle traps between the time of preparation and the beginning of the exposure period? ____________________

2. Which location showed the greatest air pollution? What do you think caused the pollution?

3. Which location showed the least air pollution? Why? ____________________

4. Why were four particle traps used at each location? ____________________

5. If no particles were trapped at a given location, is it correct to assume that air at that location is not polluted? Explain your answer. ____________________

6. Why did you stir each water sample before conducting your tests?

7. Before completing the procedure, which water sample would you have predicted to be clean and safe for human use? Did this experiment change your evaluation of that sample? Explain your answer. ____________________

Name ______________________ Class __________ Date __________

Critical Thinking and Application

1. In what type of weather would you expect air pollution to be the worst? Explain your answer.

2. How might pollutants be removed from water supplies?

3. What are some ways in which you can help reduce air and water pollution in your community?

4. Explain how the statement "Think Globally, Act Locally" applies to the problems of air and water pollution.

Going Further

Conduct a study of the pollution in your community. Use the air and water pollution tests from this investigation to determine air and water quality, and make a visual survey to evaluate land pollution. Report your findings on a map of the area. Be sure to identify the major landmarks; roadways; and residential, commercial, and industrial areas of your community on the map. Your map key should explain any colors or symbols used in reporting the data.

Name ______________________ Class ____________ Date ____________

The Effects of Acid Rain on Seed Germination and Plant Growth

Pre-Lab Discussion

One serious result of air pollution is acid rain. Acid rain is the term used to describe any type of precipitation that has a pH below 6. Remember that the pH scale ranges from 1 to 14, with 7 being neutral. Substances with a pH from 1 to 6 are acidic; substances with a pH from 8 to 14 are alkaline (basic).

Acid rain forms when water vapor in the atmosphere combines with pollutants released by the burning of fossil fuels in industrial plants and in automobiles. When fossil fuels are burned, sulfur and nitrogen oxides are released. These pollutant gases combine with water vapor to form drops of sulfuric acid and nitric acid. Acid rain can damage buildings, plant and animal life, and soil. Dying forests and lifeless lakes in the northeastern United States, Canada, and Europe are being blamed on acid rain.

In this investigation, you will observe the effects of acid rain on seed germination and plant growth.

Problem

How does acid rain affect the germination of seeds and the growth of plants?

Materials *(per group)*

10 bean seeds
Empty beaker
Paper towels
Glass-marking pencil
Medicine dropper
Rain solutions
Half a petri dish
Seedling plants
Metric ruler
Forceps
pH paper

Safety

Put on a laboratory apron if one is available. Put on safety goggles. Handle all glassware carefully. Always use special caution when working with laboratory chemicals, as they may irritate the skin or cause staining of the skin or clothing. Never touch or taste any chemical unless instructed to do so. Note all safety alert symbols next to the steps in the Procedure and review the meanings of each symbol by referring to the symbol guide on page 10.

Procedure

Part A. The Effects of Acid Rain on the Germination of Seeds

1. From your teacher, obtain a beaker containing the rain solution you will be using during this investigation. Use pH paper to determine the pH of this rain solution. Record the pH in Data Table 1.
2. Loosely pack an empty beaker with paper towels.
3. Obtain 10 bean seeds that have been soaked in the rain solution assigned to you. Using forceps, place the bean seeds between the paper towels and the edge of the beaker so that you can easily observe the seeds each day. See Figure 1.

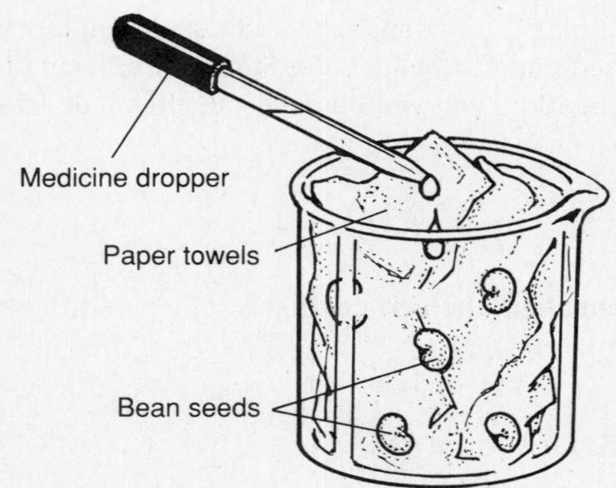

Figure 1

4. Use the medicine dropper to moisten the paper towels with your assigned rain solution.
5. Cover the beaker with half of a petri dish and place the beaker in a well-lighted location.
6. Observe the seeds daily for signs of germination or other evidence of change. Record your observations in Data Table 1.
7. Check the moisture in the paper towels each day. If necessary, moisten the paper towels with your assigned rain solution.
8. Report your data to the class. As each group reports their data, complete Data Table 2.

Part B. The Effects of Acid Rain on Seedlings

1. Obtain a seedling plant from your teacher.
2. Label the plant with your name and the pH of your assigned rain solution.

Name ______________________ Class ________ Date ________

3. Describe the appearance of your seedling plant in Data Table 3. Your description should include quantitative and qualitative data. Quantitative data might include the height and width of the plant and the number of leaves. Qualitative data might include the overall appearance, color, odor, and texture of the plant.

4. Place your plant in the sink or other area designated by your teacher. To simulate daily rainfall, sprinkle your plant's leaves and stem with 2 full medicine droppers of your rain solution. Be sure to moisten the soil each day with the rain solution.

5. Repeat step 4 every day for 5 days. Record you observations in Data Table 3.

Observations

Data Table 1

pH of Rain Solution ________

Day	Number of Germinating Seeds	Other Observed Changes
1		
2		
3		
4		
5		

Data Table 2

Day	Number of Seeds Germinating				
	pH ______	pH ______	pH ______	pH ______	pH ______
1					
2					
3					
4					
5					

Data Table 3

pH of Rain Solution ____________

Day	Appearance of Seedling Plant
1	
2	
3	
4	
5	

Analysis and Conclusions

1. Would you classify the rain solution used by your group as acid rain? Why or why not?

____

____

____

2. In which solution did the greatest number of seeds germinate? The least number of seeds germinate? ____

____

____

3. Which solution had the greatest effect on the seedling plant's growth and development? The least effect? ____

____

____

4. Why is it necessary to observe the seeds and the seedling plant over a 5-day period?

____

____

____

Name ______________________ Class ____________ Date ____________

5. By adding an alkaline substance such as lime to soil, farmers can help neutralize the effects of acid rain on soil. Based on the results of this investigation, will adding lime to soil protect crops from the effects of acid rain? Explain your answer. ______________________

____

Critical Thinking and Application

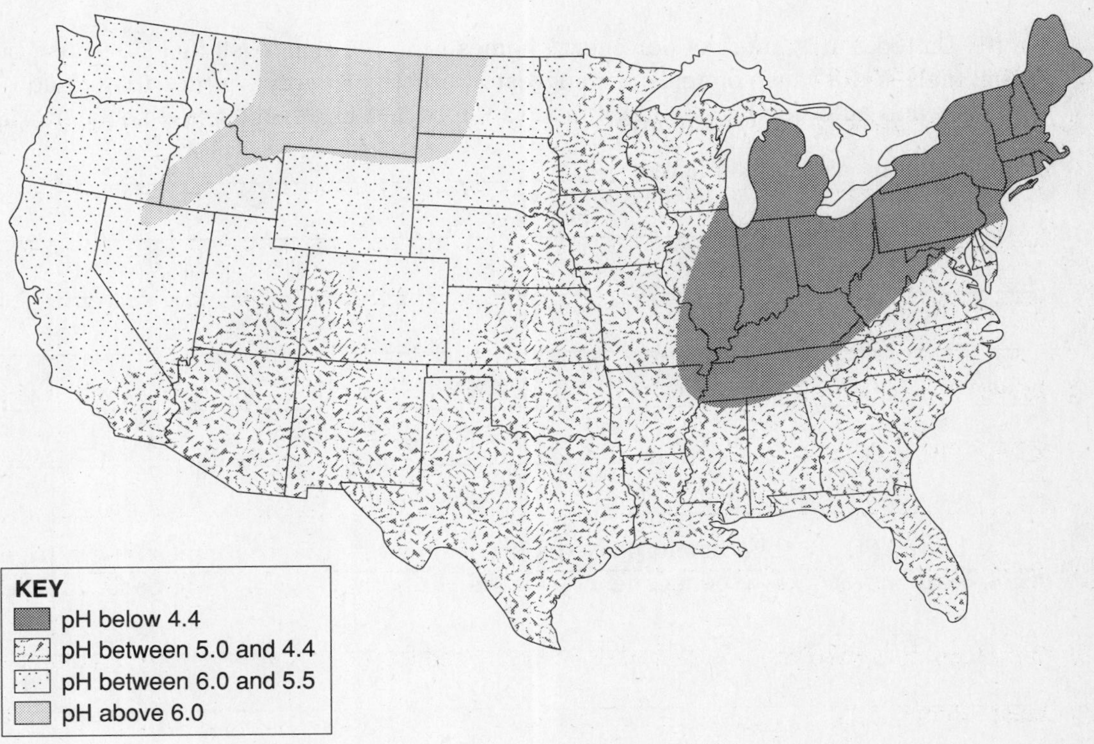

Figure 2

Use the map in Figure 2 to answer questions 1 through 3. The map identifies the acidity of rainfall across the United States.

1. What is the pH of the rain in your area? Would it be considered acid rain?

____

____

2. What happens to the acidity of rain as you move from the east coast to the west coast?

3. How can you explain your answer to question 2?

4. In the United States, most of our energy comes from burning fossil fuels. Because burning fossil fuels is the cause of acid rain, suggest alternative energy sources that could be used by industry and automobiles. Identify the advantages and disadvantages associated with each type of energy source you listed.

5. If the United States government passed a law prohibiting the use of fossil fuels, predict the ecological and economic consequences that might ensue. Would these consequences be positive or negative?

Going Further

Design an experiment that would test the possible effects of different types of pollutants on seed germination and plant growth and development. Pollutants that you might use in your experiment include detergents, table salt, and motor oil.

Name ______________________ Class ____________ Date ____________

Classifying Garbage

Pre-Lab Discussion

Throughout history, people have produced waste material, or garbage. And waste disposal has always been a problem that people have had to solve. Early civilizations had the option of moving to new locations when too much waste material accumulated near their homes. In addition, the garbage produced by early human societies was biodegradable; living organisms in the soil decomposed the waste material, thus solving the garbage problems.

Today, garbage is a major concern of our society. In addition to biodegradable materials, our garbage contains huge amounts of nonbiodegradable materials, which create more challenging waste disposal problems. And the problems of waste disposal increase with each passing year. Today, the average American produces 1000 kilograms of garbage annually. What happened to the 1000 kilograms you produced last year?

A common means of disposing of garbage is to add it to landfills. Landfills, however, are rapidly reaching capacity levels. In fact, many landfills are already full. Because of the mounting volume of garbage, new methods of disposal are needed. In planning for new methods, scientists must take into account the type of garbage, the effect of the garbage on the environment, and the cost of new technologies of disposal.

In this investigation, you will determine what types of garbage are biodegradable.

Problem

What types of garbage are most effectively disposed of in landfills?

Materials *(per group)*

Garbage
Garden soil
4 petri dishes
Scissors
Forceps
Newspaper
Medicine dropper
Water
Glass-marking pencil
Masking tape
Metric ruler
Trowel or scoop

Safety

Put on a laboratory apron if one is available. Handle all glassware carefully. Be careful when handling sharp instruments. Note all safety alert symbols next to the steps in the Procedure and review the meanings of each symbol by referring to the symbol guide on page 10.

Procedure

1. With a glass-marking pencil, label the bottoms of the four petri dishes as follows: Dish 1, Dish 2, Dish 3, and Dish 4. Include your name and the date as well.
2. From the garbage provided by your teacher, use the forceps to select 4 different types of garbage. Be sure to select both food and nonfood items.
3. Use the scissors and forceps to cut or pull apart the samples so that they are approximately 2 square centimeters in size. You will need 3 small pieces of each item of garbage you select.
4. Use the forceps to place the 3 small pieces of each type of garbage into a labeled petri dish.
5. In the Data Table, record a brief description of the items of garbage in each petri dish. Indicate size, color, texture, odor, and other important features of each garbage sample.
6. Place the petri dish on newspaper and use the trowel or scoop to carefully fill each dish with garden soil. With the medicine dropper, add enough water to each dish to make the soil moist, but not soggy.
7. Place a cover on each dish and tape each dish closed with masking tape. Carefully turn each dish over so that the samples can be seen through the glass. Leave the dishes in a warm place.
8. In the appropriate place in the Data Table record your prediction of how each sample of garbage in each petri dish will look after one week.
9. Examine each dish after one week. Use the forceps to remove the samples of garbage. Record any changes in the size, color, texture, and odor of the samples.
10. Follow your teacher's instructions for the proper disposal of all materials.

Observations

Data Table

Dish	Type of Sample	Initial Description Date ________	Predicted Change	Final Description Date ________
1				
2				
3				
4				

Name ______________________ Class __________ Date __________

1. Which sample(s) showed the greatest amount of change? Describe the kinds of changes that occurred. ______________________

2. Which sample(s) showed the least amount of change? Describe the kinds of changes that occurred. ______________________

Analysis and Conclusions

1. Based on the results of this investigation, identify the samples you studied as biodegradable or nonbiodegradable. ______________________

2. Which materials might be recycled? ______________________

3. What substances in the petri dishes caused the decay of the materials that showed evidence of change? ______________________

Critical Thinking and Application

1. What types of waste material should be disposed of in landfills? ______________________

2. Many communities are charging additional fees if people do not separate their garbage into recyclable and nonrecyclable bundles. What are the environmental benefits of this policy?

3. How can you reduce the amount of trash you produce each day?

Going Further

Design an experiment to determine the amount of garbage you produce in a day. Use your daily figure to calculate the amount of garbage you produce in a year. You may use a calculator or a computer for your computations. Prepare a graph or chart that illustrates the different types of materials found in the garbage you produce.